Wranglers and physicists

WRANGLERS AND PHYSICISTS

Studies on

Cambridge physics in

the nineteenth century

EDITED BY

P.M. HARMAN

Manchester
University Press

Copright (c) Manchester University Press 1985

Whilst copyright in the volume as a whole is vested in Manchester University Press, copyright in the individual chapters belongs to their respective authors, and no chapter may be reproduced whole or in part without the express permission in writing of both author and publisher.

Published by Manchester University Press
Oxford Road, Manchester M13 9PL, UK
and 51 Washington Street, Dover
New Hampshire 03820, USA

ISBN 0 7190 1756 4 *cased*

BRITISH LIBRARY CATALOGUING IN PUBLICATION DATA
Harman, P. M.
Wranglers and physicists; studies on Cambridge physics in the nineteenth century.
1. University of Cambridge--History. 2. Physics --Study and teaching (Higher)--England--Cambridge (Cambridgeshire)--History.
I. Title
530.7'1142659   QC47.G73C3

LIBRARY OF CONGRESS CATALOGUING IN PUBLICATION DATA
Main entry under title:
Wranglers and physicists.
Papers presented at a conference held at the University of Cambridge.
Bibliography: p. 242.
Includes index.
1. Physics--England--Cambridge (Cambridgeshire)-- History--Congresses. 2. Mathematics--England-- Cambridge (Cambridgeshire)--History--Congresses.
3. Physics--Study and teaching (Higher)--England-- Cambridge (Cambridgeshire)--History--Congresses.
5. Maxwell, James Clerk, 1831-1879--Congresses,
I. Harman, P. M. (Peter Michael), 1943-
QC9.G7W73 1985   530'.09426'59   85-1485

Printed in Great Britain
by Unwin Brothers, Old Woking

Contents

		Page
	Preface	vii
1.	Introduction P.M. HARMAN (*University of Lancaster*)	1
2.	The educational matrix: physics education at early-Victorian Cambridge, Edinburgh and Glasgow Universities DAVID B. WILSON (*Iowa State University*)	12
3.	Geologists and mathematicians: the rise of physical geology CROSBIE SMITH (*University of Kent*)	49
4.	Mathematics and mathematical physics from Cambridge, 1815–40: a survey of the achievements and of the French influences I. GRATTAN-GUINNESS (*Middlesex Polytechnic*)	84
5.	Integral theorems in Cambridge mathematical physics, 1830–55 J.J. CROSS (*University of Melbourne*)	112
6.	Mathematics and physical reality in William Thomson's electromagnetic theory OLE KNUDSEN (*University of Aarhus*)	149
7.	Mechanical image and reality in Maxwell's electromagnetic theory DANIEL M. SIEGEL (*University of Wisconsin*)	180
8.	Edinburgh philosophy and Cambridge physics: the natural philosophy of James Clerk Maxwell P.M. HARMAN (*University of Lancaster*)	202
9.	Modifying the continuum: methods of Maxwellian electrodynamics JED Z. BUCHWALD (*University of Toronto*)	225
	Bibliography	242
	Index	257

Preface

This volume consists of essays based on papers presented at a conference on 'Cambridge Mathematical Physics in the Nineteenth Century'. Physicists educated at or teaching at the University of Cambridge made important contributions to the conceptual development of physics in the nineteenth century. The essays in this book discuss the way in which the Mathematical Tripos at Cambridge shaped the physics of men such as William Thomson (later Lord Kelvin) and James Clerk Maxwell.

Cambridge mathematical physics is discussed in relation to the stress on 'mixed mathematics', which included mechanics, hydrodynamics, and physical astronomy and optics, in the Mathematical Tripos. Special attention is paid to the work of Thomson and Maxwell, but the cultural and educational context in which mathematical physics was pursued is also considered; and the relation between Cambridge 'mixed mathematics' and the analogical physics taught in Scottish universities (both of which may well have shaped the work of Thomson and Maxwell) is discussed.

Without aiming to provide an exhaustive study of Cambridge physics in this period or to systematically demarcate its scope, the essays in this book explore important mathematical, physical, philosophical, cultural and educational dimensions of Cambridge physics in the nineteenth century. The breadth of coverage of documentary sources and historical problems provided by a volume containing chapters by different authors will, it is hoped, stimulate future work on these topics.

The conference at which the papers on which these chapters are based was held under the auspices of the British Society for the History of Science in March 1984. It is a pleasure to thank Robert Fox and Jack Morrell, successive Presidents of the Society, and John Brooke and the late Bernard Norton for their advice and support.

I am grateful to the participants at the meeting for contributing to the discussion of the papers, especially to Fabio Bevilacqua, John Brooke, David Gooding, Martin Klein and Paul Theerman, who made formal contributions to the conference as commentators on the papers; and to John Heilbron for presenting a paper to the meeting.

The Wordsworth Hotel, Grasmere, provided a splendid location for the conference in the magnificent environment of the Lake District, which contributed greatly to the relaxed fostering of informal discussions, and Mr Robin Lees and his staff are to be thanked for their efficiency and courtesy.

An international scholarly meeting of this kind could not be held without financial support. I am extremely grateful to the Council of the Royal Society and the Division of History of Science of the International Union of the History and Philosophy of Science for their generous support of the meeting. I am especially grateful to Erwin Hiebert, President of the Division of History of Science of the IUHPS, for his kind advice and help.

The publication of this book owes much to our editor, Alec McAulay. I am very grateful to him for his support of the project, and to Robert Fox for first encouraging his interest.

P.M.H.

1

Introduction

P.M. HARMAN

In his classic *History of the Theories of Aether and Electricity* Sir Edmund Whittaker referred to the '"Cambridge school" of natural philosophers' (Whittaker 1951-53, 1:153), which included George Green (1793-1841), George Gabriel Stokes (1819-1903), William Thomson, later Lord Kelvin (1824-1907), James Clerk Maxwell (1831-79), and Joseph Larmor (1857-1942). The 'Cambridge school' includes most of the major physicists active in Britain *c.* 1820-1900 (two significant exceptions being Michael Faraday and James Prescott Joule, who were experimental scientists). The members of the 'Cambridge school' were all graduates of the Mathematical Tripos at Cambridge; and although most of these men became skilled experimentalists, mathematical physics formed the core of their scientific achievement. The general impact of the Mathematical Tripos in physics education can be gauged from the fact that in the second half of the nineteenth century, nearly half the chairs in physics in British universities were held by wranglers, men who had obtained first-class degrees in the Cambridge Mathematical Tripos (Forman, Heilbron and Weart 1975, 32).

The notion of a 'Cambridge school' in the nineteenth century needs to be regarded with some care. Whittaker employed the term to reflect the educational role of the Mathematical Tripos in providing the techniques of advanced mathematics and in fostering a commitment to the mathematisation of physical phenomena. However, the members of the 'Cambridge school' had diverse educational backgrounds and careers, and for most of the period the University of Cambridge did not have a teaching or career structure for physics which would give the notion of a 'school' of physics a more precise meaning. Qualifications such as these are necessary, but there is a kernel of truth in Whittaker's characterisation. 'Cambridge' physics in this period may be understood to mean the physics of men educated, if not also teaching at, the University of Cambridge in the nineteenth century, without the implication that this physics can be characterised as solely 'Cambridge' in character. The distinctive feature of Cambridge physics in the nineteenth century was its mathematical character, reflecting the educational impact of the Mathematical Tripos.

A special feature of the Mathematical Tripos was its emphasis on physical topics, which muted attempts to encourage abstract analysis.

P.M. Harman

The Tripos was reformed in the late 1840s under the influence of William Whewell (1794–1866), its analytical focus being replaced by a renewed emphasis on 'mixed mathematics', embracing mechanics, hydrodynamics, astronomy and planetary theory, and physical and geometrical optics, despite the removal of theories of electricity, magnetism and heat from the syllabus (Becher 1980a). As a Tripos examiner in the late 1860s, Maxwell was instrumental in broadening the Tripos to encompass problems of current concern in physics, by including a few questions on heat, electricity and magnetism. These developments led to moves to establish a professorship of physics and a laboratory to provide instruction in these topics (Sviedrys 1970). The lectures of the professor — Maxwell himself was appointed in 1871 — were subject to the Board of Mathematical Studies, and the role of the laboratory was envisaged as a means of broadening the teaching for the Mathematical Tripos. The Mathematical Tripos, rather than the Natural Sciences Tripos (which emphasised experimental physics taught in consort with other sciences) was overwhelmingly dominant as the educational path for prospective physicists. Only in the 1890s did physics education through the Natural Sciences Tripos become the major educational route for tyro physicists (Wilson 1982). Physics at Cambridge in the nineteenth century was therefore 'mixed mathematics'.

The distinctive content of the Mathematical Tripos, emphasising 'mixed mathematics', geometry, and the ability to solve problems, rather than the pursuit of analytical rigour in the study of the foundations of mathematics, had the consequence that the research emphasis of the most able graduates (Cayley and Sylvester being obvious exceptions) tended to be towards the cultivation of 'mixed mathematics' (Becher 1980a). The wranglers of the Mathematical Tripos — Stokes, Thomson, Maxwell and Larmor — became the 'Cambridge' physicists who pursued physical theory as 'mixed mathematics'.

The word 'physics' acquired its modern connotations, as the science of mechanics, electricity, optics and heat, and as employing a mathematical and experimental methodology, in the first half of the nineteenth century. This encompassed quantification and the search for mathematical laws; the emergence of a unified physics based on the programme of mechanical explanation, which sought to explain physical phenomena in terms of the structure and laws of motion of a mechanical system; and the development of the concept of energy, which provided the phenomena of physics with a new and unifying conceptual framework within the mechanical view of nature (Harman 1982a).

Mathematical physicists educated at Cambridge contributed substantially to the establishment of these aims and the realisation of this programme of research. These physicists emphasised mathematisation and quantification, and the special status of mechanics — the science of matter in motion — as standing at the summit of physics and as providing the paradigm for physical theory. They stressed a commitment to the mechanical view of nature, which supposed

Introduction

that matter in motion was the basis of all physical phenomena and that mechanical explanation was the programme of physical theory. The explication of physical science within a mathematical framework of analysis has been described as 'the Cambridge programme', especially as enunciated by Whewell, who emphasised the cultural benefits of mathematical reasoning in providing evidence of design in nature by establishing mathematical laws, and hence confirming the status of natural knowledge as 'objective, value-transcendent, and tinged with the Divine' (Morrell and Thackray 1981, 481).

The writings of John Herschel (1792–1871) have been seen as especially important in shaping the outlook of Cambridge mathematical physicists (Wilson 1982). In his *Preliminary Discourse on the Study of Natural Philosophy* (1830) Herschel emphasised the role of mathematics as the key to scientific knowledge, and placed dynamics, the science of force and motion, 'at the head of all the sciences' and as capable of providing 'a certainty no way inferior to mathematical demonstration' (Herschel 1830, 96). In Herschel's view, knowledge of the hidden processes of nature would depend on the successful formulation of dynamical theories, the realisation of the programme of the mechanical worldview. This programme of research was fundamental to the work of Stokes, Thomson and Maxwell and their successors. In the 1830s and 1840s the concept of the luminiferous ether, which was envisaged as providing a dynamical basis for optical theory, was actively developed by Green and Stokes. In the 1840s Stokes elaborated hydrodynamical theories, which he applied to the theory of the ether (Wilson 1972). In the 1840s and 1850s Thomson sought to establish dynamical theories of heat and magnetism, emphasising the programme of establishing physics in terms of matter and motion (Knudsen 1976; Smith 1976a, 1978). Hydrodynamical theories of the ether played an important role in his work (Knudsen 1972); and his theory of atoms as vortices in a fluid had an important influence on Larmor's theory of the ether and electromagnetism of the 1890s. The search for a dynamical theory of electricity and magnetism, and a dynamical theory of gases and molecular physics, shaped the influential work of Maxwell from the 1850s to the 1870s (Brush 1976; Heimann 1970a, 1970b; Klein 1970, 1972; Siegel 1981). In characterising the physical sciences for the *Encyclopaedia Britannica* in the 1870s, Maxwell identified the scope of physics with the programme of mechanical explanation, and emphasised that dynamics, the theory of the motion of bodies as affected by forces, was fundamental to physics (Harman 1982a).

It is therefore appropriate to consider the work of Stokes, Thomson and Maxwell in the context of a 'Cambridge school' of mathematical physics, but such a denotation should not be interpreted as an exclusive demarcation. Although all but one of the major figures — the exception being Thomson who pursued a vigorous academic, technological and entrepreneurial career in Glasgow — spent all or part of their professional careers at Cambridge, the education of individuals, the breadth of their interests, and the development of career patterns cannot be satisfactorily characterised solely in 'Cambridge' terms. In the nineteenth century, physics was international in character, and the work of Fourier on heat theory and Fresnel on optics and the luminiferous ether had a major impact on Cambridge physics (Whittaker 1951–53, 1:128–69; Wise 1979, 1981). That many of the objectives which have been identified as characteristic of Cambridge physics were typical of British physics in the nineteenth century can in part be explained as a consequence of the dominance of Cambridge's mathematical physicists; but other factors may need to be invoked (Wise 1982). Thomson and Maxwell were Scots and educated at Scottish universities before attending Cambridge, and certain features of their work, including an emphasis on the unification of phenomena and the stress on the role of physical and mathematical analogies, may reflect themes promulgated in the natural philosophy classes of Scottish universities. For this reason it has been claimed that the emergence of mechanistic physics in nineteenth-century Britain rested on the fusion of the conceptual unity stressed by Scottish physicists with the 'mixed mathematics' of Cambridge (Smith 1976*b*).

The chapters in this book, based on papers presented at the conference on 'Cambridge Mathematical Physics in the Nineteenth Century', provide studies that illuminate these themes. Without attempting to provide an exhaustive account of Cambridge physics, or to systematically demarcate its scope, these essays explore important aspects of the educational, mathematical, physical, philosophical and cultural dimensions of Cambridge physics in the nineteenth century. Emphasis is placed on the formative role of Cambridge 'mixed mathematics' and the development of dynamical theories of physics in the work of Thomson and Maxwell, but attention is also given to more general cultural issues, as well as to the important relations between Scottish analogical physics and Cambridge 'mixed mathematics'.

In his chapter on 'The educational matrix' David Wilson contrasts physics education at Cambridge with that at Scottish universities. Wilson is concerned with the dominance of Scots, notably Thomson and Maxwell, among the Cambridge mathematical physicists. By drawing upon manuscript lecture notes and university archives, Wilson is able to reconstruct the patterns of physics education at Cambridge, Edinburgh and Glasgow in the first half of the nineteenth century, contrasting the Cambridge Mathematical Tripos with Scottish natural philosophy courses as the sources of instruction in physics at Cambridge and the Scottish universities.

At Cambridge Wilson places special emphasis on the competition to secure a high place in the list of 'wranglers', and on the role of the mathematics coaches, notably William Hopkins (1793–1866), who tutored Stokes, Thomson and Maxwell. Wilson notes that Hopkins placed special emphasis on the physical subjects in the Mathematical Tripos, especially physical optics, a topic central to Cambridge 'mixed mathematics'. Although J.D. Forbes at Edinburgh admired

Introduction

the Cambridge mathematical method, his course maintained the traditional broad physical range of Scottish teaching of natural philosophy, with its basis in experimentation. In reconstructing the content of Forbes's course in the period 1833–56, Wilson notes the primacy of mechanics and Forbes's commitment to the view that 'mixed mathematics' would subsume all physical science, as well as his emphasis on analogies in physics. In his course at Glasgow the Cambridge wrangler Thomson provided a course that reflected more accurately the current state of physics, but similarly placed an emphasis on the unity of nature grounded on the science of dynamics.

In remarking on the stress in Scottish natural philosophy on analogies, connections and the unity of nature, features which were striking components of the thermodynamics and field physics of Thomson and Maxwell, Wilson contrasts their work with that of Stokes, whose physics remained largely confined to established topics in 'mixed mathematics', hydrodynamics and ether theory. Wilson thus sees the innovatory strength of Thomson and Maxwell as deriving from the Scottish tradition of natural philosophy; but he notes the educational impact of Cambridge 'mixed mathematics' on Forbes (who taught Maxwell at Edinburgh) and Thomson. The Scottish tradition of analogical physics played a formative role in shaping British physics, though Scottish physics was itself transformed by Cambridge attitudes.

Whereas David Wilson draws attention to the educational impact of Cambridge 'mixed mathematics', Crosbie Smith, in his chapter on 'Geologists and mathematicians', explores the way in which mathematics served as the framework for the broadening of the physical sciences in Cambridge. Smith emphasises the importance of the cultural values associated with mathematical science, especially physical astronomy. The search for mathematical laws was envisaged by Hopkins and Whewell as the uncovering of divine laws, and they sought to elevate geology to a similar status for the benefit of natural theology. Just as astronomy was grounded on dynamical principles, so geology should be based on geological dynamics. Hopkins attempted to chart the geometrical laws of geology and then to establish their physical causes; there was an analogy with astronomy, where the establishment of the laws of planetary motion by Kepler was prior to the explication of their causal and dynamical basis in Newton's theory of gravity. The Cambridge programme of 'mixed mathematics', stressed by Hopkins as a mathematics teacher in focusing on physical optics as well as physical astronomy, had important scientific applications.

Mathematical analysis could therefore be applied to geology as well as to physical astronomy. The establishment of mathematical laws exemplified the cultural benefits of the study of science, in revealing the divine framework of natural laws. The application of the method to new domains thus had important cultural and educational as well as scientific implications. Smith indicates that Cambridge mathematical physics should be viewed in relation to the community of Cambridge theologians and mathematicians, and their interests, values and educational attitudes (see also Cannon 1964). Smith stresses the theological and cultural benefits of mathematical science as perceived by Cambridge academics.

In his chapter on 'Mathematics and mathematical physics from Cambridge, 1815–40', Ivor Grattan-Guinness surveys the development of Cambridge 'mixed mathematics' in the period, noting the influence of French mathematics and mathematical physics in shaping Cambridge science. He emphasises the role of the Analytical Society, founded in 1812 by Herschel, Babbage and Peacock, in transforming attitudes to the calculus by bringing continental analysis to Cambridge, but notes that Cambridge mathematicians continued to emphasise Lagrange's approach in which a derivative is defined in terms of a Taylor series, and failed to keep up with Cauchy's approach based on limits, which sought more adequate foundations for the calculus. Under Whewell's influence, analysis was subordinated to geometric arguments, and 'mixed mathematics' took precedence over abstract analysis. Grattan-Guinness emphasises the role of George Biddell Airy (1801–1892), whose *Mathematical Tracts*, first published in 1826, became a major textbook of Cambridge 'mixed mathematics'. The 1826 edition of Airy's book was concerned with mechanics and astronomy, but the 1831 edition added a discussion of the wave theory of light. These topics remained fundamental to the work of Cambridge mathematical physicists in the 1830s and 1840s. Herschel provided a major review of the wave theory of light, and Green and Stokes developed theories of the luminiferous ether, following the work of Fresnel and Cauchy. Surveying the research publications of Cambridge-educated mathematicians in the period 1815–40, Grattan-Guinness concludes that physical optics and physical astronomy were especially important, just as they were in the Mathematical Tripos itself. In particular he draws attention to the work of Airy, as a major contributor to mathematical physics.

Grattan-Guinness's chapter provides a detailed taxonomy of journals and the contributions of individuals. The pattern that emerges, which supplements previous accounts of Cambridge mathematics teaching (Becher 1980a), provides clear evidence that Cambridge 'mixed mathematics' had its research as well as its teaching emphasis in mechanics, hydrodynamics, physical astronomy and physical optics. The downgrading of abstract analysis, coupled with the high status of 'mixed mathematics', had the consequence that mathematical physics became the dominant field for Cambridge mathematicians.

In his chapter on 'Integral theorems in Cambridge mathematical physics, 1830–55', Jim Cross reviews the development of mathematical theorems concerned with the relations between lines and surfaces. Theorems of this type, concerned with the relation between quantities measured in relation to lines and quantities measured in relation to areas (theorems concerning line-and surface-integrals), were to be of fundamental importance in the physics of Thomson and Maxwell. Cross surveys the development of integral theorems from the early work of Gauss, Ampère and Poisson to the application of integral theorems to

Introduction

heat conduction and the distribution of electric charge by the Cambridge mathematician Robert Murphy (1806–1843). Green and Stokes applied these methods to hydrodynamics, and Thomson applied theorems of this type to electricity and magnetism. In reviewing these mathematical developments, Cross notes the importance of the physical contexts in which integral theorems were deployed, the close connection between problems in mathematical physics and the development of appropriate mathematical techniques. By the 1840s Cambridge mathematicians were developing mathematical methods applicable to hydrodynamics, ether theories, and the mechanics of continuous media, the main teaching and research areas of Cambridge 'mixed mathematics'. These mathematical methods were to be important in later field and ether theories, in which Thomson and Maxwell were to broaden the scope of Cambridge 'mixed mathematics'. One of the most important integral theorems, establishing a relation between a surface integral over a surface and line integral round its boundary, deployed in Maxwell's *Treatise on Electricity and Magnetism* (1873) to express relations between electromagnetic quantities, and known as Stokes's theorem, was first stated by Thomson in a letter to Stokes of 2 July 1850, and then set by Stokes in his Smith's Prize examination paper at Cambridge in 1854 (where Maxwell was placed equal Smith's Prizeman). This episode illustrates the close connection between mathematics education and the development of mathematical physics at Cambridge, and the close personal relations between members of the 'Cambridge school'.

Ole Knudsen's chapter on 'Mathematics and physical reality in Thomson's electromagnetic theory' provides a systematic analysis of Thomson's work on electromagnetism and ether theory; though Thomson's early work has been discussed in some detail (Wise 1979, 1981), Knudsen's account extends across Thomson's entire career. Knudsen argues that Thomson's attitudes were shaped by his early outlook, which emphasised analogical reasoning as a means of establishing the relation between the mathematical formalism of a theory and its physical content. This was established in Thomson's early work on the analogy between heat flow and electric flux, where he developed the analogy to the solution of integral equations; and in his demonstration of the mathematical equivalence of theories based on particle-action and field concepts, by establishing the analogy between the concept of lines of electric force and the notion of the flux of heat between particles. Knudsen shows how Thomson's various ether theories were based on mathematical and physical analogies with elastic solids and with fluids. Thomson was thus able to broaden the conceptual framework of Cambridge 'mixed mathematics'. In his 'Baltimore Lectures' of 1884, Thomson aimed to link hydrodynamical and elastic solid analogies as a means of explaining electromagnetism and light, seeking to explicate the relation between light and matter.

Knudsen stresses that these analogies were merely dynamical 'illustrations', falling short of the complete mechanical theory of physics that Thomson hoped to achieve. Thomson envisaged a three-tier approach to the explication of physical reality: mathematical analogy; dynamical illustration based on physical analogies; and a complete dynamical theory of nature. This programme of physical explanation is characteristic of Cambridge 'mixed mathematics'. Knudsen argues that Thomson's construal of the relation between mathematics and physical reality and of the role of analogies explains his attitude to Maxwell's field theory and electromagnetic theory of light. In Thomson's view, Maxwell's theory failed to provide adequate dynamical foundations; introduced unwarranted physical hypotheses which were not based on analogies; and failed to explain the interaction of light and matter dynamically.

Knudsen's construal of Thomson's approach to mechanical explanation bears upon the well-known criticism by Pierre Duhem, in his *Aim and Structure of Physical Theory* (1906), of the 'English school' of mathematical physics. Duhem contrasted the phenomenological and deductive approach of French mathematical physicists such as Ampère with the intuitive, model-building style of Thomson and Maxwell, claiming that Thomson did not aim to be 'furnishing an explanation acceptable to reason, and that he has only wished to produce a work of imagination' (Duhem 1954, 85). Duhem's criticism was deliberately polemical, but Knudsen's chapter shows that Thomson was deeply preoccupied with questions of consistency and the status of dynamical models: by application of the method of mathematical and physical analogy, allied to hydrodynamical and elastic solid analogies, Thomson sought a unified and comprehensive dynamical theory of nature.

These issues, which were fundamental to the dynamical programme of Cambridge 'mixed mathematics', are further pursued by Dan Siegel in his chapter on 'Mechanical image and reality in Maxwell's electromagnetic theory'. Siegel seeks to analyse the extent to which Maxwell's mechanical models were conceived as literal representations of reality. Maxwell's transition from an essentially geometrical and analogical theory of lines of force to a realistic physical model of lines of force in terms of a theory of molecular vortices, followed by a retreat to a more general dynamical theory of the field, has received much discussion (Heimann 1970a; Klein 1972), but Siegel explores the issue further by a detailed analysis of the status of the molecular vortex model of the ether in Maxwell's paper 'On physical lines of force' (1861/2). Siegel charts the development of the successive stages of this four-part paper, and places the paper in the context of Maxwell's other work on electromagnetism. He distinguishes between Maxwell's continued commitment to a theory of molecular vortices (which was grounded on a physical phenomenon, the rotational effect of magnets on polarised light), and the specific mechanism Maxwell employed in the paper to represent vortex motion, which he always considered to be illustrative and provisional.

Siegel emphasises the explanatory power of Maxwell's physical model. The vortex model explained the relationships between the electric current and the

Introduction

magnetic field, which were expressed mathematically in terms of relations between linear and rotational quantities, and mechanically in terms of relations between the rotational (angular) motion and the linear displacement of a vortex medium. The development of the theory of charge and of the theory of light as transverse waves in the ether, as Maxwell's paper progressed, illustrates the explanatory power of Maxwell's vortex model. In contrast to Duhem, Siegel views Maxwell's appeal to mechanical models as essentially consistent, grounded on Maxwell's fundamental commitment to the view that reality is mechanical, as reflected in his later qualified commitment to a mechanical representation of the field by molecular vortices.

Duhem's thesis about the character of the 'English school' of mathematical physics in the nineteenth century has been interpreted in terms of the impact of Scottish philosophy on British physicists: that the emphasis on models, analogies and geometrical arguments by British physicists echoed the treatment of the role of analogies in the work of Scottish philosophers (Olson 1975). In particular, Maxwell's work has been seen as constrained by metaphysical precepts acquired from Scottish philosophy.

In his chapter on 'Edinburgh philosophy and Cambridge physics', Peter Harman aims to refute claims that Maxwell's physics was derived from the metaphysics of William Hamilton, whose lectures he attended at Edinburgh University. Harman shows that Maxwell deployed philosophical argument merely as a means of justifying his mathematical and physical world view; and moreover that his philosophical arguments also manifest the influence of Whewell's writings. Harman shows that Maxwell's use of geometrical and physical arguments, and his treatment of the relation between mathematics and physical reality, owed much to the tradition of Cambridge 'mixed mathematics'. He argues that Maxwell's use of physical–geometrical theory is characteristic of the work of his mentors, Stokes and Thomson, and that Maxwell does not appeal to geometrical foundations for the fluxional calculus, an argument that was characteristic of the Scottish tradition of mathematics and philosophy. Harman thus seeks to refute claims that Maxwell's stress on geometrical argument shows his indebtedness to distinctively Scottish traditions.

Harman sees Hamilton's influence on Maxwell as exemplified by Maxwell's continued concern to justify his physics by appeal to more general philosophical argument, and the main part of Harman's paper is concerned with discussion of the distinctive style of Maxwell's natural philosophy, which embraced philosophical, mathematical and physical arguments. Harman seeks to characterise Maxwell's construal of the relation between mathematics and physical reality, showing that Maxwell's use of physical analogies and a physical–geometrical style of argument enabled him to express the link between theory and reality. Maxwell's concept of dynamical explanation in his *Treatise on Electricity and Magnetism* (1873) is fundamental to these questions. Harman argues that Maxwell did not conceive the generalised dynamical equations of the electrom-

P.M. Harman

agnetic field as merely providing a symbolic representation of the field, but that Maxwell expressed these equations in terms of physical principles of energy and momentum, while avoiding the supposition of any theory about the structure of matter constituting the field. Harman contends that Maxwell thus sought to embody a link between the mathematical formalism and the physical reality depicted, while avoiding the specification of a mechanical model. Rejecting claims that Maxwell's mode of dynamical explanation was grounded on prior metaphysical principles, Harman charts the physical and mathematical basis of Maxwell's theory, and the status of his appeal to more general philosophical principles as a means of justifying his physical–mathematical worldview.

In their chapters Knudsen, Siegel and Harman portray different aspects of the dynamical programme of Cambridge 'mixed mathematics' as applied to a wider range of physical phenomena in the work of Thomson and Maxwell. In his chapter on 'Modifying the continuum', Jed Buchwald discusses the way in which Maxwell's dynamical theory of the electromagnetic field shaped mathematical physics in Cambridge in the 1880s and 1890s. Buchwald is concerned with a group of 'Maxwellian' physicists, most of whom attended Cambridge and nearly all the rest being taught by Cambridge-educated physicists, who sought to develop and modify Maxwell's electromagnetic field theory. Buchwald discusses the development of 'Maxwellian' dynamical physics, especially by Larmor, which sought to explain all electromagnetic processes in terms of energy changes governed by the least action principle.

Buchwald traces the development and modification of the Maxwellian theory that the seat of electromagnetic processes was a continuous ether governed by the laws of dynamics, to its ultimate transformation by Larmor into a theory of the field based on models of material bodies built out of charged particles, 'electrons'. Larmor's work bears all the hallmarks of Cambridge 'mixed mathematics': graduating as Senior Wrangler in 1880, hydrodynamics and mechanics were fundamental to his research. His papers on the theory of the electromagnetic field in the 1890s followed Maxwell's *Treatise* in emphasising dynamical foundations: the Lagrangian formalism of dynamics would permit the details of the mechanism of the ether to be ignored. The electromagnetic properties of the field were explained in terms of a dynamical theory of the ethereal plenum; this was a realisation of the Cambridge programme of mechanical explanation, the attempt to explain all physical phenomena in terms of mechanical principles. Larmor's theory of the ethereal plenum unified the electromagnetic and dynamical properties of the ether, and his work falls squarely into the tradition established by Stokes, Thomson and Maxwell. Succeeding Stokes as Lucasian Professor of Mathematics at Cambridge, Larmor continued the tradition of Cambridge 'mixed mathematics'.

The distinctive character of nineteenth-century 'Cambridge' physics thus owed much to the educational impact of the Mathematical Tripos. An

Introduction

alternative tradition of Cambridge physics began to develop in the 1870s, the experimental physics of the Cavendish Laboratory. First under Maxwell, and then under his successors Lord Rayleigh (1842–1919) and Joseph John Thomson (1856–1940), both of whom were distinguished mathematical physicists and successful graduates of the Mathematical Tripos (and whose contributions to mathematical physics alone are of great importance), the Cavendish Laboratory established itself as a teaching and research institution in Cambridge (Sviedrys 1976). By means of curriculum reform, introducing a greater mathematical content into the Natural Sciences Tripos, J.J. Thomson made the Natural Sciences Tripos rather than the Mathematical Tripos the educational route for prospective physicists. The increasing success of the Natural Sciences Tripos in attracting students in the 1890s, providing a physics education grounded on experimental physics taught together with other sciences, and the close links between the Natural Sciences Tripos and the Cavendish Laboratory, were undoubtedly important in shifting the emphasis to experimental physics at Cambridge (Wilson 1982). In a paper presented at the conference, John Heilbron drew attention to the decline in vitality in research at the Cavendish in the early 1890s as indicated by numbers of research students and publications, and offered certain intellectual and institutional factors as causes. Removal or relaxation of these factors conditioned the rapid rise in the strength of Cavendish physics after 1895 by incorporation of non-Cambridge elements.[1] By the mid-1890s the Cavendish Laboratory began to dominate physics at Cambridge, and Cavendish physicists and graduates of the Natural Sciences Tripos established the Cambridge tradition of experimental physics of the twentieth century.

Note

1. Professor Heilbron plans to develop his argument further before publishing it.

= 2 =

The educational matrix: physics education at early-Victorian Cambridge, Edinburgh and Glasgow Universities

DAVID B. WILSON

Early-Victorian Cambridge University graduated many important mathematicians, astronomers and mathematical physicists. They were among the 'wranglers', or top students, taking the 'Mathematical Tripos', an examination required of candidates for 'honours' degrees and taken in an undergraduate's fourth year. The vast majority of high wranglers were Englishmen who attended English schools before entering Cambridge. However, of the six strongest mathematical physicists to graduate from Cambridge during this time, the formative period of the 'Cambridge school' of physics, none had this conventional background.

George Green (Cambridge BA1837), the only Englishman of the group, was a self-taught forty-year-old who had already published some of his best work before he became an undergraduate in 1833. An Irishman, George Gabriel Stokes (BA 1841) entered Cambridge from Bristol College. Born in Ireland, William Thomson (BA 1845) for several years attended Glasgow University, where his father was professor of mathematics, before entering Cambridge. A Scot, P.G. Tait (BA 1852), attended the Edinburgh Academy and for one year Edinburgh University before coming up to Cambridge. An Irishman, W.J. Steele (BA 1852), entered Cambridge from the University of Glasgow where he was an outstanding student of the new professor of natural philosophy, William Thomson. Like Tait, James Clerk Maxwell (BA 1854) attended both the Edinburgh Academy and Edinburgh University, though he spent three years at the latter.

As four of these six — including the two major figures, Thomson and Maxwell — attended Scottish universities as well as Cambridge, this chapter examines physics education at Cambridge, Edinburgh, and Glasgow Universities in the early-Victorian period. Though useful, the term 'physics education' is somewhat anachronistic. In Scotland, students took a class in natural philosophy. At Cambridge they prepared for the Mathematical Tripos. These were hardly equivalent courses of study, and the differences between them

The educational matrix

reflected overall differences between university education in Scotland and Cambridge.[1]

Traditionally, the 'liberal' education found at Scottish universities emphasised ancient languages and modern philosophy, studied at a general level. A broad course in natural philosophy occupied one year in the four-year curriculum and usually required only elementary mathematics. Each Scottish university had one professor of natural philosophy, who taught the course. Enrolments for the course at Glasgow and Edinburgh averaged around 60 per year at the former, 120 at the latter. General courses suited Scottish 'democratic' ideas, which encouraged students to enter at various ages, with varying levels of preparation, and for various periods of time. Few took degrees.

At Cambridge, a 'liberal' education was an intense grounding in elementary and advanced mathematics designed to train minds to think clearly and rationally when encountering subjects more cluttered than mathematics. Including satisfactorily mathematised physical subjects in the curriculum made the educational point that clarity was possible outside pure mathematics. Students took the demanding, several-day Mathematical Tripos during their fourth year, and those who passed received honours degrees. Around 120 each year did so. Unlike Glasgow and Edinburgh, Cambridge possessed a large collection of professors, college lecturers and private tutors to prepare students for the Tripos.

The curriculum at Cambridge and Edinburgh during the early-Victorian period was relatively stable. A committee at Cambridge which formally defined the contents of the Tripos in 1850, for example, stated that they were mainly codifying long-standing practice. James David Forbes, the new professor of natural philosophy at Edinburgh, had introduced his educational ideas by the mid-1830s, and he held his professorship until 1860. Only at Glasgow was there substantial change. Appointed professor of natural philosophy in 1803, William Meikleham gave way to substitutes because of illness during the early 1840s and was replaced after his death in 1846 by William Thomson.

This chapter, then, is primarily a comparative study of physics education at Cambridge, Edinburgh and Glasgow from about 1835 to about 1855. It discusses not only which branches of physics were taught, but also broader issues, like the place in physics education of both mathematics and the philosophy of science. It especially deals with the distinct, but often related, questions of the similarities, connections and unities perceived to exist between different branches of the physical sciences. The point here is not whether such views originated in these universities at this time, but rather which university represented which point of view within that spectrum of viewpoints from mere (or no) similarity to total unity.

We can summarise the differences in the respective courses of study. Meikleham's course at Glasgow was the most philosophically oriented and probably the least detailed scientifically. Cambridge was the strongest in mathematics and optics, but the least philosophically inclined. Forbes's course at Edinburgh was stronger in more physical subjects than either Cambridge or Meikleham's Glasgow. Meikleham's and Thomson's courses at Glasgow laid the greatest stress on the unity of the physical sciences, Cambridge the least. In the late 1840s and early 1850s, the most modern physics education was offered in Glasgow by William Thomson. But while Thomson presided in Glasgow, the best student of the day, Maxwell, was attending Edinburgh and Cambridge.

This question of the education of individual students leads the chapter, finally, into that thicket of speculation concerning the respective roles of Scotland and Cambridge in shaping the research of the 'Cambridge' physicists. Without pretending that their physics education was the only influential factor involved, one can still point out noticeable correlations between their respective educational backgrounds and their respective careers of research.

Whatever the three universities' individual degrees of importance in shaping Victorian physics, Cambridge's educational system was the largest, most prestigious, and least susceptible to change. Because Cambridge influenced Edinburgh and Glasgow, not vice versa, this chapter treats the universities in the order opposite to that in which they were attended by the Cambridge physicists.

Cambridge

The best mathematical students at early-Victorian Cambridge were part of a high-level nearly professional milieu. It consisted not only of an educational system, but also of a scientific society, two scientific journals, and employment for those excelling in the educational system. This system culminated in two examinations: the Mathematical Tripos and the Smith's Prize examination, which immediately followed the Tripos and in which only the very best students competed. A student who was placed as a top wrangler in the Mathematical Tripos could count on being elected to a college fellowship and thus staying in Cambridge to help train future students for the Tripos. The four examiners per year for the Tripos, for example, were drawn from the ranks of college fellows. For some — like Stokes and Thomson — a college fellowship served until a professorship in mathematics or natural philosophy came open, either in Cambridge or elsewhere. All six Cambridge professorships in mathematics and physical sciences were held by tripos graduates. The most effective teaching for the Tripos during this period, however, was done by private coaches, themselves Tripos graduates. By the mid-1830s, this rather large group of coaches, fellows and professors had created a substantial institutional framework for their mathematical studies. Founded in 1819 to promote research in the natural sciences generally, the Cambridge Philosophical Society and its *Transactions* were dominated by the disciplines of the Mathematical Tripos by the mid-1830s. Also in the mid-1830s, Tripos graduates founded the *Cambridge Mathematical Journal*. It published papers mainly by Cambridge wranglers, some of the

The educational matrix

articles being directed towards undergraduates preparing for the Tripos. Under Thomson's editorship, it broadened into the *Cambridge and Dublin Mathematical Journal* in the mid-1840s.

Among the many people responsible in one way or another for the Tripos, there were four or five key figures. William Whewell had helped introduce Continental mathematics into the Tripos around 1820, and he continued to have an influence on the content of the Tripos. In the 1830s he succeeded in getting heat, electricity, magnetism and the wave theory of light included, and in the 1840s he was instrumental in making tripos mathematics more intuitive and geometrical than it had been (Becher 1980*a*). Whewell's suggestion led Robert Murphy, third wrangler in 1829, to publish his textbook on electricity 'designed for the use of students in the university' in 1833 (Murphy 1833*c*, v.) From 1835 to 1855 Whewell did not instruct students for the Tripos, but as the Master of Trinity College he was, beginning in 1842, one of the four examiners for the Smith's Prizes. George Biddell Airy's *Mathematical Tracts* had influenced the Tripos during the late 1820s, and his professorial lectures 'had the effect of raising in a very important degree the standard of *information* among our students, and exciting their interest in classes of phenomena till then little known in the University'.[2] Among the new subjects was the undulatory theory of light, which Airy added to the second edition of his *Mathematical Tracts* in 1831, making it possible for questions on the subject to appear in the tripos. It remained the standard reference on the subject for students during the early-Victorian period. In 1836 Airy was replaced as Plumian professor by James Challis, whose lectures constituted an important source of instruction for physical subjects on the Tripos. His lectures on astronomy, hydrodynamics and optics were 'attended by a class varying from twenty to thirty consisting chiefly of the most advanced students'.[3] Like Whewell, Challis examined for the Smith's Prizes but not for the Tripos. When Stokes became Lucasian professor in 1849, he took over Challis' lectures in hydrodynamics and optics and joined Challis and Whewell as Smith's Prize examiners. The most important instructor for the Tripos, however, was William Hopkins, foremost of the many private coaches. He trained Stokes, Thomson, Tait, Steele and Maxwell. Consequently, to understand what students studied for their Cambridge examinations, one must concentrate on the writings of Airy, the professorial lectures of Challis and Stokes, and the instruction of Hopkins, as well as on the content of the Mathematical Tripos and the Smith's Prize examination.

In addition to mathematics, the early-Victorian Tripos focused on several areas of mathematical science: mechanics, hydrostatics, hydrodynamics, observational astronomy, gravitational theory and geometrical optics. Students faced questions on the laws of motion and the experiments confirming them, on levers and pulleys, and on different combinations of forces. Geometrical optics overlapped with astronomy in questions on various kinds of telescope. Moreover, occasional questions on gas laws, steam engines, dew, the speed of sound, and thermometers involved the subject of heat. Mathematical treatments of electricity and magnetism were included in one or two questions each year (of a total of some 175 to 200 questions). The undulatory theory of light was the subject of four or five questions each year.

Heat, electricity and magnetism were only marginal Tripos subjects. Not only did they barely appear on the Tripos itself, but the questions that were asked did not deal with the possible connections or analogies between heat, electricity and magnetism. There were no mathematical lectures on the subjects given in Cambridge.[4] Hopkins ignored them.[5] When the Board of Mathematical Studies was created at mid-century, its members wanted to define allowable Tripos subjects. In their first report in 1849 they stated: 'Taking into consideration the great number of subjects which now occupy the attention of the candidates, and the doubt which exists as to the range of subjects from which questions may be proposed, the Board recommends that the Mathematical Theories of Electricity, Magnetism, and Heat, be not admitted as subjects of examination'.[6] Only underscoring the tenuous position these subjects had had in the Tripos, the Board members clearly felt that their various recommendations were mainly 'giving a definite form to that which has for some years past been the practice in the University Examinations'. They were not introducing 'violent change'.[7]

Optics, on the other hand, occupied a strong position in the Tripos. The many questions on geometrical optics led progressively to a few on physical optics. Together, geometrical and physical optics constituted a substantial and predictable examination subject. Moreover, Hopkins' opinion guaranteed that his students would give the subject its due:

> It is only when the student approaches the great theories, as Physical Astronomy and Physical Optics, that he can fully appreciate the real importance and value of pure mathematical science, as the only instrument of investigation by which man could possibly have attained to a knowledge of so much of what is perfect and beautiful in the structure of the material universe, and in the laws that govern it. (Hopkins 1841, 10.)

Hopkins' students were, therefore, presumably among the ten to fifteen who each year satisfactorily answered the advanced questions on physical optics.[8]

Airy introduced his tract on the undulatory theory by distinguishing between the theory's 'geometrical' and 'mechanical' parts. The former, like the theory of gravitation, was 'certainly true'. However, 'the *mechanical* part of the theory, as the suppositions relative to the constitution of the ether, the computation of the intensity of reflected and refracted rays, &c., though generally probable, I conceive to be far from certain'. (Airy 1842, v–vi.)

Airy used sound as the best of various analogies for supporting weak portions of the mechanical part of the theory. The theory of sound, for example, showed that the resultant of combining waves was simply the sum of the disturbances which each wave would have produced by itself. The same could therefore be

The educational matrix

supposed true for light, Airy stated, because of similarities between air and any other medium in which undulations were propagated (ibid. §§10–12). Sound also helped explain the degree to which light would spread out after passing through an aperture of a given size (ibid. §28), and sound provided an 'analogical illustration' of the reflection of undulations of light (ibid. §31). However, this 'analogy which has guided us in several instances' failed to remove the 'obscurity in the undulatory theory' in regard to dispersion. Happily in this case, water waves did behave like light waves. Hence, 'in this respect there is a fair analogy between the relations of the velocity to the length of the wave, in glass or other refracting substances (as regards luminiferous waves), and in water (as regards waves of water)' (ibid. §38). Moreover, Airy attempted to remove the 'vagueness' in the undulatory theory's account of reflection by appealing to the behaviour of differently sized elastic bodies striking one another (ibid. §§36 and 128). For Airy, therefore, sound and other mechanical analogies helped make certain points plausible which were otherwise vague or obscure.

In addition to differential and integral calculus, Hopkins lectured to his students on mechanics, gravitational theory, hydrostatics, hydrodynamics, sound, light, and astronomical instruments. Force, matter and motion were the key concepts, with basic truths like the laws of motion being established by indirect induction rather than direct experiment. In astronomy, for example, the proper process was to 'assume' the truth of Newton's laws of motion and his theory of gravity, to calculate the consequent planetary motions, and finally to compare the calculated results with observations. 'The entire accordance in this case of the results of calculation and observation establishes at once the truth of the laws of motion and of the theory of gravitation.'[9]

Like Airy, Hopkins used sound to assist in explaining the undulatory theory. In his lectures, he discussed sound before light in order to introduce, for example, the five 'fundamental hypotheses of the undulatory theory'. As recorded by Stokes, three of the hypotheses were:

(1) That every luminous body is in a state of isochronous vibration similar to that of a sonorous body producing a musical note.
(2) That a perfectly elastic medium pervades all space, (whether occupied by other matter or not) and by means of w^h the undulations proceeding from the luminous body are transmitted to all parts of surrounding space, in a manner precisely analogous to that in w^h sonorous vib^{ns} are transmitted by the atmosphere.
(3) That the sensation of vision is produced by the mechanical action of the particles of this medium on the organ of vision in a manner analogous to that in w^h the sensation of sound is prod^d by the action of air on the organs of hearing.[10]

The lectures did not employ water waves or colliding bodies as analogies. Hopkins did follow Airy, however, in imagining a mechanical system that would transmit the transverse vibrations required by the phenomena of polarised light. He described properties germane 'to any medium such as the luminiferous ether':

> In any system of particles acting on each other with forces w^h are functions of their mutual distances there are three directions at $r^t \angle$ s to each other along w^h if a particle can be displaced the force of restitution will act in the same direction.[11]

Rather than an analogy, this was perhaps an 'assumed' mechanical system awaiting indirect inductive verification.

Challis, in his professorial lectures, also discussed sound before light, stating that 'many of the laws of aerial vibrations serve to illustrate the theory of Light' (Challis 1838, 10). The object of physical optics was 'to refer the various phenomena of light to some mechanical action of which a distinct conception can be formed' (ibid. 22). According to the undulatory theory, Challis said, 'light is an effect of the action of a very elastic medium, as sound is produced by means of the air'. And, like Hopkins, Challis compared the 'mechanical action' of ethereal vibrations on 'the constituent atoms of the eye' to the effects of aerial vibrations on the ear (ibid.). He thought it might be possible to treat the ether 'en masse', as one did air in acoustics, rather than as a medium constituted of particles.[12]

Tripos questions on physical optics from 1835 to 1855 asked about many aspects of the undulatory theory — interference, diffraction, double refraction, polarisation. They occasionally asked about specific difficulties facing the theory, comparisons with the corpuscular theory, and descriptions of experimental results (e.g. *Cambridge Calendar* 1835, 389; 1841, 333; 1848, 362; 1851, 375; 1853, 393). Despite Airy's, Hopkins', and Challis' discussions of the similarities of sound and light, however, only one question during those two decades asked for such a comparison (ibid. 1849, 339). Moreover, only one question explicitly asked for an analogy to help justify an explanation by the undulatory theory (ibid. 1835, 389).

Questions on the Smith's Prize examinations deserve separate attention. Examiners for it were free not only to pose questions more demanding than those for the Tripos, but also to cover subjects not usually found on the Tripos. However, the distribution of questions among physical subjects from 1835 to 1855 did, in fact, approximate that on the Tripos. Physical optics received a fair amount of emphasis, but heat, electricity and magnetism very little. In Whewell's fourteen examination papers (containing about twenty questions each) from 1842 to 1855, for example, there was only one question on heat (on the conduction of heat) and none on electricity.[13] From 1847 to 1855, Whewell usually asked one or two questions each year on the form of the curved lines around a magnet and on related issues in terrestrial magnetism. The papers set by the other two regular examiners from the mid-1830s to the mid-1850s, Challis and George Peacock, seldom ventured beyond optics to heat, electricity and magnetism. In his examination questions from 1850 to 1855, Stokes always

The educational matrix

included physical optics, but never asked about electricity or magnetism and only twice about heat. The subjects of electric currents and of connections between branches of physical science were confined to a mere five questions during the entire two decades. These were usually set by substitute examiners, notably W.H. Miller and Samuel Earnshaw.[14]

Consequently, on the basis of his undergraduate preparation for the Mathematical Tripos and the Smith's Prize examination, a high wrangler could be expected to be a technically skilful mathematician not only adept at mechanics, hydrodynamics and gravitational theory, but also thoroughly familiar with the intricacies of optical theory. He would know little about mathematical theories of heat, electricity and magnetism, and even less about any similarities or connections between them. Even the much-discussed analogy between sound and light was not an important examination topic, though Airy's textbook did uphold the analogy's potential significance for optical research. Hopkins' students would have heard his view of scientific knowledge and seen the view applied to speculations on the structure of the luminiferous ether. The physical natures of heat, electricity and magnetism, however, were not examination topics. Moreover, despite Hopkins' methodological statement and the philosophical writings of Whewell, methodological issues were not a significant part of Cambridge examinations. In short, the Cambridge mathematical student would know much about Laplace, Young and Fresnel, and little about Oersted, Ampère, and Faraday. He would know much about Newtonian gravitation, and little about Newton's rules of philosophising. His education differed greatly from that of students in Edinburgh and Glasgow.

Edinburgh

In a time of criticism and reform of British science, the self-proclaimed critic and reformer, James David Forbes, combined Cambridge and Edinburgh traditions into an all-inclusive course on natural philosophy. To understand the thrust of Forbes's actions, consider his situation in the early 1830s, before being elected professor of natural philosophy in January 1833.

On the one hand, he had enjoyed a highly successful early career, featuring close connections with the two leading Scottish natural philosophers, John Leslie and David Brewster. He had twice taken Leslie's natural philosophy course at Edinburgh and each time won Leslie's prize for the top student. Brewster's *Edinburgh Journal of Science* had provided a means for Forbes to publish several papers while still an undergraduate, and eventually he became Brewster's protégé. Proposed by Brewster for membership of the Royal Society of Edinburgh, Forbes was elected as soon as he became eligible, on his twenty-first birthday in 1830.

Despite such institutional and intellectual support rendered him in Edinburgh, however, Forbes saw much to criticise. He had not learned the higher analysis, for example, and with Brewster's encouragement had had to begin studying it on his own during the summer of 1830. Moreover, when, in the autumn of 1830, Forbes carefully read Leslie's major work on heat, he was not overly impressed, dismissing one of Leslie's key experiments as 'monstrously fallacious' and one of his chapters as 'singular and desultory'.[15] He later wrote that Leslie had unfairly rejected or ignored newer ideas and 'with few exceptions ... carried his scientific views of 1804 with him to the grave' (Forbes 1860, 946). Though Forbes had made the momentous decision to abandon the law for a career in science, he could rightly see that, as he wrote to Leslie in September 1830, he had 'a deplorable prospect' of finding a suitable position.[16] Indeed, Forbes regarded science in Edinburgh and Scotland from the perspective of Charles Babbage's *Decline of Science in England*, published in 1830. In June 1830, he and John Robison, son of Professor Robison, praised Babbage's book to one another and hoped to apply its lessons to the Royal Society of Edinburgh, Forbes writing that he was glad to see that Robison, too, was 'inclined to think that the Society wants jogging'.[17] In September he wrote to Leslie: 'I doubt not you have seen Babbage's book which is making some little stir in the torpid region of science. It was certainly much required and if it had been extended to Scotland, so much the better'.[18] The sorry state of Scottish science was a recurrent theme in Forbes's correspondence of the 1830s, and as late as 1839, trying to encourage a closer relationship between himself and the newly elected professor of mathematics at Edinburgh, Philip Kelland, he wrote to Kelland:

> Before you have lived much longer here, you will find that there are scarcely any persons in Edinburgh or in Scotland who understand or care one *iota* about subjects such as occupy your papers and my experiments (polarized heat &c) — that, in short, unless you have a very stoical pleasure in the solitary contemplation of beautiful Theorems you will be glad to find anyone to converse with on such subjects, like myself .... [19]

Hence, even though Babbage had criticised science education at Cambridge in his book, Forbes found much to admire there during his well-known visit in 1831. Not only was there a group of people in Cambridge who cared about the things Forbes did, but, through a system of written examinations, Cambridge fostered the study of analysis. Whereas Scotland's great optician Brewster (who was himself no mathematician despite his encouragement of Forbes's studies) continued to endorse the old corpuscular theory of light, the Cambridge-educated Whewell, Airy and John Herschel supported the modern wave theory, the full development of which had depended heavily on the higher analysis. Babbage reinforced the importance of analysis, writing to Forbes in 1832 that 'it is the great key to Nature and every great discovery renders it more necessary'.[20] There naturally followed Forbes's consultation with members of the Cambridge group, especially Whewell, as he refashioned the Edinburgh natural philosophy course. His was the same spirit of reform found both in Babbage's *Decline* and in the 1831 founding of the British Association for the Advance-

The educational matrix

ment of Science (which Forbes invited to Edinburgh in 1834 for its first meeting outside England).[21]

Forbes's aim, however, was not simply to imitate the Mathematical Tripos, and the resulting course was still grounded largely in the Edinburgh tradition. In the first place, Forbes regarded even the introduction of analysis as the continuation of an existing Edinburgh tradition, albeit at a higher level. Coexisting with the well-known emphasis at Edinburgh on the great pedagogical value of elementary geometry was the view of Forbes's professorial predecessors that the natural philosophy course should contain mathematics useful for the study of natural philosophy. Unfortunately, students displayed, in Leslie's words, 'a woeful want of mathematics'.[22] Justifying his own plans for reform, Forbes wrote to Henry Moseley in 1833: 'I need only refer you to the text books respectively published and employed by my three immediate predecessors Professors Robison, Playfair, and Leslie, proving that they considered the department broadly called mixed Mathematics as the most important part of their course.'[23]

In the second place, Forbes's course remained much broader than the Mathematical Tripos. He insisted that both mathematics and experimentation were essential to natural philosophy and that neither should be excluded.[24] His course ranged over such topics as heat, electricity and magnetism which were barely present in the Tripos. Reflecting Forbes's conviction that there existed an 'intimate and reciprocal connection' between natural philosophy and 'the Mechanical Arts, of which we take Civil Engineering to represent the department most cognate to that of Natural Philosophy' (Forbes 1860, 7), the course also included principles of machines and the theory of steam engines. Indeed, Forbes vehemently defended his prerogative to teach these technological subjects when a new professorship of technology was proposed at Edinburgh in the 1850s,[25] even adding a set of lectures on civil engineering in the 1855–56 session. (See Table 1.)

The administrative stroke that allowed Forbes to include advanced mathematics in a general course was his organisation of the class into divisions, each division being determined by competency in mathematics. He apparently incorporated some such division from the outset, because his first list of class prizes in 1834 distinguished between an 'Advanced Division' and a 'Junior Division' (*Edinburgh Almanack* 1834, 203). However, he instituted a more formal system beginning with the 1838–39 session, dividing the class into three divisions:

> In the junior Divison, a knowledge of Elementary Plane Geometry and Trigonometry, and the first principles of Algebra, will alone be expected: in the middle Division, Algebra, in its more ordinary branches, its application to Geometry and to the working of Problems, with some notions of Analytical Trigonometry: in the first Division, some knowledge of both the Differential and Integral Calculus will be required.[26]

Table I Contents of J.D. Forbes's natural philosophy course, 1833–56[a]

	LDBG,[b] JAL 33–4	JAL 34–5	JAL 35–6	WJMR 36–7	WS, WJMR 37–8	JAB, AS 38–9	39–40	40–1	JW 41–2	42–3[c]	44–5	BS 45–6	46–7	PGT, JCM 47–8	JCM 48–9	49–50	50–1	54–5	55–6
Introduction and general	18	7	6	2	5	7	4	6	3	6	2	2	4	4		5	7	8	5
Introduction and properties of bodies								2							16				
History of mechanics																			
System of the world												13							
Properties of bodies	14	6	9	9	11		10		12		12								11
Statics	30	21	27	23	21	25	22	21	25	30	24	24	34	31	32	25	29	24	16
Hydrostatics		4		6	7		3			5	7		4		3	6		7	6
Hydrostatics and pneumatics								14				7					9		
Dynamics	24	22	20	23	21	21	21	23	20	18	17	23	20	21	20	18	20	18	16
Electricity							9						15				18	5	
Magnetism																			
Hydrodynamics	9	5	11	5		10	5			4		12							15
Barometer																			
Steam engine	5	5	8	6	7	2	9	9	3	10	4	11	3	3	3		4	6	3
Pneumatics																		6	5
Meteorology						12										9			
Acoustics	5	8	6	6		11		14		14				11		9			
Pneumatics, heat, and steam engine				17															
Heat and steam engine																			
Heat	4	13				20			20		20		22		17	18		19	
Optics		22	21		11	15		22	17	22	7	19		21	21	17		11	
Electromagnetism[d]	19	7	15	15	19		6		13				6				17		13
Astronomy			2				5							37					
Miscellaneous																			
Civil engineering																			12
Total regular lectures	[128]	120	123	114	102	111	106	111	113	109	102	111	117	104	112	107	104	104	102
Glaciers									5		9		9						
Extra lectures	20	23	31	25	19	27	21	22	14	8	2		2		12	21	n.r.[f]	n.r.	n.r.
Examinations[e]	23	24	22	25	18	24	23	22	12	16	14	11	5	5	6	6	n.r.	n.r.	n.r.
Total hours of instruction	[171]	167	176	164	139	162	150	155	144	133	118	[122]	124	109	130	134	n.r.	n.r.	n.r.

[a] Source: Record of lectures delivered to class, 1833–57, St. Andrews University Library, Forbes Papers, Box IX, vol. 5. The numbers and descriptions of the topics are Forbes's.
[b] Initials are those of prominent students in the course that session. LDBG = Lewis D.B. Gordon, JAL = James A. Longridge, WJMR = William J. Macquorn Rankine, AS = Archibald Sandeman, JW = John Welsh, BS = Balfour Stewart, PGT = P.G. Tait, JCM = James Clerk Maxwell, WS = William Swan, JAB = John A. Broun.
[c] Omitted sessions are those of Forbes's absence due to illness, i.e. 1843–44 and 1851–54.
[d] Forbes's description for 1842–43 was: Electricity, galvanism, magnetism, electromagnetism.
[e] These were the total number of hours devoted to both oral and written examinations. During 1846–47 and 1847–48 there were an unspecified number of oral examinations conducted during the regular lectures.
[f] n.r. = not recorded.

The educational matrix

First-division students used the calculus in examinations over their textbooks in mechanics and optics. They usually read Poisson's *Mechanics*, Whewell's *Mechanics*, or Thomas Jackson's *Mechanics*, plus Airy's tract on the undulatory theory. Forbes awarded his class prizes to students who gained the highest marks on these examinations. Though Forbes's main course of some 110 lectures (which students from all divisions attended) was not highly mathematical, he gave extra lectures for first-division students which were more mathematical and of assistance in preparation for the examinations. As these extra lectures seldom dealt with electricity and magnetism, it appears that the mathematical portion of Forbes's course confined itself to the Cambridge subjects represented in the students' textbooks. The subject of heat is a possible exception, for Forbes did often devote extra lectures to the areas of his own research on radiant heat.[27]

Over the years Forbes's basic scheme varied in two ways. First, the course's content differed from year to year because the session was too short to cover all branches of natural philosophy. In the *Edinburgh Almanack* for 1833, Forbes announced that he would lecture on electromagnetism in 1833–34 and replace it with optics in 1834–35 (page 36). His detailed record of lecture topics allows reconstruction of the numbers of lectures devoted to each subject for each session of his professorship. Table I provides a summary. Secondly, as shown in Table I, after Forbes's serious illness which caused him to miss the 1834–35 session, he gave considerably fewer extra lectures. In the ten years before 1843–44 he averaged twenty-one extra lectures per year, whereas in the next decade he gave an appreciable number of such lectures in only two sessions. In the 1847–48 session, when both Tait and Maxwell were in the class (though only Tait was in the first division), Forbes gave no extra lectures, noting in his record book that 'some extra lectures [were] given from 12 to 1 but were not wanted'. Moreover, the extra lectures for 1848–49 and 1849–50 did not include mathematical theories of light, astronomy, electricity or magnetism. In both sessions, there were a few extra lectures on mechanics and, in 1849–50, several on instruments and on radiant heat. Undoubtedly, therefore, after 1843–44 Forbes lectured much less on mathematical physics, and in 1855 he stated that higher mathematics was best learned through 'the labour of private study and ardent self-exercise', that is, as he had done.[28]

Forbes's course presented a Baconian methodology as represented by Playfair and Herschel but not Whewell. In an essay for his moral philosophy class, which drew heavily on Playfair's account of Bacon, Forbes called Bacon a singular instance, because the right methodology could only be invented once. And 'the system of Bacon ... was adapted to the true constitution of the human mind and is indeed the triumph of that modesty which belongs to genuine science over the self-conceited semi-ignorance of former times'.[29] Like Playfair and Herschel, Forbes thought the inductive method should be as quantitative as possible. In 1841 he examined his third division on Herschel's *Preliminary*

Discourse, and the questions, all keyed to specific sections of Herschel's book, highlighted the dependency of progress in science on the establishment of highly quantitative laws derived from careful observation.[30] In the session of 1845–46, Forbes asked his students to evaluate the respective philosophies of Whewell and John Stuart Mill.[31] In his lectures, he considered but did not endorse Whewell's version of induction. Regarding the laws of motion — Whewell's strongest case for inductive truths established *a priori* — Forbes told his class: 'The science of dynamics is founded upon three fundamental axioms called the laws of motion. These are all derived and proved to be true from experience and observation. Indeed it is doubtful whether any of these could be proved *a priori*'.[32]

As disclosed by Table I, mechanics was the centrepiece of Forbes's course. Divided into statics and dynamics, mechanics constituted nearly half of the course in every session. Mechanics, lectured Forbes, was the science of force.

> The question then occurs What is force? It is brought in in mechanics as an additional element along with figure and space & quantity. Mechanics constitute the most stable part of Natural Philosophy. It is not founded upon conjecture or hypothesis. It is permanent while the other parts are progressive. What is force then? It is that which produces or tends to produce motion and which indeed will always produce it unless it is counteracted by an antagonist force tending in an opposite direction. This gives rise to the very simple division of Mechanics into statics and dynamics.[33]

The *application* of dynamics to specific areas, however, was progressive: 'So many variables and so many modifications of the circumstances attending motion take place that the science is really without limit and is a progressive science depending greatly upon mathematics and becoming enlarged in proportion as our mathematical investigations are more comprehensive'.[34]

Forbes's emphasis on quantitative laws and mathematical dynamical theories reflected his view that scientific progress would follow a pattern established by the heroes of seventeenth-century science. Kepler and Newton put Baconian methodology into practice, the first formulating his three laws of planetary motion and the second utilising them to establish his dynamical theory of universal gravitation. As Forbes wrote to Moseley while planning his course:

> Any doubt as to the propriety of viewing Mixed Mathematics as belonging to a Natural Philosophy Class is at this moment peculiarly untenable. For the whole progress of *General Physics* is happily so fast leading to a subjection to Mathematical Laws of that department of Science, that in no very long time, Magnetism, Electricity and Light may be expected to be as fully the objects of dynamical reasoning as Gravitation is at the present moment.[35]

Forbes regarded Gauss as the Kepler of terrestrial magnetism and Fresnel as the Newton of optics.[36] His lectures, though predominantly a description of empirical phenomena, nevertheless also expressed the ideal of discovering underlying dynamical systems responsible for observed phenomena. The phenomenon of Newton's rings, he declared, 'can only be fully explained upon the undulatory theory' and that of thin plates was 'quite conclusive for the theory of undulations'.[37] Going further, in a lecture on the dynamics of collisions, he explained that 'although the doctrine of impact might seem at first sight to be of little use yet it bears very closely upon dynamics and also the propagation of vibrations in elastic media to which sound and probably light are referable'.[38] And suggesting a particular dynamical explanation in optics, he told his students: 'In the theory of undulations also the velocity in the refracting medium is held to be less than in vacuum because the particles of ether [are] trammelled. They have more work to do and therefore take longer time to do it.'[39] In magnetism, Forbes supported the two-fluid theory which accounted 'for most of the phenomena' and which was 'conformable to the truth'.[40]

As Olson has argued, Forbes's Baconianism included an emphasis on analogical reasoning (Olson 1975, 226). In both his research and teaching, Forbes pointed to the importance of analogies and connections between branches of natural philosophy in the progress of science. Just as analogies between sound and light had been instrumental for Young's research in optics, for example, so also in the early 1830s similarities between light and radiant heat were guiding Forbes's own research. 'The importance of analogies in science has not perhaps been sufficiently insisted on by writers on methods of philosophizing. A clear perception of *connexion* has been by far the most fertile source of discovery.' (Forbes 1836a, 147n.) In outlining his natural philosophy course in the *Edinburgh Almanacks* for 1833 and 1834, Forbes included introductory lectures 'On the connexion of physical sciences with each other and with general knowledge' and discussed 'Analogies of heat, light, and electricity' (pages 35, 170). Under electromagnetism he included 'Analogies of magnetism and electricity' and the discoveries made by Oersted, Seebeck and Faraday concerning various connections between electricity, magnetism and heat (page 36). His unpublished lists of lecture topics regularly included subjects like the heating, luminous and chemical effects of electricity.[41]

The actual *unity* of different branches of natural philosophy, however, received only selected notice from Forbes. One example was Forbes's argument that the difference between light and radiant heat was no more than the difference between different colours of light. Light and heat were undulations in the same medium of ethereal particles (Forbes 1836b, 247–8; 1860, 956). He also accepted with Faraday that electrical action and chemical affinity were 'one and the same force' (Forbes 1860, 978). However, he did not think electricity and magnetism were the same, labelling Ampère's theoretical reduction of magnetism to electricity, for example, as 'arbitrary and improbable'.[42] Moreover, regarding the recent transformation of Joule's admittedly important work on the science of thermodynamics — an extension with obvious implications to one *looking* for unity — Forbes cautioned that 'a larger induction is still required to justify all the conclusions which the zealous promulgators of this comparatively new "mechanical theory of heat" have advanced'.[43]

Significantly, as Playfair had done earlier, Forbes stressed caution as another crucial feature of Baconianism: 'The uncompromising lesson of cautious induction which Newton thus practically taught, was better worth knowing than even the law of gravity'.[44] Playfair had warned against both the propensity 'to find in nature a greater degree of order, simplicity, and regularity, than is actually indicated by observation' and also the urge to extend a favourite subject to other areas:

> Thus, electricity has been applied to explain the motion of heavenly bodies; and, of late, galvanism and electricity together have been held out as explaining not only the affinities of Chemistry, but the phenomena of gravitation, and the laws of vegetable and animal life. It were a good caution for a man who studies nature, to distrust those things with which he is particularly conversant, and which he is accustomed to contemplate with pleasure. (Playfair 1860, 571–2.)

Glasgow

Glasgow differed significantly from both Cambridge and Edinburgh. Whereas Edinburgh had one professor of natural philosophy in the early-Victorian period, Glasgow, in effect, had three: William Meikleham from 1803, David Thomson, his replacement, from 1840 to 1845, and William Thomson beginning in 1846. Unlike Glasgow before 1846, both Edinburgh and Cambridge boasted prominent physicists who published important original research and actively participated in the reform of early-Victorian scientific institutions. Meikleham's and David Thomson's lack of publications and prominence obviously increases the difficulty of ascertaining their views. Nevertheless, one can still trace much of the changing character of Glasgow's natural philosophy course from Meikleham to David Thomson to William Thomson.

Meikleham had been a much-honoured undergraduate at Glasgow, winning a dozen prizes including one 'for the best explanation of Sir Isaac Newton's Rules of Philosophising' in his final session, 1792–93 (Addison 1902, 40–53). He became the assistant to the professor of natural philosophy, John Anderson, who was noted for drawing Glasgow artisans to the natural philosophy course with a non-mathematical experimental course separate from the mathematical course (Anderson 1837, 10; Muir 1950, 7–10). As professor, Meikleham retained the division, telling a Royal Commission in 1827 that of his seven hours of lectures per week, 'four of them are given upon a popular experimental course, a general view of Natural Philosophy, without the mathematical illustrations; three hours are devoted to those mathematical illustrations'. The experimental course could be taken by itself, but evidently the mathematical course could not: 'The mathematical lectures form a supplementary course to that which is called the experimental course.' ('Evidence' 1837, 117.)

His commitment to this arrangement affirmed Meikleham's belief in the

The educational matrix

importance of mathematics to natural philosophy, but also allowed him to uphold the Scottish ideal of openness in 'a town distinguished by its mechanical arts' and in a university not distinguished by undergraduates of high skill in mathematics. Meikleham testified to the Commissioners that he wanted his students to have as much preparation in mathematics as possible and at least to have taken the junior mathematics course, which included the elements of arithmetic and algebra, the first six books of Euclid, with mensuration, plane trigonometry and surveying. However, he also noted the importance of making the course 'accessible to all persons who have not perhaps made all the preparation desirable' ('Evidence' 1837, 120). A student's notes on Meikleham's mathematical lectures during the 1812–13 session are dominated by geometrical diagrams, many very intricate, dealing with the composition and resolution of forces, simple machines, and gravitational theory including the three-body problem.[45] A contemporary evaluation of Meikleham's course, presumably of its mathematical component, was 'that most of the students could not profit by his lectures on account of their ignorance of mathematics' (Murray 1927, 263).

Turning to material surviving from the sessions of 1818–19 and 1821–22, we find students' lecture notes and their essays summarising lectures. They are not very mathematical but are much concerned with the foundations and extent of natural knowledge. Though the course ranged over most branches of natural philosophy, Meikleham apparently confined the subjects of heat, electricity, magnetism and optics to the fifth month of the five-month session.

Meikleham defined the scope of his course. Whereas natural history involved describing phenomena and 'distinguishing the diversified objects of nature', in natural philosophy one 'labours for an acquaintance with the qualities or causes which generate these diversities. In short, the first is a knowledge of effect; the second of cause'.[46] Furthermore, whereas chemical change involved a body's *internal* alterations, mechanical change involved the motion of the whole body. 'Physical philosophy therefore or that branch of natural science which forms the object of our present course, contemplates the mechanical motions of bodies, or the forces by which these motions are produced'.[47] The study of forces and motion included statics and dynamics, this distinction depending on whether the forces were in equilibrium with one another or not.[48]

Like Forbes, Meikleham praised the inductive methodology and its introduction by Bacon, with whom 'the period arrived when she [science] was no longer to be held in bondage by her servants'.[49] According to Meikleham, 'in contemplating the diversified phenomena of nature; their motions, qualities, and mutual relations, we find that almost the only way of discovering causes, is by attending to effects'.[50]

However, in making such inductive inferences about causes, one had to be critically aware of the limited extent of human knowledge. In the case of the best-established cause, gravitation: 'That Gravitation is a power which exists in all matter may be asserted with a great degree of probability. About its nature we are less interested. The futility of speculations on this subject appears now very evident from the unsatisfactory results to which Philosophers have come.'[51] In addition to *gravitation*, Meikleham thought that terms like *attraction* and *inertia*, 'employed to indicate qualities observable only by their effects are generally of an arbitrary kind, that is they convey to us no notion of the real nature of the qualities or agents represented by them'.[52] Knowledge of universal gravitation was, in reality, knowledge only of the universal tendency of bodies to move towards one another.[53] The limitation of current knowledge also included magnetism ('that unknown property' or 'mysterious sympathy') and electricity (for which there was 'no satisfactory explanation').[54]

Meikleham was willing, however, to accept the probable truth of theories of nature's hidden realm, especially regarding nature's material side as opposed to its qualities, powers, or forces. Meikleham explained friction, for example, in terms of the composition of matter: 'Philosophers regard matter as composed of ultimate indivisible particles, indefinitely small; which never come into absolute contact.' Hence, 'matter is universally porous, which experiment has proven to be the case', and the resultant 'protuberances' and 'cavities' at one surface would interact with their counterparts on a second, producing friction.[55] Meikleham's concept of matter was also germane to his views in optics, where — though there was 'no *certain* knowledge' — Newton's particle theory of light 'is now the generally received opinion', because if light travelled like sound in undulations, there would be no shadows.[56] 'The doctrine of the porosity of matter' had now disproved the old idea that reflection was caused when particles of light actually 'impinged' against matter in the reflecting object. Meikleham favoured the view that light particles were reflected because of a repulsive force surrounding the particles of matter in the reflecting body's surface. He proposed a way of testing the validity of this view, but admitted that currently it 'is only a Hypothesis'.[57]

Analogies and the possible unity of the physical sciences were also part of Meikleham's course. His view that creative analogical thinking must be restrained by careful Baconian induction was implicit in his judgement on Kepler: 'This intense Philosopher had been from his youth remarkable for a fondness to descry analogies in nature's operations, and the many fanciful ones which he thought he found and which a fertile imagination abundantly furnished him are now justly esteemed truly ridiculous.'[58] He used analogies in two ways. As had Newton in his third rule of reasoning, Meikleham used analogy to justify the universality of gravitation.[59] As Forbes was later to do, Meikleham used analogy in discussing similarities between different areas of natural philosophy, especially electricity and magnetism.[60] Finally, going beyond analogy to the question of unity, he concluded a discussion of the Leyden jar with the statement:

> From these various facts, from the velocity of operation, and from many other attributes disclosed by late investigators, electricity, it seems probable, is the same as

The educational matrix

magnetism. And it is daily becoming more probable that the wonderfully varied effects of electricity, fire, heat, magnetism, and gravitation, may be at some future period, assigned to one common orgin.[61]

Meikleham did tell the Commissioners five years later in 1827 that in his course 'I consider the affections of heat, the doctrines of electricity and magnetism, which of late years have become particularly interesting' ('Evidence' 1837, 115).

At the beginning of the 1839–40 session, before illness forced his retirement, Meikleham was able to give enough lectures to provide an overview of his course.[62] He offered a mathematical course on mechanics, astronomy and optics, and an experimental treatment of heat, electricity and magnestism. Though Newton, Laplace and Lagrange were the ones who had 'chiefly enriched Natural Philosophy', Meikleham also called attention to more recent developments. Mentioning Davy and Faraday, he spoke of the 'great and numerous discoveries' made in electricity and magnetism during the last three decades. He cited the French physicist Biot as a source for 'recent discoveries on the polarization of light' and declared that 'Airy has given the best account of the undulatory theory of light'. Others acknowledged included the Frenchmen Poisson and Poinsot and the Cambridge graduates, Whewell (on mathematics and dynamics), Herschel (sound and light), H. Coddington (geometrical optics), and J. Pratt (mechanics). Meikleham stressed the importance of mathematics, telling his students that they might have difficulty in the natural philosophy course because of a lack of mathematics, including differential calculus. He continued to emphasise relations between branches of natural philosophy: 'Heat has a very obvious connection with light. Cohesion and repulsion ⟨are⟩ and chemical affinities are very similar to gravitation ... Also that by electricity or galvanism all substances may be put into such a state as to be affected by the magnet. And hence magnetism and electricity are modifications of the same ⟨substance⟩ agent.' His scepticism was present, too. Though heat, light, electricity and magnetism constituted the category of imponderable substances (as distinct from ordinary solids and fluids), Meikleham declared that it was actually 'incorrect' to call them imponderable, 'as we know them, not as substances, but as states of bodies'. Moreover, 'with regard to the nature of light we know nothing', and, indeed, at least in these few lectures, he did not endorse either the particle or undulatory theory, instead citing works by supporters of each (i.e. Biot and Airy).

Hence, at the end of his career, Meikleham was imparting a philosophical caution, an awareness of the importance of high-level mathematics, a knowledge of the older masters of mechanics and astronomy, and an introduction to the connections and unities in the world of 'imponderables' disclosed by nineteenth-century research.

Little evidence survives concerning David Thomson's conduct of the natural philosophy class from 1840 to 1845. Thomson had won many honours as a Glasgow undergraduate before going to Cambridge with high expectations —

expectations undermined by ill health, which prevented his reaching wrangler status in the Mathematical Tripos. It is likely that he followed the basic course he had learned from Meikleham, whom he was now 'assisting'. However, there was at least one change and perhaps a second. Poisson's *Mécanique* and Samuel Earnshaw's *Mechanics* replaced Newton's *Principia* and Laplace's *Mécanique céleste* in the 1844 *Calendar* as the required books for the examination leading to a degree with highest distinction. Also, Thomson may have emphasised Faraday's work more than Meikleham had. William Thomson credited David Thomson with introducing him to Faraday's ideas, presumably during the session of 1840–41 when William assisted David in experiments for the course (Thompson 1910, I:178–9).

In the Glasgow mathematics classes, William's father introduced changes in the 1830s. Appointed professor of mathematics in 1832, James Thomson attempted to raise the level of mathematical studies by introducing written examinations and offering class prizes to encourage additional work. Hence, in addition to the usual list of prizewinners for 'eminence' or 'proficiency' in the day-to-day work of the classes, Thomson awarded prizes for 'the solution of problems not previously known to students' and for 'vacation exercises'.[63] Moreover, his calculus textbook, published in 1831, urged advanced students of mathematics to explore physical subjects, especially mechanics and physical astronomy: 'These sciences, indeed, merit, in the highest degree, the subsequent attention of the Student; as they display, in the most striking manner, the power of the Modern Analysis, and manifest its triumphs in numerous discoveries of the most profound and valuable nature.' (J. Thomson 1831, iii.)

During the 1830s and early 1840s, therefore, both the mathematics and natural philosophy classes held up French mathematicians and mathematical physicists as an ideal. Under both Meikleham and David Thomson, the experimental side of the natural philosophy course evidently included a significant amount of modern work and focused on connections and unities in the physical world.

Following his father's example, William Thomson changed the natural philosophy course from what it had been under Meikleham. It was still divided into mathematical and experimental parts with students able to take the latter alone. In similar fashion to Meikleham, Thomson distinguished between natural history and natural philosophy, called mechanics the science of force, and divided it into statics and dynamics. The mathematics course still concentrated on topics like the resolution and composition of forces in a variety of circumstances. Not yet sufficiently mathematical, heat, electricity and magnetism fell to the experimental course. Many things were different, however. By varying the content of the experimental course from year to year, for example, Thomson was able to treat those subjects in vastly greater detail than Meikleham had in his one month per year.[64] Thomson utilised the Glasgow distinction between public and private students to allow stronger students to

The educational matrix

stay in the class a second year at a reduced fee to do more advanced work. He introduced written examinations and offered a variety of class prizes to reward advanced work. The laboratory he established allowed interested students to perform experimental research ancillary to his own research. His own research interests also influenced the topics proposed for university prize essays.[65]

The undergraduate years of W.J. Steele illustrate Thomson's innovations. As a public student during Thomson's first session, 1846–47, Steele paid a fee of four guineas and was placed third in the list of students noted for 'ability, diligence, and proficiency'. He also won two other prizes, one for excelling on monthly written examinations and another for excelling on written examinations on the mathematical theory of magnetism. At the end of the session he won a university medal as 'the most distinguished student in the examination for the degree of MA in mathematics and physics'. For the session of 1847–48, Steele was a private student, paid two guineas, and won three prizes: one for excelling on written examinations on the mathematical theory of electricity, a second 'for exercises performed by private students during the session', and a third 'for solutions of problems performed during last summer'.[66] In the spring of 1848 Steele calculated tables 'at my [Thomson's] request, of a new thermometric scale' and wrote a paper on Forbes's research on underground temperatures.[67]

Moreover, this was the very period in which Thomson's research was explicating analogical relations between branches of physical science and groping for an underlying unity. It was also the time of his major paper in 1851 on thermodynamics. As early as the session of 1852–53, Thomson was presenting such considerations to his students. He lectured not only on Joule's mechanical equivalent of heat, but also linked heat with electric currents and chemical activity. Heat was not caloric but the result of matter in motion. Both light and radiant heat consisted of vibrations in a medium resembling an incompressible elastic solid. Thomson explained Clausius' formulation of the second law of thermodynamics and declared that the 'vital' power is not a source of energy but 'a mere power of direction'.[68]

Thomson's influence on Glasgow students showed up in their writings. A prize essay for 1849, after discussing the similarities of radiant heat and light, states:

> Lastly, the undulatory theory, which is now generally considered as true, explains equally well the phenomena of all. These facts are sufficient to show, if we reflect at the same time on the simplicity and centralising affinity of nature's laws, where science and intellect can grasp them, that there is at least a strong hypothesis, for the apriori supposition, that heat, light, and electricity, are different modifications of one universal permeating substance ....[69]

An 1851 prize essay on 'The motive power of heat' discussed Carnot's work which had been 'rescued from oblivion and put forward with a most beautiful demonstration' by Thomson in his paper on the subject. This student discussed Oersted, Ampère and Seebeck, conjecturing 'that all the various phenomena of Heat (and among the rest 'Motive Power') may be developed from Electricity alone'.[70] An 1855 essay on 'Mechanical energy of electricity in motion' discussed the first and second laws of the 'Dynamical Theory of Heat'.[71]

John Ferguson took Thomson's course in the 1859–60 session and was appointed professor of chemistry at Glasgow in 1874. Later, he recalled Thomson's lectures:

> His impulse was to correlate phenomena and arrive at the principle underlying them, and this gave him a certain impatience with branches of science which were still in the observational stage, and had not yet come under mechanical laws. Hence the most brilliant and weighty part of his course was at the end, when he summed up his teaching and generalised upon energy and the correlation of the physical forces, showed us Faraday's experiments on the conversion of electricity and magnetism, and Joule's conversion of work into heat. (Ferguson 1908, 281.)

Hence, pursuing pedagogical policies suggested by his father's career, Thomson changed the environment for natural philosophy in Glasgow. Though some accounts seriously underestimate the level of Meikleham's course (Thompson 1910, I:183), Thomson's was undoubtedly a more detailed and up-to-date portrayal of the current state of both experimental and theoretical natural philosophy than Meikleham's had been. Thomson valued openness as much as Meikleham had, but contrived more ways than he for both ordinary and talented students to thrive in the same class. On the other hand, Meikleham's course had been richer than Thomson's in philosophical reflection on the foundations of natural knowledge.

Thomson's course also contrasted with Forbes's. Forbes, unlike Meikleham, had been part of the reforming spirit of the early 1830s, and he had designed his course accordingly. Thomson's course, however, better represented the reformed state of science at mid-century. As active in the new institutions of science as Forbes, Thomson more clearly reflected the ideal of original research and the ideas of mathematical physics. Whereas Forbes's students themselves only occasionally made use of the experimental apparatus of the natural philosophy class, Thomson formed an ongoing team of undergraduate research assistants. By 1850 Forbes was giving fewer lectures on mathematical physics, and the mathematical part of his course still emphasised mechanics, astronomy and light. By 1850 Thomson was placing area after area of the mathematisation of heat, electricity and magnetism before advanced Glasgow undergraduates. Whereas Forbes never accepted the new science of thermodynamics, Thomson was, of course, one of its principal creators.

Forbes and Thomson differed most over the unity of nature. Forbes looked forward to the formation of mechanical theories in the *different* branches of natural philosophy and proclaimed that in a few specific instances the unity of

The educational matrix

supposedly separate aspects of nature had been demonstrated. Thomson, stimulated in large measure by Faraday's experiments on the rotation of polarised light in a magnetic field, was as early as 1846 seeking a highly unified conception of physical nature, envisaging an elastic solid as a *single* mechanical system whose various motions or strains could 'represent' phenomena of electricity, magnetism, light and radiant heat. Such thoughts — crucial to Thomson's deepest speculations for decades — lay behind his short paper, 'On a mechanical representation of electric, magnetic, and galvanic forces', a copy of which he sent to Forbes at the end of 1846.[72] It elicited little enthusiasm from Forbes, who merely thanked Thomson for the paper, noting that 'though as you say it is in its present form somewhat too concentrated'.[73] They still differed in the 1850s. In his history of modern science, Forbes carefully qualified and limited the significance of Faraday's experiment on polarised light (Forbes 1860, 981–2). By contrast, Thomson used it in yet another deep probe for nature's dynamical unity, one which directly influenced Maxwell's developing ideas on electricity and magnetism (Thomson 1856; Siegel 1981, 244–6).

Scotland and Cambridge

What kinds of influence did these three educational systems have on the research of the early-Victorian Cambridge physicists? To suggest answers, this section summarises differences and similarities among the systems, next discusses the careers of the three universities' second-level graduates, and then turns to the major figures themselves. Was theirs truly 'Cambridge' physics, or — given the importance of Maxwell and Thomson — was it, more properly, 'Scottish' physics?

On the surface, mathematics would appear to be an area of great difference in the educational systems. However, Meikleham did expect his advanced students to read Laplace's *Mécanique céleste*, and Forbes organised his course in a way that would promote mathematical physics. James Thomson taught calculus at Glasgow during the 1830s and 1840s and was replaced in 1849 by a high wrangler, Hugh Blackburn. At Edinburgh, a senior wrangler, Philip Kelland, replaced William Wallace in 1838. Recently, scholars have emphasised the similarity, not the difference, between the mathematics taught at Cambridge and that at Edinburgh (Becher 1980a, 46–7; Wise 1982, 175–8). Both Thomson and Maxwell published mathematically sophisticated papers before coming to Cambridge, and Harman's chapter argues that there was no disjunction in Maxwell's mathematical studies as he moved from Edinburgh to Cambridge. Even Rankine, who did not go on to Cambridge and who studied at Edinburgh before Kelland arrived, was a strong enough mathematician to be regarded as a mathematical physicist.[74] Green was a mathematical physicist before coming to Cambridge. Meikleham, Forbes and Hopkins would all have endorsed John Herschel's dictum that numerical precision was 'the very soul of science and its attainment affords the only criterion, or at least the best, of the truth of theories, and the correctness of experiments'.[75]

Meikleham, Forbes and Hopkins appeared also to agree that it was possible to discover mechanical properties of nature at the microscopic level. Meikleham spoke of corpuscles of light and of matter's ultimate particles, while Forbes and those at Cambridge regarded the ether as an actually existing mechanical medium. None was a naive realist in these matters, all evidencing some degree of scepticism and all acknowledging the usefulness of tentative or hypothetical explanations. Meikleham, for example, regarded his theory of reflection as hypothetical, and Forbes regarded Ampère's electrical theory of magnetism, like Newton's optical theory of fits, as a useful hypothesis (Forbes 1860, 974).

Clearly, Scotland and Cambridge included different branches of natural philosophy in their respective curricula. If Cambridge students analysed optics more extensively than the Scottish, the latter paid more attention to heat, electricity and magnetism. Another difference lay in the Scottish emphasis on analogies, connections and unities within a context of critical reflection on the possibility and scope of human knowledge. Meikleham and Forbes may not always have agreed, but these issues were a regular feature of their courses — a feature sustained in his own way by Thomson in the late 1840s and 1850s. To be sure, even though the Tripos largely ignored such issues, there was much discussion in Cambridge of one analogy — that between sound and light. Note the difference, however. Forbes cited this as having been a fruitful analogy for Thomas Young, and viewed himself as creatively pursuing the similar analogy between light and radiant heat. On the other hand, in Cambridge, except for Airy's textbook, the sound-light similarity had become an analogy of pedagogy, not discovery. Once acoustics was used to help explain optical concepts, Tripos questions asked about light, not about the relationship between sound and light.

There is one obvious fact about both the top wranglers at Cambridge and the prize-winning students in the Edinburgh and Glasgow natural philosophy classes: the great majority of them were destined to be neither mathematicians nor natural philosophers. At Glasgow and Cambridge, more students went into the church than into any other career. About seventy per cent of the Glasgow students and about forty per cent of those at Cambridge became clergymen. Unlike their Glasgow counterparts, however, nearly half of the Cambridge clergymen divided their careers between the church and either secondary education or a college fellowship. No career besides the church claimed a large number of Glasgow students, but many Cambridge students became college fellows (thirty per cent) or secondary teachers (seventeen per cent), or went into the law (nineteen per cent). Only about five per cent of the Glasgow students became professors, whereas nearly fifteen per cent of the Cambridge students did so.[76] Similar information on Edinburgh's top students is not easily attained.

Table II lists several Cambridge and Scottish students. Though they were of secondary importance in the history of physics, they were the élite students in

Table II Second-level students: Glasgow, Edinburgh, and Cambridge[a]

Student	NP class	FRS	FRSE	Career[c]	Math.	Astr.	Mech.	Hydr.	Opt.	Snd.	Heat	Elec.	Mag.	Other
(a) Glasgow (1833–34 to 1854–55; 10 of 167 students)[b]														
D. Thomson[e]	1833–34			Prof. nat. phil.										15
J.D. Hooker	1835–36	1847	1883	Kew Gardens										
J.R. McClean	1835–36	1869		Engineer										1
J. Thomson	1839–40	1877	1875	Eng./Prof. eng.			2	2			1			
J. Napier	1839–40		1877	Shipbuilder										
J. R. Napier	1839–40	1867	1874	Shipbuilder										3
J. Kerr[f]	1845–49	1890		Education										
D. M'Farlane[g]	1847–48			W. T.'s assist.										
W. Craig	1853–54		1875	Medicine										
F. McClean[h]	1853–55	1895		Engineer										
(b) Edinburgh (1833–34 to 1854–55; 13 of 87 students)[i]														
J.A. Longridge[j]	1833–35			Engineer										
T. Cleghorn	1835–36		1856	Law										
E.C. Batten	1835–36		1857	Law										
W.J.M. Rankine	1836–38	1853	1850	Eng./Prof. eng.	1		1				4			3
J. Turnbull	1837–38		1879	?										
W. Swan	1837–38		1848	Prof. nat. phil.					4					
J.A. Broun	1838–39	1853		Private observatories				1					10½	1½
G.E. Day[k]	1839–40	1850		Med./Prof. med.										1
W. Welsh	1841–42	1857		Kew Observatory							2		1	3
W. Turner	1844–45	1877	1861	Prof. anat.										1
B. Stewart	1845–46	1862	1878	Prof./Kew Observatory	1	1			1		2			1
J. Sime[l]	1847–48		1870	Education										
R.M. Ferguson[m]	1851–52		1868	Education										

Student	NP class MT	FRS	FRSE	Career	Math.	Astr.	Mech.	Hydr.	Opt.	Snd.	Heat	Elec.	Mag.	Other
(c) Scotland–Cambridge (7 of 7 students)[n]														
A. Smith	1831–32 (G) sw 1836	1856		Law	14	1	7	1	4					
D.F. Gregory	1832–33 (E)[o] 5w 1837			College[p]	21		2				1			
A. Sandeman	1838–39 (E) 4w 1846			Prof. nat. phil. & math.										
W.A. Porter	1844–45 (G) 3w 1849			Law/Educ.										
J. Porter	1846–47 (G) 9w 1851			College					1					
C.A. Smith	1852–53 (G) 2w 1858			Government										
W. Jack	1852–53 (G) 4w 1859		1875	Prof. math.								$\frac{1}{2}$	$\frac{1}{2}$	
(d) Cambridge (1830 to 1854; 41 of 242 students)[q]														
C. Pritchard	4w 1830	1840		Educ./Prof. astr.										
S. Earnshaw	sw 1831			Coll./Church										
T. Gaskin	2w 1831	1839		Coll./Educ.				2	5					
G. Budd	3w 1831	1836		Prof. med.										
G.E. Paget	8w 1831	1873		Med./Prof. med.										
J.H. Pratt	3w 1833	1866		Church	1	1	1	1	$7\frac{1}{2}$	$\frac{1}{2}$	3			
P. Kelland[x]	sw 1834	1838	1839	Prof. math.	1		8							
R. Main	6w 1834	1860		Royal Obs.		5								

Student	MT	FRS	FRSE	Career	Math.	Astr.	Mech.	Hydr.	Opt.	Snd.	Heat	Elec.	Mag.	Other
(d) continued: Cambridge (1830 to 1854; 41 of 242 students)[q]														
S.S. Greatheed	4w 1835			Church	1	1								
W. Walton	8w 1836	1837		College	16	1	4							
J.J. Sylvester	2w 1837	1839	1874	Prof. math.	16			1						
A.J. Ellis	6w 1837	1864		Philologist					1					
M. O'Brien	3w 1838			Prof. nat./phil.	1	$3\frac{1}{2}$	$2\frac{1}{2}$	1	5					
R. Potter[r]	6w 1838			Prof. nat. phil.		1		2	8					
P. Frost	2w 1839	1883		College	4									
G.W. Hearn	6w 1839			Professor	7	2	1		1				1	
R.L. Ellis	sw 1840			College	19	2	1						1	
H. Goodwin	2w 1840			Coll./Church	9		2	2	3					
A. Cayley	sw 1842	1852	1865	Law/Prof. math.	113		6		3					
F. Fuller	4w 1842		1867	Coll./Prof. math.										
J.C. Adams	sw 1843	1849	1849	Coll./Prof. math.		9								
F. Bashforth	2w 1843			Church/Prof.	1		1							
T.M. Goodeve	9w 1843			Law/Prof.										
E. Walker[u]	8w 1844	1869		Coll./Law										
S. Parkinson	sw 1845	1870		College		1								
R. Peirson[v]	3w 1845			Coll./?										
F.W.L. Fischer	4w 1845	1855		Prof. nat. phil.		1								
H. Blackburn	5w 1845		1850	Prof. math.										2
J.B. Cherriman	6w 1845			Prof. nat. phil.	1									
W.P. Wilson	sw 1847			Prof. math.					1					
I. Todhunter	sw 1848	1862		College										
W. Scott	3w 1848			Prof. astr./Educ.		2								
H.M. Jeffery	6w 1849	1880		Education	1									
W.H. Drew	8w 1849			Educ./Prof. math.										
W.H. Besant	sw 1850	1871		College			3							
H.W. Watson	2w 1850	1881		Educ./Church										
J. Wolstenholme	3w 1850			Coll./Prof. math.	4									
R.B. Hayward	4w 1850	1876		Education										
N.M. Ferrers	sw 1851	1877		College	21									
E.J. Routh	sw 1854	1872		College	3	1	2							
B.W. Horne	4w 1854			College	1									

a This table includes students who became fellows of either the Royal Society of London or Edinburgh, who became professors of mathematics or natural philosophy at a British university, or who were publishing articles in mathematics or natural philosophy. Use of the term *second-level* is not intended to imply, of course, that J.D. Hooker, for example, was a second-level botanist. Sources: Addison 1898; Addison 1902; Addison 1913; *Catalogus togatorum 1836–54*; *Glasgow University prize and degree lists, 1834–1863*; *Edinburgh Academy Register* 1914; *Edinburgh University Almanack 1833–34*; class competitions in the natural philosophy class for 1839, 1841, 1842, 1845, 1846, and 1850 (Edinburgh University Library); Edinburgh University class lists for natural philosophy, 1833–34 to 1854–55 (Edinburgh University Library); *The Scotsman*, 23 April 1836, 22 April 1837, 28 April 1838; Tanner 1917; Venn 1940–54; *Record of the Royal Society* 1940; Royal Society of London 1867–79; lists of fellows in the *Transactions of the Royal Society of Edinburgh*.

b The total of 167 includes all the prize-winners in the natural philosophy classes (173 students) minus relevant students in the Scotland–Cambridge list (4) and minus W. Thomson and W.J. Steele. There was no distinction in the Glasgow prize lists between students who knew calculus and either those who did not (as at Edinburgh) or those who knew it very well (as at Cambridge, by implication). There was a total of some 1300 students in the class during this twenty-two year period.

c The student's main career is listed, though two careers are given for some, for example, James Thomson who was an engineer and then a professor of engineering.

d These are the papers published within twelve years after the student left the natural philosophy course or, in the case of a wrangler, after he took the Mathematical Tripos. The assumption is that this is about the length of time when the student's university education would be most influential on his choice of research topics. In fact, including papers over a longer period would seldom alter the distribution of papers among the topics. Papers which discuss two of the topics have been divided between them, one-half to each.

e Thomson was a senior optime in the Mathematical Tripos of 1839. He published an encyclopaedia article on acoustics later in his career.

f Kerr's famous work on the 'Kerr effect' was not published until the 1870s.

g M°Farlane did publish two papers on heat in the early 1870s.

h McClean was 27th wrangler in the 1859 Tripos.

i The total of 87 includes Forbes's 'advanced' prize-winners for 1833–34 (eight students), plus all the prize-winners from 1834–35 to 1837–38 (38), plus the first-division prize-winners from 1838–39 to 1854–55 (44), minus Sandeman, Tait and Maxwell. There were probably around 180

j total of 2636 students in the class during this twenty-two-year period. Lewis Gordon won no prize in his year, 1833–34, but became FRSE in 1846 and was professor of engineering at Glasgow.

j Longridge attended Cambridge later but did not take a degree. In the 1850s he published papers on ventilation in mines and in 1858 won the Telford medal.

k Day was 23rd wrangler in the 1837 Tripos.

l Sime assisted Forbes during the 1854–55 session.

m Ferguson later published papers and a book on electricity and electrical technology.

n This list includes all those who were prize-winners in the natural philosophy class at Edinburgh or Glasgow from 1833–34 to 1854–55 and who were also placed among the top ten wranglers at Cambridge. It also includes all the top ten wranglers from 1830 to 1854 who also attended a Scottish university. There evidently were no students from Aberdeen or St. Andrews among the top wranglers during this period. The addition of Thomson, Steele, Tait, and Maxwell would make a total of only eleven students in this exceedingly select category. I have found ten or eleven other prize-winners from Glasgow or Edinburgh who attended Cambridge, but who were not placed in the top ten in the Tripos.

o According to R.L. Ellis's obituary of Gregory, this is the session which Gregory attended the University of Edinburgh (Ellis 1845b). Neither Ellis nor the *Dictionary of National Biography* states (and I have not been able to determine) whether he attended the natural philosophy class. If so, he was not among the prize-winners. (*Edinburgh University Almanack* 1833, 138.) He was, therefore, the only one of this group of eleven not to win a prize in a natural philosophy class. His chief teacher at Edinburgh was the professor of mathematics, William Wallace.

p That is, a fellow of a Cambridge college.

q The total of 242 includes the top ten wranglers from 1830 to 1854 (253 students — there were ties for tenth), minus the relevant students in the Scotland–Cambridge list (five), and minus Green, Stokes, Thomson, Steele, Tait and Maxwell. During this twenty-five-year period, there were 952 wranglers and 2979 altogether who took honours in the Mathematical Tripos (i.e. wranglers, senior optimes, and junior optimes).

r Several wranglers wrote textbooks on Tripos subjects, which are not noted here.

s Kelland also wrote a book on heat (Kelland 1837b).

t Potter published several papers before entering Cambridge, which are not listed.

u Walker won the Adams prize in 1865 with an essay on magnetism.

v Peirson's one publication entry is his winning Adams prize essay of 1850.

The educational matrix

the educational matrix along with Stokes, Thomson and Maxwell. They reflected differences between Cambridge and the Scottish universities. One can see, for example, the closer ties of Glasgow and Edinburgh to engineering and medicine. Whereas no Scottish minister appears in the list, there are Cambridge clergymen, no doubt encouraged in their research by that overlapping career structure of college, church, secondary school and university. Most obvious is the greater length of the Cambridge list and the greater proportion of those on it who were publishing — products of the specialised, professional Cambridge setting described earlier.

Moreover, as we might expect, Scottish students usually published on different subjects than Cambridge graduates. Whereas during this period apparently not a single student from Glasgow or Edinburgh published substantial work in pure mathematics, many Cambridge graduates did. There were no Scottish counterparts to Cayley and Sylvester, and this was the period when Cambridge graduates began taking over Scottish chairs of mathematics. On the other hand, of the Cambridge men, only Kelland published anything of note in heat, electricity or magnetism, and his work was confined to heat.[77] Ellis's one paper on magnetism, for example, discussed a particular mathematical problem more then magnetism itself (Ellis 1845a). The most important Scottish work was concentrated in exactly those areas. Rankine began publishing his far-reaching physical ideas on heat in the late 1840s, and Stewart his in the late 1850s. The one item by James Thomson recorded under heat in Table II was his basic paper on pressure lowering the freezing point of ice. Broun's and Welsh's papers presented the results of their important work in magnetic and meteorological observatories. The Scots, however, could not match in quantity or quality the Cambridge work in astronomy, mechanics and optics.

Unlike the other categories in Table II, the Scotland-Cambridge section lists all the students in its category. Of the several thousand students at Cambridge, Edinburgh and Glasgow during this period, only eleven did exceptionally well both in Scotland and at Cambridge. The list is so small that one hesitates to generalise too much. Nevertheless, it is a fact that all but two or three of these students met the criteria for inclusion in the élite groups in the other categories — a vastly higher proportion than in those other categories. (See note *a* to Table II for the criteria.) Also, there obviously was no simple deterministic connection between a high-level, combined Scottish–Cambridge educational background and later research; these seven were not attempting research similar to Thomson's and Maxwell's. Only Smith and Gregory published significantly, and, even though Smith later did important research on magnetism, their early research careers — unlike Thomson's and Maxwell's — emphasised typically Cambridge subjects, especially pure mathematics. Gregory's one article listed under heat, for example, concerned a mathematical point found in Fourier's work on heat, rather than the physics of heat (Gregory 1839).

Returning to the major Cambridge physicists mentioned at the outset of this chapter, what can we now state about any connections between their educational background and their later research?[78] First, there were broad differences in research between the Scots, Thomson and Maxwell, and the non-Scots, Green and Stokes. Moreover, placing these major figures in their educational context underscores the pivotal importance of Thomson's ideas, the uniqueness of Thomson and Maxwell, and the subsidiary character of Tait's early research.

Before coming to Cambridge, the self-educated Green published on electricity and magnetism, drawing on work by Poisson and Laplace. However, after coming to Cambridge, he focused his few remaining papers on Cambridge subjects: hydrodynamics, sound and light. In a manner reminiscent of Airy's textbook treatment of light and sound, Green published on the reflection and refraction of sound before publishing a similar paper on light. As the editor of his collected papers explained, the second paper was intended 'to do for the theory of light what in the former paper has been done for that of sound' (Green 1871, vii).

Stokes entered Cambridge from Bristol College, whose principal was a Cambridge graduate and whose mathematics curriculum followed 'the plan at present pursued in Trinity College, Cambridge'. In a student's first two years at Bristol, he studied mathematics, including differential and integral calculus, statics and dynamics, and the first three sections of Newton's *Principia*. In the third year, 'he will be occupied with the principles of Hydrostatics and Optics, and with the remainder of the first book of the *Principia*, as well as with Spherical Trigonometry and Physical Astronomy' (*Outline* 1830, 9–10). By the time Stokes took his degree in 1841, therefore, he had spent seven years with the Cambridge cluster of physical and mathematical subjects. In 1841 he published the first in a series of papers on hydrodynamics that led to papers in the mid-1840s on elastic solids and on the nature of the luminiferous ether. In addition to hydrodynamics, his papers over the next few years dealt with sound, the figure of the earth, and topics in optics, including his discovery of fluorescence in the early 1850s. Stokes's early career contrasted with Thomson's.

However, the contrast was not in their use of mechanical explanations. It was only two years before Thomson's 'mechanical representation' paper of 1847 that Stokes extended his hydrodynamical research into his 'glue-water' concept of the luminiferous ether, according to which the ether possessed both fluid and solid properties (Wilson 1972). Stokes thought Thomson was employing the same concept in his 1847 paper: 'Perhaps the jelly-like fluid that we once spoke of may be made in your hands to explain the law of the mutual action of electric currents and the phenomena of the induction of these currents.'[79] The wave theory of light was well enough established that Stokes could regard his glue-water theory as approximating the physical reality of the ether. No similar theory in electromagnetism was so well established, but Thomson hoped his mechanical 'representation' would lead to one.[80]

The educational matrix

Obviously, Stokes and Thomson did research in different areas. Though they corresponded about one another's research and though they wrote occasional papers in the other's area, Stokes pursued the Cambridge subjects while Thomson investigated heat, electricity and magnetism. As Thomson replied to Stokes' letters on certain mathematical issues:

> I have been for a long time thinking on subjects such as those you write about, and helping myself to understand them by illustrations from the theories of heat, electricity, magnetism, and especially galvanism; sometimes also water. I can strongly recommend heat for clearing the head on *all* such considerations, but I suppose you prefer cold water.[81]

A deeper difference lay in Thomson's enduring search for a unified theory of physical phenomena. Stokes attempted nothing of the kind. Of course, one would like to have a more definitive record of Meikleham's views on the unity of natural phenomena. But it does appear that though Meikleham and Forbes both saw themselves as cautious Baconians, Meikleham's course in the 1830s emphasised physical unity more than Forbes's. Meikleham seemed to interpret experimental results as indicating unity, but not as sufficient to reveal the underlying nature of heat, electricity and magnetism. Forbes's interpretation seemed to be that without a much better, experimentally supported understanding of their underlying nature, one could not know whether they were unified or not. Meikleham's course certainly emphasised unity more than the Cambridge curriculum. To the degree these views, associated with Meikleham's course, influenced Thomson, they would help explain Thomson's enthusiasm for Faraday's ideas, as presented to him first by David Thomson and later by Faraday himself.

Maxwell, as Knudsen's, Harman's and Siegel's chapters point out, was following a Thomsonian programme in his research on electricity and magnetism. Though he had not formally studied electricity and magnetism at either Edinburgh (see Table I) or Cambridge, Maxwell completed his paper 'On Faraday's lines of force', read to the Cambridge Philosophical Society in two parts in December 1855 and February 1856. His method of treating Faraday's lines of force obviously owed much to two of Thomson's papers, one on the analogy between statical electricity and the conduction of heat and the other on the 'mechanical representation' of electrical and magnetic phenomena. The imaginary fluid invoked by Maxwell to interpret Faraday's lines 'was confessedly adopted from the analogy pointed out by Thomson in 1843 [*sic*] between the steady flow of heat and the phenomena of statical electricity' (Tait 1879, 334). Maxwell argued that by showing that the mathematical laws of electrical attraction were 'identical in form' to those of a phenomenon not caused by attractive forces, Thomson had called into question the prevailing view that electrical phenomena were caused by attractive forces. Similarly, Maxwell used the flow of an imaginary fluid to foster Faraday's theory of lines of force as an alternative to the accepted use of attractive forces to explain electromagnetic phenomena. Like Thomson's paper of 1847, Maxwell's sought a unified mechanical representation of electricity and magnetism.

What about Forbes's influence on Maxwell? Though Forbes had not lectured on electricity and magnetism while Maxwell was in his class, one would expect Maxwell to have picked up much of Forbes's thinking as a result of their close friendship. In that regard, electricity and magnetism were, at least, subjects important to the Edinburgh curriculum, in contrast to Cambridge. However, despite Forbes's praise for Faraday's experimental demonstrations of connections between electricity and magnetism, he apparently had little regard for Faraday's theory of lines of force, which would be the basis of Maxwell's unified view of electromagnetism. He did not mention them, for example, either in his lectures in 1845–46 or in his extended discussion of Faraday in his *Dissertation* on the history of recent science.[82] Forbes did lecture on heat and light to Maxwell's class in the 1847–48 and 1848–49 sessions. However, to judge from Forbes's titles for the lectures, he did not discuss radiant heat, the polarisation of heat, or the wave theory of light.[83] Once again, one must assume that Maxwell knew Forbes's views on the unity of light and radiant heat. But Forbes evidently did not emphasise such views in the lectures Maxwell heard. Forbes's lectures for 1847–49 thus contrast interestingly with Thomson's 'mechanical representation' paper of 1847, and it is a contrast that makes it all the more correct to place Maxwell in a Thomsonian research programme, rather than in one defined by Forbes, or by William Hopkins.

Tait's highly derivative early publications eventually seemed to advocate a wedding of William Rowan Hamilton's mathematics to Thomson's and Maxwell's physical conceptions. Tait went on from Cambridge to become professor of mathematics in Queen's College, Belfast, and in 1860 to replace Forbes at Edinburgh. Tait's earliest papers were written jointly with Thomas Andrews, professor of chemistry at Belfast, whom he assisted with experiments on ozone. By 1860 Tait had heartily endorsed the conservation of energy, making it the centrepiece of his inaugural lecture that year. The transformation of nature's forces into one another in equivalent amounts afforded 'a strong argument for the ultimate identity of these forces' (Tait 1860, 25). He also endorsed Hamilton's quaternions, whose simplifying power would 'render progress easier' in natural philosophy (ibid. 36). In 1859 Tait published the first of a series of papers demonstrating quaternions' usefulness to modern optics, electricity and magnetism. Though he did not explicitly discuss Thomson's and Maxwell's work in electricity and magnetism in his inaugural lecture, two years later he related quaternions to Thomson's 'mechanical representation' paper and to Maxwell's recently published paper, 'On physical lines of force' (Tait 1862, 623). In 1863 he declared that 'it may now be asserted that the next grand extensions of mathematical physics will, in all likelihood, be furnished by quaternions', and he cited Thomson's paper of 1847:

The educational matrix

> Thomson has shown ... that the forces produced by given distributions of matter, electricity, magnetism, or galvanic currents, can be represented at every point by displacements of such a solid producible by external forces. It may be useful to give his analysis, with some additions, in a quaternion form, to show the insight gained by the simplicity of the present method. (Tait 1863, 117.)

In the light of the early research careers of these five physicists, in the light of the careers of the several dozen 'second-level' students, and in the light of the differences in physics education from Glasgow to Edinburgh to Cambridge, what is the historically accurate label for this early-Victorian school of physics? The answer, it seems to me, is to recognise the presence of *two* principal schools. First, there was the large *Cambridge* school of mathematics and mathematical physics, which was shaped by Whewell, Airy and Hopkins, which drew the forty-year-old Green into its sphere, of which Stokes was the central figure, and to which were attracted the mathematically minded Scots, Smith and Gregory. Secondly, there was the much smaller *Glasgow* school of mathematical natural philosophy, which was shaped by Meikleham, of which William Thomson was the central figure, and to which were attracted the physically minded Edinburgh students, Maxwell and, eventually, Tait. Perhaps its only other members, in 1850, were Thomson's brother James, and maybe Thomson's students, Steele and Kerr.

Such an analysis suggests that Forbes's Edinburgh constituted a third school, in the event one of lesser importance. One could see its ancestors as Leslie and Brewster and its main beneficiary as Balfour Stewart. Like Leslie and Brewster before him and Stewart after, Forbes gained fame as an experimental, not mathematical, physicist. Moreover, his conceptual conservatism — which brings to mind the conservatism of Leslie and Brewster criticised by the young Forbes — marked the school at a key juncture in the history of physics. Unwilling to follow Thomson's jumps to conclusions, Forbes interpreted Faraday's and Joule's findings with great circumspection and saw unity chiefly where it was demonstrated by his own overwhelming experimental evidence. In fairness, however, we should recognise the possibility that Forbes's conservatism might have carried the day. In the early 1850s, it was not unreasonable to predict that Thomson's brilliant theories about heat, electricity and magnetism would be as short-lived as his brilliant theory of vortex atoms later turned out to be. In that case, this book might have placed Forbes's Edinburgh school of physics at centre stage, relegating the impulsive Thomson and Maxwell to the wings.

Recognising three schools leads to further insights. Despite the high degree of professionalism at Cambridge in the 1830s and Forbes's promotion of professionalism at Edinburgh, it seems that it was the obscure Meikleham's viewpoint which lay closest to the methodology that revolutionised British physics. Consider Stokes's and Thomson's different approaches to different branches of physics in the 1840s. Now, instead of seeing them as two members

David B. Wilson

of the same Cambridge school who happened to be interested in different subjects, we can see them as members of two different physical traditions who happened to take the same Tripos. One's monumental efforts went into elaborating what had been a chief British research programme in physics during the preceding two decades; the other's into developing his tradition into what was to be the chief research programme for several subsequent decades. Consider Maxwell's career. We can now see that around the time he moved geographically from Edinburgh to Cambridge, he may have been moving in his field of natural philosophy from Edinburgh to Glasgow. Given his long association with Forbes, it may have seemed to him a long trip. But it would have seemed even longer for someone with a strictly Cambridge education, like Stokes. Indeed, no Cambridge-only graduate during this period took up research in electricity and magnetism, much less tried to construct a unified theory of physical nature. Finally, we can better appreciate how great a transformation of early-Victorian institutional–conceptual patterns was wrought by the acceptance of Thomson's and Maxwell's ideas. For example, Forbes's long-time student and assistant, Stewart, became a leading populariser of thermodynamics, and Tait, Forbes's successor, made energy physics crucial to the Edinburgh natural philosophy class for the last four decades of the century. Moreover, Stokes's position at Cambridge changed dramatically. Cambridge officially embraced the new theories in the 1870s and 1880s, and the Cavendish Laboratory — headed successively by Maxwell, Lord Rayleigh and J.J. Thomson — became the late-century focus for Cambridge physics education (Wilson 1982). As Buchwald's chapter shows, the late-century Maxwellians were mostly Cambridge wranglers. Still lecturing on the old Cambridge subjects, Stokes, the pre-eminent physicist resident in Cambridge in the 1850s, had by the 1880s and 1890s declined to a distinctly secondary role in Cambridge physics education.

Conclusion

This chapter, in the first instance, has been a comparative study of physics education at Cambridge, Edinburgh and Glasgow Universities. It discusses not only similarities, but also significant differences among the universities, thus supporting the general importance of recognising regional and institutional variation within the history of scientific ideas. Specifically, the chapter shows that we should not let the early-Victorian 'Cambridge' physicists' common Tripos experience obscure the great differences in their respective educations in physics. Educational influence was obviously not the only factor shaping the thought of these physicists. Nevertheless, specific patterns of influence are revealed by detailed study of their educational matrix. Postulating the existence of three schools of physical thought, the chapter points to correlations between the institutional–regional diversity, on the one hand, and variation within the research of the 'Cambridge' group of physicists, on the other. The contrast

The educational matrix

between Stokes and Thomson in the 1840s thus emerges as a central conflict between traditions, a conflict heralding the imminent ascent of one tradition over the other, previously supreme. Consequently, not only does the educational matrix merit historical investigation in its own right, but the resultant study provides an introduction and backdrop for the remaining chapters in this volume.

Acknowledgements

The chapter has benefited from comments by Peter Harman and several participants at the Grasmere Conference, especially Paul Theerman, John L. Heilbron, Daniel M. Siegel and Keith Hutchison. I am grateful to Dr Peter Swinbank and Professor J.M.A. Lenihan for assisting me with my research in Scotland, and to the University of Glasgow for making me an honorary research fellow.

My research has been financially supported by the National Science Foundation, the American Philosophical Society, the Royal Philosophical Society of Glasgow, Iowa State University, and the Mechanical Engineering Department of Iowa State University. My attendance at the Grasmere Conference was made possible by a travel grant from the American Council of Learned Societies.

Notes

1. For standard histories of the universities, see Winstanley 1955 and Davie 1961. Anderson 1983 questions parts of Davie's thesis.
2. Report of the Board of Mathematical Studies, 19 May 1849, Cambridge University Archives; Airy 1826.
3. Report of the Board of Mathematical Studies, 19 May 1849. Hopkins regretted that Challis' 'admirable lectures' did not have a larger audience (Hopkins 1841, 22).
4. The professor of chemistry, James Cumming, gave non-mathematical lectures on heat, electricity and magnetism. Stokes attended Cumming's lectures in 1844, three years after he had taken the Tripos. (Stokes, Notes on Cumming's lectures, February to May 1844, Cambridge University Library, Stokes Collection, Add. MS 7656, PA29 to PA31.)
5. The most complete set of notes by a Hopkins student that I know of are Stokes's, and they include nothing on heat, electricity and magnetism. (Stokes, Notes on Hopkins' lectures, Cambridge University Library, Stokes Collection, Add. MS 7656, PA2 to PA24.)
6. Report of the Board of Mathematical Studies, 19 May 1849, Cambridge University Archives.
7. Report of the Board of Mathematical Studies, 3 June 1850, Cambridge University Archives.
8. Report of the Board of Mathematical Studies, 20 May 1852, Cambridge University Archives.
9. Stokes, Notes on Hopkins' lectures on 'Mechanics, machines with friction, dynamics part 1', Cambridge University Library, Stokes Collection, Add. MS 7656, PA6.
10. Stokes, Notes on Hopkins' lectures on 'Sound and light I', Cambridge University Library, Stokes Collection, Add. MS 7656, PA20.
11. Stokes, Notes on Hopkins' lectures on 'Sound and light I'. See Airy 1842, §103, and, for a discussion of this tradition of research on the ether, Buchwald 1980.
12. Stokes, Notes on Challis' lectures, no date, Cambridge University Library, Stokes Collection, Add. MS 7656, PA35.
13. Smith's prize examinations are printed in the *Cambridge Calendar* following the questions for the Mathematical Tripos.
14. Peacock was second wrangler in 1813 and the Lowndean professor of astronomy and geometry from 1837 to 1858. W.H. Miller was fifth wrangler in 1826 and professor of mineralogy from 1832 to 1880. Earnshaw was senior wrangler in 1831. (See Table II.)

15. Forbes, 'Analysis of Sir John Leslie on heat', 6 November 1830, St. Andrews University Library, Forbes Papers, Box VI, vol. 11. Leslie 1804.
16. Forbes to Leslie, 13 September 1830, St. Andrews University Library, Forbes Papers.
17. Forbes to Robison, 16 June 1830, St. Andrews University Library, Forbes Papers.
18. Forbes to Leslie, 13 September 1830, St. Andrews University Library, Forbes Papers.
19. Forbes to Kelland, 20 February 1839, St. Andrews University Library, Forbes Papers.
20. Babbage to Forbes, 25 January 1832, St. Andrews University Library, Forbes Papers.
21. On the influence of Whewell on Forbes, see Shairp, Tait and Adams-Reilly 1873, 93–120; Davie 1961, 169–75; and Smith 1976b, 25–8. On Forbes and the British Association, see Morrell and Thackray 1981, 430–4.
22. Leslie to J. Brown, 6 December 1817, Edinburgh University Library, Dc. 2. 57, letter 202.
23. Forbes to Moseley, 8 August 1833, St. Andrews University Library, Forbes Papers.
24. As Forbes advised his successor, P.G. Tait: 'But what I would wish earnestly to recommend to your consideration is the extreme undesirableness of separating systematically the Mathematical and Experimental (or popular) departments of Nat. Philosophy. It seems to me to be characteristic of the Scottish System that this has not been done by its best teachers.' (Forbes to Tait, 17 January 1861, St. Andrews University Library, Forbes Papers.)
25. Forbes to J. Lee, principal of Edinburgh University, 10 September 1855, National Library of Scotland. He had earlier urged William Thomson to make a similar protest when Lewis Gordon's lectures as new professor of engineering at Glasgow appeared to overlap with the natural philosophy course: 'I think you ought to make a mild but decided remonstrance against the Invasion which would confine Natural Philosophy to Physical Astronomy, Optics, and Electromagnetism'. (Forbes to Thomson, 27 November 1848, St. Andrews University Library, Forbes Papers, Cambridge University Library, Kelvin Collection, Add. MS 7342, F198.)
26. Natural philosophy class competitions for 1839, 1841, 1842, 1845, 1846, and 1850. Edinburgh University Library.
27. Natural philosophy class competitions, Edinburgh University Library; Forbes, Record of lectures delivered to class, 1833–57, St. Andrews University Library, Forbes Papers, Box IX, vol. 5; and Balfour Stewart, Notes on Forbes's lectures, 1845–46, Edinburgh University Library. These are the only students' notes on Forbes's lectures that I know of.
28. 'Professor Forbes on English and Scotch University Reform', *The Scotsman* (10 November 1855), copy in Edinburgh University Library, Lyell Papers, 1/1479.
29. Forbes, 'On the inductive philosophy of Bacon, his genius and achievements', p. 31, in Moral philosophy essays, St. Andrews University Library, Forbes Papers, Box V, vol. 4. He acknowledges his reliance on Playfair's 'full' discussion of Bacon on page 12. Playfair 1860, 584.
30. Natural philosophy class competitions for 1841, Edinburgh University Library; Playfair 1860, 586–8; and Herschel 1830, *passim*.
31. B. Stewart, Notes on Forbes's lectures, 1845–46, lecture 21. Forbes offered a prize for the best essay 'On the best method of arriving at truth in physical investigations'.
32. Ibid. lecture 41.
33. Ibid. lecture 8.
34. Ibid. lecture 39.
35. Forbes to H. Moseley, 8 August 1833, St. Andrews University Library, Forbes Papers. This letter is mistakenly identified as one from Forbes to Whewell in Shairp, Tait and Adams-Reilly 1873, 98.
36. Stewart, Notes on Forbes's lectures, 1845–46, lecture 66; Forbes 1860, 920.
37. Stewart, Notes on Forbes's lectures, lectures 100 and 105.
38. Ibid. lecture 63.
39. Ibid. lecture 105.
40. Ibid. lecture 49.
41. Forbes, Record of lectures delivered to class, 1833–57, St. Andrews University Library, Forbes Papers, Box IX, vol. 5.
42. Forbes 1860, 974. See Stewart, Notes on Forbes's lectures, lecture 49.
43. Forbes 1860, 942. See Shairp, Tait and Adams-Reilly, 1873, 491.

The educational matrix

44. Forbes, 'The history of science; and some of its lessons', reprint from *Fraser's Magazine*, March 1858, in Edinburgh University Library.
45. Anonymous, 'Notes taken from the lectures of Dr Meikleham', 1812–13, Glasgow University Library, MS Gen. 44.
46. R. Pollok, 'Statement of the subjects treated in the natural philosophy class', 7 November 1821, Glasgow University Library, MS Gen. 1355/56.
47. See note 46.
48. Ibid. Anonymous, 'Notes taken from the lectures of Dr Meikleham', lecture 9.
49. R. Pollok, 'Of induction', 12 November 1821, Glasgow University Library, MS Gen. 1355/57; Anonymous, 'Notes taken from the lectures of Dr Meikleham', lecture 3.
50. Pollok, 'Of induction'.
51. W. Sommerville, 'Exercises in natural philosophy prescribed by Dr Meikleham Prof. of Nat. Phil. in the University of Glasgow during the Session of 1818–19': 'On tides', Glasgow University Library, MS Gen. 608.
52. R. Pollok, 'Of inertia', 22 November 1821, Glasgow University Library, MS Gen. 1355/60; Pollok, 'Notes taken in the natural philosophy class of the University of Glasgow — W. Meikleham Professor', 12 November 1821, Glasgow University Library, MS Gen. 1355/104.
53. Pollok, 'Of inertia'.
54. Sommerville, 'Exercises in natural philosophy prescribed by Dr Meikleham': 'On the directive property of the magnet'; Pollok, 'Notes taken in the natural philosophy class', 14 March 1822; Pollok, 'Of the electrical charge', 14 March 1822, Glasgow University Library, MS Gen. 1355/65.
55. Sommerville, 'Exercises in natural philosophy prescribed by Dr Meikleham': 'Of friction'; Pollok, 'Notes taken in the natural philosophy class', 12 November 1821; Pollok, 'Solidity', 15 November 1821, Glasgow University Library, MS Gen. 1355/58.
56. Sommerville, 'Exercises in natural philosophy prescribed by Dr Meikleham': 'On the laws of the propagation of light'.
57. Pollok, 'Notes taken in the natural philosophy class', 6 March 1822; Sommerville, 'On the laws of the propagation of light'.
58. Sommerville, 'Exercises in natural philosophy prescribed by Dr Meikleham': 'On the evidence which the laws of Kepler afford in support of the system of gravitation'.
59. See note 58.
60. Sommerville, 'On the directive power of the magnet'; Pollok, 'Of magnetism', 22 March 1822, Glasgow University Library, MS Gen. 1355/66; Pollok, 'Notes taken in the natural philosophy class', 14 March 1822.
61. Pollok, 'Of the electrical charge', 14 March 1822, Glasgow University Library, MS Gen. 1355/65.
62. W. Thomson, Notes on the natural philosophy class, 1839–40, Cambridge University Library, Kelvin Collection, Add. MS 7342, NB9. Angle brackets ⟨ ⟩ used in quotations from manuscripts indicate deletions.
63. Addison 1902, 33–57; *Glasgow prize and degree lists*.
64. Shortly after beginning his first course of lectures, Thomson wrote to Forbes: 'In the experimental course, of which there have been five lectures already, we have only got through part of the subject of magnetism. At this rate we might get through magnetism and common electricity in one ⟨year⟩ session, galvanism and electro magnetism in another, and get something of heat &c, in a third.' (Thomson to Forbes, 22 November 1846, St. Andrews University Library, Forbes Papers.)
65. Thompson 1910, 1:239–51; *Glasgow University prize and degree lists*; W. Smith, Notes on W. Thomson's lectures, 1849–50, Glasgow University Library, MS Gen. 142; W. Jack, Notes on W. Thomson's lectures, 1852–53, Glasgow University Library, MS Gen. 130.
66. *Glasgow University prize and degree lists*; 'Students' entrance book', 1846–81, Department of Natural Philosophy, Glasgow University.
67. W. Thomson, Journal and research notebook, 1845–56, Cambridge University Library, Kelvin Collection, Add. MS 7342, NB34.
68. Jack, Notes on Thomson's lectures, 1852–53.
69. R.G. Weldon, 'The chemical properties of light', 1849, Glasgow University Archives, 41128.

David B. Wilson

70. R. Lockhart, 'An essay on the motive power of heat', 1851, Glasgow University Archives, 40981.
71. T. Logan, 'Mechanical energy of electricity in motion', 1855, Glasgow University Archives, 40986.
72. Thomson 1847; Thomson, Journal and research notebook, entries for 31 October 1846, 28 November 1846, and 29 November 1846, Cambridge University Library, Kelvin Collection, Add. MS 7342, NB34 (printed in Thompson 1910, I:197–8); Thomson to Faraday, 11 June 1847, in Thompson 1910, I:203–4; Thomson to Forbes, 29 December 1846, St. Andrews University Library, Forbes Papers, Cambridge University Library, Kelvin Collection, Add. MS 7342, F167.
73. Forbes to Thomson, 31 December 1846, St. Andrews University Library, Forbes Papers, Cambridge University Library, Kelvin Collection, Add. MS 7342, F168.
74. On Rankine, see Hutchison 1981 and Channell 1982.
75. Herschel 1830, §23. There remains the possibility that it was not the content, but the intensity of Cambridge mathematical studies which aided the Cambridge graduates, by giving them greater facility in the use of mathematics. (Daniel Siegel suggested this possibility at the Grasmere Conference.) This view would be supported by the fact that the major 'Cambridge' physicists did, after all, graduate from Cambridge and by the fact that Rankine, who did not, was not really so strong a mathematician as he may have appeared to his contemporaries. (This is Keith Hutchison's view of Rankine, expressed to me at Grasmere.) On the other hand, there is the view that, for Thomson at least, the Cambridge examination contributed little. (Crosbie Smith made this suggestion at Grasmere.) Several considerations support this view. Thomson and Maxwell were doing high-level mathematics before coming to Cambridge. Rankine not only did not attend Cambridge, he attended Edinburgh before Kelland became professor of mathematics. Possibly he would have been a better mathematician had he studied with Kelland. Much of the well-known intensity at Cambridge came from relentless preparation to answer examination questions, which was evidently not the kind of preparation Thomson and Maxwell went in for. They did both finish second, not first. Green was a successful mathematical physicist before he attended Cambridge. Therefore, it may be that the Scots were attracted to Cambridge because of its reputation, but that Cambridge mainly confirmed abilities and knowledge already present, without adding a great deal to them. It may be that the kind of intensity that was important was the continual working with mathematics, rather than the specific Tripos preparation itself. In any case, whether the Cambridge intensity was significant or not, on the assumption that Stokes, Thomson, and Maxwell all knew the same mathematics and had more or less equal mathematical abilities, mathematics would seem not to afford an explanation of the *differences* between them which this chapter is concerned with.
76. This information is drawn from the sources for Cambridge and Glasgow listed in note *a* to Table II. The figures for Cambridge clergymen does not include those who took holy orders without making a career in the church.
77. Kelland 1836*a*, 1836*b*, 1837*a*, 1837*b*, 1840, 1841.
78. As Steele died before he was able to publish any research, he is not discussed here.
79. Stokes to Thomson, 13 March 1847, Cambridge University Library, Kelvin Collection, Add. MS 7342, S323.
80. According to Thomson: 'When I wrote the paper I had some hope, w^h I still retain, that a satisfactory physical theory of all those agencies, including besides light, is approachable.' (Thomson to Stokes, 4 February 1849, Cambridge University Library, Stokes Collection, Add. MS 7656, K28.)
81. Thomson to Stokes, 7 April 1847, Cambridge University Library, Stokes Collection, Add. MS 7656, K19.
82. Daniel Siegel first suggested to me that Forbes had a low opinion of Faraday's concept of lines of force.
83. Forbes, Record of lectures delivered to class, 1833–57, St. Andrews University Library, Forbes Papers, Box IX, vol. 5.

= 3 =
Geologists and mathematicians: the rise of physical geology

CROSBIE SMITH

Nor is that Fellow-wanderer, so deem I,
Less to be envied, (you may trace him oft
By scars which his activity has left
Beside our roads and pathways, though, thank Heaven!
This covert nook reports not of his hand)
He who with pocket-hammer smites the edge
Of luckless rock or prominent stone, disguised
In weather-stains or crusted o'er by Nature
With her first growths, detaching by the stroke
A chip or splinter — to resolve his doubts;
And, with that ready answer satisfied,
The substance classes by some barbarous name,
And hurries on; or from the fragments picks
His specimen, if but haply interveined
With sparkling mineral, or should crystal cube
Lurk in its cells — and thinks himself enriched,
Wealthier, and doubtless wiser, than before!
(Wordsworth, 1814)

Geology and geometry: the Cambridge style of geological science
William Wordsworth's scornful indictment of 'the brethren of the hammer' in Book III of 'The excursion' is an appropriate way to open a discussion of the special relationship between Cambridge geology and Cambridge mathematical physics in the first half of the nineteenth century. For not only was the symposium which was the stimulus for this chapter set in the heart of Wordsworth's beloved lakes and mountains in his adopted village of Grasmere, but two of my three central characters, Adam Sedgwick (1785–1873) and William Whewell (1794–1866), were born not thirty miles away, at Dent near Sedbergh, and at Lancaster, respectively. Furthermore, the path from Cambridge to the Lakes in the 1820s and 1830s was a well-trodden one. William's brother, Dr Christopher Wordsworth (1774–1846), was the celebrated Master of Trinity until succeeded by Whewell in 1841, and Sedgwick, elected fellow of

Crosbie Smith

Trinity in 1810 and Woodwardian professor of geology in the University in 1818, explored the geology of the Lake District with William's guidance in the years 1822–24. From Sedgwick's own account, Wordsworth 'joined me in many a lusty excursion, and delighted me (amidst the dry and sometimes almost sterile details of my own study) with the outpourings of . . . the beauteous and healthy images which were ever starting up within his mind during his communion with nature, and were embodied, at the moment, in his own majestic and glowing language' (Clark and Hughes 1890, 1:246–9). And in 1842 William Hopkins (1793–1866), last of my trio, illustrated the principles of his new physical geology by reference to the valleys and lakes of the region (Hopkins 1842*b* and 1848; Davies 1969, 244–5).

Three aims run in parallel through this chapter. First, I want to examine the emerging characteristics of a distinctive Cambridge school of geological science which cannot be as easily subsumed under a more general English school of geology, centred on Oxford, as a recent study suggests (Rupke 1983). The term 'school' in this context, however, is not employed in the later, more rigorous sense of an academic 'Cambridge school of geology' aimed at training professional research geologists (Porter 1977). Rather, it is a convenient label for a particular group of Cambridge scholars, inspired by Sedgwick, and sharing a characteristic geometrical style and method in their approach to the subject both in theory and in practice. Secondly, I hope to show by an analysis of the writings of three Cambridge wranglers (Sedgwick, Whewell and Hopkins) how closely their approach to geological science was tied to the intellectual goals of Cambridge University, and to draw attention to the significance of this study for understanding the subsequent mathematical style of such well-known Cambridge wranglers as William Thomson, G.G. Stokes and James Clerk Maxwell. Thirdly, I shall try to elucidate the nature of the very important interactions between the demands of natural theology and geological science, especially through the doctrine of inorganic progression.

Adam Sedgwick was the father of Cambridge geological science, and as such provides the obvious starting point for this study. Although fifth wrangler in 1808, he had no enthusiasm for mathematics as such. Leading 'a life of dull uniformity' as a fellow of Trinity, neither vacation reading parties with his mathematical pupils nor the prospect of ordination as an Anglican clergyman stirred him from intellectual torpor. Only an enthusiasm for outdoor activity seemed to preserve his mind and body from total atrophy. When, therefore, the Woodwardian Chair became vacant, it was not from any devotion to geological science, but from a desire 'for active exertion in a way which will promote my intellectual improvement' which persuaded him to become a candidate. Thus 'so soon as I was seated in the Woodwardian Chair I gave away my dogs and my gun, and my hammer broke my trigger'. In the succeeding years, Sedgwick's hitherto-latent intellectual energies surfaced as he devoted himself to the learning and promoting of his subject.

Geologists and mathematicians

To begin with, the new professor pursued geological science by frequent and intensive excursions such as that to the Lakes in 1822–24. Sedgwick's other travels took him to the Isle of Wight with J.S. Henslow in 1819 and to the West Coast of Scotland with Roderick Murchison in 1827, and in 1829 he visited Germany, Austria and Switzerland, accompanied again by Murchison. Perhaps most famous of all in retrospect, however, was the excursion in North Wales in 1831 when young Charles Darwin joined him. A similar expedition in North Wales a year later set William Hopkins along the path to promoting the mathematisation of geological science with which this chapter will ultimately be concerned. Sedgwick's principal aim during these excursions was to follow the course of geological strata, to collect specimens for display in Cambridge, and above all to determine the geological structure of Wales. These 'professional pursuits', as he called his activities, were in keeping with his threefold hope for the Professorship:

> First, that I might be enabled to bring together a Collection worthy of the University, and illustrative of all the departments of the Science it was my duty to study and to teach. Secondly, that a Geological Museum might be built by the University, amply capable of containing its future Collections; and lastly, that I might bring together a Class of Students who would listen to my teaching, support me by their sympathy, and help me by the labour of their hands (Clark and Hughes, 1890, 1:165).

Thus it was that Sedgwick's own intellectual vitality contributed in no small way towards moving Cambridge from the undoubted intellectual stagnation of previous decades into a long period of academic pre-eminence and renown. Sedgwick's particular approach to geology, and his inspiration of a Cambridge school of geological science, are what principally concerns us here, however, and it is to a closer examination of his approach that I shall now turn.

Rival geological theories of the Wernerians and Huttonians had tended to dominate late-eighteenth century studies of the earth. Where the 'Neptunist' disciples of Werner would permit only aqueous causes of major geological change, the 'Vulcanist' disciples of Hutton emphasised igneous causes (Gillispie 1951; Porter 1977, 170–6; 184–202). Mineralogy, the classification and analysis of rocks according to their mineral constituents, was characteristic of the Wernerians, and these 'barbarous' classifiers probably formed the object of Wordsworth's poetic attack in 1814. Indeed, Sedgwick himself remarked on 'the barbarous nomenclature of Geology' in 1820–22 (Sedgwick 1822, 92). Encouraged by the investigations of strata and their characteristic fossils carried out by, among others, William Smith — whom Sedgwick termed 'the father of English geology' (Clark and Hughes 1890, 1:284; Rupke 1983, 191–3) — the Cambridge professor soon reacted against a theoretical, speculative study of the earth's crust. The ancient speculations on the 'Theory of the earth' were inadequate principally because they had been formed, first, 'without an extended acquaintance with the laws of nature', and secondly, 'without any knowledge of the interior structure of the earth' (Sedgwick 1821). His severely antihypothetical, 'inductive' methodology thus shaped, as we shall see, much of his (and his Cambridge contemporaries') subsequent work.

A more systematic understanding of Sedgwick's distinctive geological style is fundamental to the aims of this chapter. An important clue occurs in an 1828 letter from Murchison suggesting that Sedgwick accompany him on a second Continental excursion: 'And here let me . . . invoke the spirit of inquiry which prevails at Cambridge, and urge you, who are really almost our only *mathematical* champion, not to let another year elapse without endeavouring to add to the stock of your British Geology some of the continental materials' (Clark and Hughes, 1:361). A further clue may be found in his biographers' remarks that Sedgwick's 1830 Presidential Address to the Geological Society of London 'hints at the true method of correlating facts, and establishing a correct induction from them — in a manner well worthy of the "mathematical geologist"' (Clark and Hughes, 1:361; Sedgwick 1830, 211–12). What, then, entitled the Cambridge professor to these epithets of 'mathematical champion' and 'mathematical geologist'?

In his forthcoming book, *Controversy in Victorian Geology. The Cambrian–Silurian Dispute*, James Secord provides an excellent analysis of the differing geological styles of Sedgwick and Murchison in relation to this famous controversy which originated in the 1830s from the attempts to classify the ancient stratified rocks of Wales. Secord shows convincingly that Sedgwick's style was characterised above all by his primary concern with geological structures and by correspondingly less emphasis on palaeontological techniques. Sedgwick's style, in short, was geometrical, and his geological language that of lines, dips, strikes and synclines. By contrast, Murchison, ex-soldier and fox-hunter, aimed for a rapid grasp of the fundamental features of a district. As an effective means to this end, he placed a consequent emphasis on palaeontology. Similar contrasts may also be drawn between Sedgwick — whose work Edward Sabine classified in 1863 as 'mountain geometry' (Clark and Hughes, 1890, 2:562) — and palaeontologically orientated contemporaries such as William Buckland and William Conybeare.

To appreciate the full significance of Sedgwick's geometrical geology, we must recognise the true nature of British geology in the first half of the nineteenth century. As Secord stresses, mainstream geology was above all a practical activity not concerned with speculative cosmological theory but rather with geology as natural history and hence with *classification*. In a taxonomic enterprise which treated of the ordering of rock strata, then, I would argue that geologists employed various investigative procedures which characterised their differing styles and methodologies. Consequently, where Wernerians would employ chemical and mineralogical procedures in classifying the strata, historical geologists such as Buckland would emphasise palaeontology, and structural geologists such as Sedgwick the importance of form. Geologists in general

Geologists and mathematicians

believed that classification offered the only road to sound 'inductive' knowledge, while Sedgwick (and Whewell and Hopkins) in particular argued that classification according to geometrical laws provided the first secure step to satisfactory geological 'theory'. A close analysis of the Cambridge classificatory procedure will thus constitute a crucial feature of the present paper.

Nowhere is the essence of Sedgwick's approach better seen than in his early paper 'On the physical structure of those formations which are immediately associated with the primitive ridge of Devonshire and Cornwall' read in 1820 to the Cambridge Philosophical Society (in the founding of which Sedgwick had played a prominent role a year earlier). Almost any passage from this lengthy paper could be used to illustrate his predominantly geometrical description of the geological phenomena, but the following extract is a particularly good example:

> In many parts of the coast of Somerset and North Devon, where we have the finest natural sections of these schistose [minerals grouped in flaky layers] masses, the strata do not preserve the same direction for many feet together; but are continually deflected, and coiled on each other; presenting in their course lines of every possible curvature. The same vertical section often exposes a series of beds, one portion of which is disposed in curves convex, and another portion in curves concave to the horizon; and so situated in respect of each other, that no force whatever, acting in a given direction, could produce the structure out of masses originally in a horizontal position. When the fundamental rock is thus constituted, the country must present a corresponding variety of surface. (Sedgwick 1822, 92.)

The strong anti-hypothetical thrust of Sedgwick's methodology, complementing his emphasis on geometrical description of the phenomena, may be seen explicitly in his 1830 and 1831 Presidential Addresses to the Geological Society of London (Sedgwick 1830, 1831). The 1831 address incorporated a well-known review of Charles Lyell's very new *Principles of Geology*, a review in which Sedgwick directed his criticism against what he regarded as Lyell's advocacy of one of the arbitrary dogmas of the Huttonian hypothesis, namely, that 'the secondary combinations arising out of the primary laws of matter have been the same in all periods of the earth' (Sedgwick 1831, 304). In particular, he condemned the assumption, implicit in this doctrine, that the geological forces of volcanic action operated with equal intensity:

> Of the origin of volcanic forces we know nothing: but we do know that they are the irregular secondary results of great masses of matter obeying the primary laws of atomic action — that they differ in their intensity — are interrupted in their periods — and are aggravated or constrained by an endless number of causes, external and purely mechanical .... To assume, then, that volcanic forces have not only been called into action at all times in the natural history of the earth, but also, that in each period they have acted with equal intensity, seems to me a merely gratuitous hypothesis, unfounded on any of the great analogies of nature, and I believe also unsupported by the direct evidence of fact. This theory confounds the immutable and primary laws of matter with the mutable results arising from their irregular

combination .... And what is this but to limit the riches of the kingdoms of nature by the poverty of our own knowledge .... (Sedgwick 1831, 301.)

The final part of this extract in fact echoes a discussion in John Herschel's *Preliminary Discourse on the Study of Natural Philosophy* (Herschel 1830, 42).

Fundamental to Sedgwick's argument against Lyell's theoretical framework was his belief in the 'utter hopelessness of bringing under the definite calculations of any mechanical law [such as the law of gravity], those mighty combinations still going on in the great laboratory of nature'. Unlike astronomy, the phenomena of geology were 'mutable and indefinite'. Geology, then, had to be 'a science of observation'. All *a priori* reasoning must be banished, and 'the language of theory can never fall from our lips with any grace of fitness, unless it appear as the simple enunciation of those general facts, with which, by observation alone, we have at length become acquainted' (Sedgwick 1831, 298–302).

In this last remark we have a revealing glimpse into what subsequently characterised the basis of the Cambridge approach to geological (an indeed all scientific) theorising. What might look here rather like naive inductivism may be read as a statement that theory and practice should not be rigidly demarcated from one another. Sound theories must not be comprised of speculative and *ad hoc* hypotheses but must, in Whewell's later phrase, be 'true Theories precisely because they are identical with the total system of the Facts' (Whewell 1840, 2:121–2). Theory and practice therefore interpenetrate, as is well illustrated in mathematical physics at this time by the techniques of Fourier's 1822 *Théorie analytique de la chaleur*, techniques that were finding much favour in Cambridge during the 1830s in preference to those of the Laplacians (Crosland and Smith 1978, 51–2; Wise 1981, 23–32). Not surprisingly, we shall see that Fourier's approach to terrestrial physics soon became a foundation stone for Cambridge geological science.

Of crucial significance for the Cambridge response to Lyell in 1831 is the recognition, in Sedgwick's words, of Lyell as 'being the first writer in our country to make known a general system of "geological dynamics"', which, however, had yet to be 'stripped of even the semblance of hypothetical assumption' (Sedgwick 1831, 303, 311). In the same year, Whewell's review of Lyell in *The British Critic* complimented Lyell on:

... the very able manner in which he has treated that part of the subject which occupies the principal portion of his work, viz. the examination of the actual operation of the causes of change which affect the earth's surface. In general, this curious province of knowlege has been very imperfectly attended to ... Writers have been satisfied with guessing at the results of such agencies, upon the suggestion of casual and common observation instead of endeavouring to ascertain, by a general and careful collection of authentic facts, the mode of action and the work done. This appendage of geology seems, however, now to be ready to receive an orderly and

Geologists and mathematicians

systematic character, which will elevate it to the rank of a separate science: a science which has for its object to classify and analyse the changes which are perpetually occurring in the inorganic portion of nature: and which we might call GEOLOGICAL DYNAMICS, since it treats of the forces which are acting to modify the face of the earth. (Whewell 1831*b*, 195.)

In the next section we shall see more of the function and character of geological dynamics with regard to Whewell's *History* and *Philosophy of the Inductive Sciences*, but for the moment let us note, first, the importance of Lyell in Cambridge eyes as a pioneer of the science, and secondly, the Cambridge emphasis on geological dynamics as a science free of all hypothetical assumption.

This latter point has far-reaching implications for any study of Cambridge physical science in the nineteenth century. The first step towards a satisfactory theory in any branch of science was the geometrisation of the phenomena. The second step was the establishment of a well-founded dynamics which itself differed but little from geometry in its freedom from hypothesis, a feature of dynamics already well expressed in Herschel's *Preliminary Discourse*:

> Dynamics, then, or the science of force and motion, is thus placed at the head of all the sciences ... it is one in which the highest certainty is attainable, a certainty no way inferior to mathematical demonstration. As its axioms are few, simple, and in the highest degree distinct and definite, so they have at the same time an immediate relation to geometrical quantity, space, time, and direction, and thus accommodate themselves with remarkable facility to geometrical reasoning. Accordingly, their consequences may be pursued, by arguments purely mathematical, to any extent, insomuch that the limit of our knowledge of dynamics is determined only by that of pure mathematics, which is the case in no other branch of physical science (Herschel 1830, 96).

Thus we shall see the importance of beginning with geometrical description — in particular, with a geometry not only of form but of motion (the science of kinematics) — before proceeding to a science of dynamics in order to account for the motion or change of geometrical form. Such an emphasis on kinematics and dynamics viewed macroscopically and hence geometrically is the leading characteristic of the mathematical physics of Stokes, Thomson, and Maxwell (Wise, 1982, 185–90), and I shall suggest in the final section of this chapter that Hopkins' approach to geological science is evidence of both a Cambridge style and a Hopkins school of mathematical physics.

In recent years, historians of science have thoroughly reappraised Lyell's *Principles* in the context of its time. They have argued in particular that Lyell's own interpretation of the history of geology failed to do justice to the existing rival synthesis of directionalism based on the best physics and chemistry of the day (Rudwick 1970; 1971; 1974; Lawrence 1978). The Cambridge response to Lyell must therefore be seen against the background of a decade of remarkable geological activity in Britain when well-known figures such as Buckland, Murchison, Poulett Scrope, Sedgwick himself, Conybeare and others either committed themselves to a directionalist position or provided support for such a position. Viewed in historical context, then, Sedgwick's 1831 address illustrates how Lyell's *Principles* crystallised the rival directionalist synthesis and brought about among British geologists a new degree of consensus centred on opposition to Lyell's steady-state theory. Sedgwick and his Cambridge contemporaries, however, were neither the first critics of Lyell's version of uniformitarianism nor the only supporters of directionalism or progression. As early as 1830, indeed, Scrope had attacked what he believed to be Lyell's proposition that 'the existing causes of change have apparently operated with absolute uniformity from all eternity' (Scrope 1830, 464). And he argued strongly in favour of 'a progressive state and a limited existence' of the surface of the globe as well as for a progressive diminution in the 'energy' of volcanic and other geological agencies (Scrope 1830, 467–9). For his part, Lyell devoted the concluding remarks of volume three of his *Principles* (1833) to defending his views against Scrope's allegation, claiming rather that he only denied 'signs of the earth's origin, or evidences of the first introduction into it of organic beings' (Lyell 1830–33, 3:383).

Sedgwick's 1831 address was nevertheless distinguished by the sustained nature of its criticism. Even Lyell himself admitted at the time that 'Sedgwick's attack is the severest' (Clark and Hughes 1890, 1:370). Its strength derived in large measure from Sedgwick's ability to portray Lyell's work as a violation of good inductive principles by appearing 'as the champion of a great leading doctrine of the Huttonian hypothesis' — that 'the secondary combinations arising out of the primary laws of matter have been the same in all periods of the earth'. Such a hypothesis had no *a priori* probability, and was only to be maintained by an appeal to geological phenomena. Yet 'the well established facts brought to light by our [geological] investigations [showed that] there is no such thing as an indefinite succession of events', as implied by the Huttonian doctrine. In particular, the earth's surface presented a definite succession of dissimilar phenomena such as to deprive phrases like 'the invariable constancy in the order of nature' of most of their meaning. Apart from the successive geological formations of the earth, Sedgwick drew attention to the fact that 'we are surrounded by animal and vegetable forms, of which there are now no living types'. In these things we had, he believed, 'some indication of change and of an adjusting power altogether different from what we commonly understand by the laws of nature'. The repeated and almost entire changes of organic types in the successive formations of the earth, then, proved convincingly that 'the approach to the present system of things had been gradual, and that there has been a progressive development of organic structure subservient to the purposes of life' (Sedgwick 1831, 303–6). Moreover, from physics came the evidence of the figure of the earth, showing by calculation the form put on by a fluid body in rotation

Geologists and mathematicians

and confirming by direct geodesic observation the spheroidal shape of the earth (Sedgwick 1831, 298).

Lyell's *Principles* also implied that there could be 'no great violations of continuity either in the structure or position of our successive formations'. But Sedgwick argued that this doctrine of equal intensity simply did not conform to the geological facts: 'we know there are enormous violations of continuity'. As evidence, he cited 'our most recent deposits' — the enormous water-worn boulders scattered over the plain of northern Europe — and the elevation of the strata of the more ancient deposits from a horizontal to a highly inclined position without intermediate gradations. More especially:

> Volcanic action is essentially paroxysmal; yet Mr Lyell will admit no greater paroxysms than we ourselves have witnessed — no periods of feverish spasmodic energy, during which the very framework of nature has been convulsed and torn asunder. The utmost movements that he allows are a slight quivering of her muscular integuments. (Sedgwick 1831, 307.)

Taken together, the inductive evidence presented by Sedgwick against both Lyell's steady-state theory and his doctrine of equal intensity was formidable. More importantly, Sedgwick presented his contrary evidence as evidence very much in harmony with a gradually cooling globe. The most powerful support for directionalism came of course from Fourier, whose doctrines received an impressive expression in Élie de Beaumont's theory of epochs of elevation in 1828–29. In this theory, a cooling, gradually shrinking earth built up compressional strains in its crust, which would then suddenly shear and buckle along lines of weakness. Linear mountain chains would thus be elevated (Davies 1969, 215; Rudwick 1971, 219–20; 1972, 186–7; Lawrence 1978, 106–10). Consequently, Sedgwick in his 1831 address not only set forth at the outset the evidence in support of an originally fluid earth in terms of its spheroidal figure, the crystalline structure of its unstratified masses, the relative position of land and water, and the nature of fossils in the ancient strata, but he subsequently gave a very warm welcome to Élie de Beaumont's theory.

Sedgwick prefaced his appraisal of Élie de Beaumont's theory with the remark that 'the very essence of philosophical discoverey ... consist[s] in bringing to a point all the scattered lights of former observations, and giving generalization to insulated phaenomena'. To begin with, Élie de Beaumont 'by an incredible number of well conducted observations of his own, combined with the best attested facts recorded by other observers, ... has proved ... that whole mountain chains have been elevated at one geological period — that great physical regions have partaken of the same movement at the same time — and that these paroxysms of elevatory force have come into action at many successive periods'. On the one hand, then, we had been advancing step by step towards the conclusion 'that different mountain chains had been elevated at several distinct geological periods'. On the other hand:

by a long series of independent observations, Humboldt, von Buch, and other great physical geographers, had proved — that the mountain chains of Europe might be separated into three or four distinct systems; distinguished from each other, if I may so express myself, by a particular physiognomy, and, above all, by the different angles made by the bearings of their component formations with any assumed meridian. All the subordinate parts of any one system were shown to be parallel; while the different systems were inclined at various angles to each other.

From these two distinct conclusions, Élie de Beaumont took 'the next great step of generalization':

> By an unlooked-for and most felicitous generalization, M. Élie de Beaumont has now proved that these two great classes of facts are commensurate to each other; and that each of these great systems of mountain chains, marked on the map of Europe by given parallel lines of direction, has also a given period of elevation, limited and defined by direct geological observations. The steps by which he reaches this noble generalization are so clear and convincing, as to be little short of physical demonstration. It forms an epoch in the history of our science; and I am using no terms of exaggeration when I say, that in reading the admirable researches of M. de Beaumont I appeared to myself, page after page, to be acquiring a new geological sense, and a new faculty of induction. (Sedgwick 1831, 308.)

Several explanatory points need to be emphasised here. First, Sedgwick's enthusiasm for Élie de Beaumont's theory is closely related to his own enthusiasm for 'mountain geometry'. Furthermore, James Secord has shown clearly in his forthcoming *Controversy in Victorian Geology* that Sedgwick's subsequent geological practice among the mountains of North Wales in the 1830s was shaped by his employment of the theory. Secondly, Sedgwick's enthusiasm related also to what Whewell later made central to his philosophy of the inductive sciences, namely, *consilience*. Consilience, as Whewell defined the term, was a 'coincidence of propositions inferred from separate classes of fact' and offered 'one of the most decisive characteristics of a true theory' (Whewell 1840, 2:446). Élie de Beaumont's 'most felicitous generalisation', then, tied together two hitherto-independent, lower-order generalisations. Thirdly, it will be evident that Sedgwick did not regard the 'theory' as epistemologically distinct from the 'facts'. Again, in Whewell's later phrase, 'the distinction between *Fact* and *Theory* is only relative' (Whewell 1840, 2:xli).

Overall, then, Sedgwick's reference to 'a new faculty of induction' may be seen in terms of what Élie de Beaumont's theory offered to the 'mathematical geologist' — a geometrical description of the phenomena, consilience of inductions, a 'theory' which emerged from and which could in turn shape geological practice, and a 'theory' which related to the possible higher-order generalisation of a gradually cooling globe in which heat was the major agent in the geological dynamics. Finally, we must recall that all this discussion took place in the context of a critique of Lyell's non-progressionist theory. Sedgwick

Geologists and mathematicians

thus concluded that he had decided in favour of Élie de Beaumont and against Lyell, because the former's 'conclusions are not based upon any *a priori* reasoning, but on the evidence of facts (Sedgwick 1831, 311). In Élie de Beaumont's approach, therefore, Sedgwick had found a near-perfect expression of the developing Cambridge style. By contrast, other geologists such as Conybeare, and of course Lyell himself, did not share Sedgwick's enthusiasm (Conybeare 1832, 401–2; Lyell 1830–33, 3:337–51; Lawrence 1978, 117–28).

In his 1832 review of Lyell's *Principles*, Sedgwick's close colleague at Trinity College, William Whewell — Cambridge professor of mineralogy since 1828 — defined the terms 'uniformitarian' and 'catastrophist' in relation to the issue of equal intensity raised by Sedgwick. Whewell agreed with Lyell that 'all the facts of geological observation are *of the same kind* as those which occur in the common history of the world'. *Uniformitarians* — as Whewell defined the term — also held, however, that the changes which lead us from one geological state to another have been, 'on a long average', uniform in their intensity, whereas *catastrophists* held that such changes 'consisted of epochs of paroxysmal and catastrophic action, interposed between periods of comparative tranquillity' (Whewell 1832, 126; Cannon 1960, 38–40).

Whewell himself sided quite decisively with the catastrophist view, reasoning that it would be rash to suppose 'as the uniformitarian does, that the information which we at present possess concerning the course of physical occurrences, affecting the earth and its inhabitants, is sufficient to enable us to construct classifications, which shall include all that is past under the categories of the present'. Given, then, the severe limitations of our knowledge, it would not be surprising if we had been left ignorant 'of some of the most important agents which, since the beginning of time, have been in action; of something, in short, which may manifest itself in great and distant catastrophes' (Whewell 1832, 126). Whewell, then, was following a nearly identical course to that of his geology colleague, while introducing the precision of clearly defined terms in a manner wholly characteristic of the future author of the *History* and *Philosophy of the Inductive Sciences*.

Whewell's definitions of 'uniformitarianism' and 'catastrophism' did not in themselves imply a firm conceptual distinction between 'non-progressionist' (steady-state), and 'progressionist' (directionalist) views of terrestrial phenomena — between, in other words, cyclical and directionalist geological dynamics. Lyell's version of uniformitarianism certainly did advocate a non-progressionist, steady-state 'system of geological dynamics'. But catastrophism, as defined by Whewell, could also have been interpreted in non-progressionist, cyclical terms in which periods of intense geological activity alternated with periods of comparative quiet. With progressionist theories of the earth grounded on central heat doctrines, however, a tendency from greater intensity in the past to lesser intensity of geological activity in the present became the accepted version of catastrophism.

Crosbie Smith

Both Whewell and Sedgwick were staunchly committed to progression in a much wider context than geology as an examination of their respective *Bridgewater Treatise* and *Discourse on the Studies of the University*, both published in 1833, clearly shows. As I have discussed elsewhere (Smith 1976a, 303-4), Whewell's conclusions in his *Bridgewater Treatise* were that 'perpetual change, perpetual progression . . . appear to be the rules of the material world' and that 'to maintain either the past or future eternity of the world does not appear consistent with physical principles, as it certainly does not fall in with the convictions of the religious man' (Whewell 1833, 203-4). That theme of consistency between inductive science (which could include natural theology) and revealed theology was echoed many times by our Cambridge wranglers along the lines of Sedgwick's remarks as early as 1825: '. . . truth must at all times be consistent with itself. The conclusions established on the authority of the sacred records may, therefore, consistently with the soundest philosophy, be compared with the conclusions established on the evidence of observation and experiment; and such conclusions, if fairly deduced, must necessarily be in accordance with each other' (Sedgwick 1825, 34; also Whewell 1845, ix, 68). To appreciate fully the significance of these remarks one must also be sensitive to their relation to the current problems of scriptural exegesis. Whewell, for example, devoted nearly thirty pages of his *Philosophy of the Inductive Sciences* to a consideration of the relation of the Sacred Narrative to the palaetiological sciences, which, as we shall see in the next section, included geology (Whewell 1840, 2: 137-65; Brooke 1979).

The relation of natural theology in general and design arguments in particular to the liberal Anglican position during the 1830s is an important one for understanding the organisation, advancement and ideology of science at the time of the founding of the British Association (Morrell and Thackray 1981, 224-45). Latitudinarian varieties of Anglicanism encouraged the progress of science in certain directions where biblical literalism (especially Scriptural Geology) and Tractarianism (the Oxford Movement) did not. Cambridge in the 1830s and beyond was, unlike Oxford, a stronghold of liberal Anglicans, and the maintenance of close ties between the ideals of mathematical science (represented by physical astronomy) and natural theology meant that a new science such as geology had only to aspire to such ideals to find favour at Newton's university. At the same time, it is worth remarking that Sedgwick's own latitudinarianism did not extend much beyond Anglicans, Presbyterians and dissenters: as his biographers noted approvingly, he retained 'a wholesome horror of the Church of Rome' (Clark and Hughes 1890, 1:88).

As we shall now see, Sedgwick's 1832 *Discourse on the Studies of the University* illustrates very forcefully the relations of the new science to both intellectual and moral goals of Cambridge. More specifically, the *Discourse* makes clear from the perspective of natural and revealed theology the major source of Sedgwick and Whewell's opposition to Lyell's steady-state world. Sedgwick's

Geologists and mathematicians

celebrated *Discourse* was originally a sermon delivered in Trinity College Chapel on the occasion of the annual Commemoration of Benefactors in 1832. Its publication owed much to Whewell's petitioning of the preacher (Clark and Hughes 1890, 1:401–2). Sedgwick divided the studies of the University into three branches: first, 'the study of the laws of nature, comprehending all parts of inductive philosophy'; secondly, the study of ancient literature; and thirdly, 'the study of ourselves, considered as individuals and as social beings' (Sedgwick 1834, 10). Of these branches only the first need concern us directly in relation to the intellectual and moral goals of the University and to the progressionist framework of Cambridge geological science.

First, let us consider briefly the intellectual and moral aims. A study of the laws of nature, following in a track 'first trodden by the immortal Newton', offered not merely intellectual reward but the training of the mind in abstract thinking 'most difficult to acquire by ordinary means' and 'of inestimable value in the business of life' (Sedgwick 1834, 11; Garland 1980, 49). Such a study was thus ideally suited to the aims of a Cambridge liberal education in which a rigorous training of the mind, most notably through geometrical reasoning, took priority over diffusion of useful knowledge. As Harvey Becher has shown, the 'analytic revolution' in Cambridge mathematics in the first two decades of the nineteenth century brought conflict within the educational system. Pure mathematics tended towards abstract, algebraic analysis, and mixed mathematics emphasised a physical, geometric approach. Central to the course of the conflict, Whewell sided largely with geometry and mixed mathematics, which he vigorously promoted as the basis of a liberal education through the Mathematics Tripos and its various reforms in the 1830s and 1840s (Becher 1980*a*). As we shall see in the next section, Whewell, like Sedgwick in geological science, regarded geometry not only as the foundation of a Cambridge liberal education but as the foundation upon which to construct the inductive sciences themselves. Nor will it be surprising to find Hopkins and his 'school of mathematical physicists' taking a very similar course.

The moral benefits of a study of the laws of nature derived above all from the way in which such a study 'teaches us to see the finger of God in all things animate and inanimate, and gives us an exalted conception of his attributes, placing before us the clearest proof of their reality; and so prepares, or ought to prepare, the mind for the reception of that higher illumination, which brings the rebellious faculties into obedience to the divine will' (Sedgwick 1834, 14). Being as he was on the evangelical wing of the Anglican church, Sedgwick's views here coincided with those of the contemporary Scottish Presbyterian theologian, Thomas Chalmers, in that natural theology was insufficient by itself for salvation but could very effectively serve 'as a harbinger to the higher lessons of the gospel' (Brooke 1979, 41–2; Smith 1979, 61–2). Elsewhere in the *Discourse*, in fact, Sedgwick spoke approvingly of Chalmers' theology (Sedgwick 1834, 56n; Morrell and Thackray 1981, 237).

Crosbie Smith

The *Discourse* highlights a second but more specific issue which we need to consider: the theological source of Sedgwick's opposition to Lyell's steady-state world. Significantly, Sedgwick also shared with Chalmers the concern that the laws of nature taken by themselves were not wholly adequate for the purposes of natural theology. Although the unchanging perfection of the laws of nature (gravity, for example) was *consistent* with the biblical conception of 'the immutable perfections of our heavenly Father', nevertheless critics could interpret that unchangeableness of the laws as a sign of nature's independence from God — 'to deify the elements themselves, and to thrust the God of nature from his throne' (Sedgwick 1834, 18–19; Smith 1979, 60–3). But, Sedgwick argued, we do not merely contemplate a succession of material changes. The external world 'proves to us the being of a God in two ways; by addressing the imagination, and by informing the reason'. The first way 'speaks to our imaginative and poetic feelings', which were 'as much a part of ourselves as our limbs and our organs of sense'. Thus 'all the touching sentiments and splendid imagery borrowed by the poet from the world without, would lose their magic power, and might as well be presented to a cold statue as to a man, were there no pre-ordained harmony between his mind and the material things around him'. Here it is difficult to escape the inference that Sedgwick (in the presence of the Master of Trinity) was implicitly alluding to the Lakes poet (Sedgwick 1834, 20–1).

It was, however, primarily the reasoning faculties of man with which Sedgwick was concerned. Although the mind may 'become bewildered among . . . the perpetual changes produced by material actions, of which we see neither the beginning nor the end', we find strong proofs of God in the study of animated nature through contrivance as a manifestation of 'intelligent super intending power'. Thus far the argument was essentially that of Paley — that contrivance proves design and hence an intelligent Designer (Sedgwick 1834, 21–4; Smith 1979, 60–4). And for Sedgwick, this proof from living beings held good whatever the nature of the *material* world:

> It is in vain that we attempt to banish an intelligent Creator, by referring all changes organic and inorganic to a succession of constant material actions, continued during an eternity of past time. Were this true, it would not touch our argument: and every clear instance of organic contrivance or material adaptation would be a phenomenon unexplained, except on the supposition of a contriver. It would only prove that, in a certain portion of space, God had thought fit to give a constant manifestation of his wisdom and power through an indefinite period of duration. *The eternity of material forms is, however, but a dream of false philosophy, unfounded in reason or analogy*; and, as far at least as organic nature is concerned, contradicted by the plainest physical records of the past world. (Sedgwick 1834, 24. My emphasis.)

Furthermore, Sedgwick claimed that the argument from the structure of living beings to the existence of 'an intelligent overruling power' also led us 'to the inevitable belief that all inanimate nature is also the production of the same

Geologists and mathematicians

overruling intelligence', through the adaptation of the organs of living beings to the condition of the material world. Thus 'by this adaptation, we link together all nature, animate and inanimate, and prove it to be one harmonious whole, produced by one dominant intelligence'(Sedgwick 1834, 24–5).

Turning to geological science itself, Sedgwick argued that the discoveries of this new science taught us 'that the manifestations of God's power on earth have not been limited to the few thousand years of man's existence. Each successive era displayed marks of skill and wisdom. Thus geology 'adds to the greatest cumulative argument derived from the forms of animated nature, by showing us new and unlooked-for instances of organic structure adjusted to an end, and that end accomplished'. Not only then did the external conditions of the earth change, but God had adapted at many successive times the organs of living beings to those changes. And so 'the great first cause continues a provident and active intelligence' (Sedgwick 1834, 25–7).

In examining the moral consequences of a sound study of the inductive sciences, Sedgwick was anxious to undermine as unphilosophical, 'as a dream of false philosophy', the eternity of material forms. The incontrovertible evidence of the life sciences showed both contrivance and adaptation in animate beings and hence that they, at least, had not been the product of eternal material forms. The incontestable evidence of geology showed the succession of material arrangements on the earth's surface, and the evidence of the successive adaptation of living beings to those changes. Once again, the living beings could not be eternal material forms. Furthermore, the geologist 'traces these changes backwards through each successive era, till he reaches a time when the monuments lose all symmetry, and the types of organic life are no longer seen. He has then entered on the dark age of nature's history; and he closes the old chapter of her records'. (Sedgwick 1834, 26.) As he summarised his stance in a letter to Hugh Miller in 1849: 'I hold (against the Huttonians) that Creation had a beginning in time, and that this conclusion is based on inductive evidence' (Clark and Hughes 1890, 2:161). Such, then, was the essence of Sedgwick's progressionist geology, linked as it was to 'inductive science', to *natural* theology, and to a consistency with the *revealed* truths of scripture.

Sedgwick concluded this key part of his *Discourse* with a brief reference to far more speculative questions connected to geological science, namely, proofs of a higher temperature in past ages as shown by older organic forms, by the crystalline structure of lower strata, and by the figure of the Earth — making probable a condition of primeval fusion and the original 'dissipated' state of the Earth through planetary space. Here, too, the lesson was important:

Speculations like these, starting at least from actual phenomena, are not without their use. For without lowering one jot the proof of a pre-ordained intelligence, they point, through a long succession of material changes, towards a beginning of things, when there was not one material quality fitted to act on senses like our own; and thus they take from nature that aspect of unchangeableness and stern necessity which has driven some men to downright atheism, and others to reject all natural religion. (Sedgwick 1834, 28.)

Lyell's uniformitarianism and non-progression *need* not be atheistic and, as we saw earlier, Lyell had protested against Scrope's charge of having advocated an eternal system (Lyell 1830–33, 3:383). Nevertheless, both Sedgwick and Whewell were fully aware of how much more consistent were progressionist doctrines with those of revealed and natural theology. And though the future might lead to 'a cold and dismal region, where our eyes behold none but the appalling forms of nature's dissolution', yet for Sedgwick God would lead us [that is, the members of Trinity College] through the valley of the shadow of death to a land where 'death and darkness have heard the doom of everlasting banishment' (Sedgwick 1834, 31).

Sedgwick had thus charted in his *Discourse* the specific moral significance of geological studies. Given that physical astronomy was 'the peculiar boast and glory of the inductive method' (Whewell 1831*a*, 397), and that Herschel in 1830 had declared that 'Geology, in the magnitude and sublimity of the objects of which it treats, undoubtedly ranks, in the scale of the sciences, next to astronomy' (Herschel 1830, 287; Cannon 1961; Wilson 1974, 84–5; Rupke 1983, 181), it was the aim of Sedgwick and his Cambridge colleagues to attempt to elevate geology to a similarly high and secure status by employing the right methodology and so avoid the fallacies of mere metaphysical speculation. Only in this way could natural theology *and* geological science truly benefit.

It was indeed significant that Whewell dedicated his *Philosophy* to Sedgwick with the remark that if 'your life had not been absorbed in struggling with many of the most difficult problems of a difficult science, you would have been my fellow-labourer or master in the work which I have here undertaken' (Whewell 1840, 1:111). We shall therefore turn to an analysis of William Whewell's methodological pronouncements, many of which post-date the *early* physical geology of William Hopkins but which will serve to illuminate several important and hitherto-unrecognised features of Hopkins' initial researches. In so doing, the social and intellectual solidarity of this Cambridge school of geological science, inspired by Sedgwick and forged amid the controversies with Lyell, will become more apparent.

Whewell: the science of geological dynamics and the philosophy of palaetiology
Whewell, like Sedgwick, held that most past geological theories had been speculative and premature. Wernerians and Huttonians alike had moved too quickly from observations of the structure and arrangement of the materials of the earth to an account of the changes that had brought about these effects. But, Whewell argued:

> it did not at once occur to them to suspect, that their common and extemporaneous judgment on such points was far from sufficient for sound knowledge; — they did not

Geologists and mathematicians

foresee that they must create a special science, whose object should be to estimate the general laws and effects of assumed causes [such as water or heat], before they could pronounce whether such causes had actually produced the particular facts which their survey of the earth had disclosed to them. (Whewell 1837, 3:546.)

In his *History*, Whewell continued to call that special science 'geological dynamics'. Later, in his *Philosophy*, Whewell substituted the term 'geological aetiology' for reasons which will soon become apparent.

Geological dynamics was, for Whewell, almost exactly analogous to the role of dynamics in astronomy. Thus:

When phenomenal astronomy had arrived at a high point of completeness ... especially by the discovery of Kepler's laws, astronomers were vehemently desirous of knowing the causes of these motions .... Men needed a science of motion, in order to arrive at a science of the heavenly motions .... Till that task was executed, all the attempts to assign the causes of cosmical phenomena were fanciful guesses and vague assertions; after that was done, they became demonstrations. The science of *Dynamics* enabled philosophers to pass securely and completely from *Phenomenal Astronomy* to *Physical Astronomy*. (Whewell 1837, 3:547.)

In like manner, Whewell explained, we needed a science of geological dynamics in order to advance from phenomenal geology to physical geology. Geological dynamics would thus be a science 'which shall investigate and determine the laws and consequences of the known causes of changes such as those which geology considers; — and which shall do this, not in an occasional, imperfect, and unconnected manner, but by systematic, complete, and conclusive methods; shall, in short, be a science, and not a promiscuous assemblage of desultory essays'. It would be a demonstrative science dealing with general cases and so would be distinguishable from theoretical or physical geology itself, 'in which we apply our principles to the explanation of the actual facts of the earth's surface' (Whewell 1837, 3:548).

Geological dynamics, however, was still so new, Whewell continued, that he believed it scarcely possible to give any historical account of its progress, or any complete survey of its shape and component parts. Indeed, its history hardly extended beyond Lyell's *Principles*. Its subject matter, in part treated of by Lyell, comprised the causes of change in the inorganic and organic world, the inorganic causes being:

the aqueous causes of change, or those in which water adds to, takes from, or transfers, the materials of the land: — the igneous causes; volcanoes, and, closely connected with them, earthquakes, and the forces by which they are produced; the calculations which determine, on physical principles, the effects of assumed mechanical causes acting upon large portions of the crust of the earth; the effect of the forces, whatever they be, which produce the crystalline texture of rocks, their fissile structure, and the separation of materials, of which we see the results in metalliferous veins. Again, the estimation of the results of changes of temperature in the earth, whether operating by pressure, expansion, or in any other way; the effects of assumed changes in the superficial condition, extent, and elevation, of terrestrial continents upon the climates of the earth; the effect of assumed cosmical changes upon the temperature of this planet .... (Whewell 1837, 3:549–50.)

The progress of geological dynamics, moreover, like the progress of all other inductive sciences, depended on the formulation and establishment of laws of greater and greater generality into which existing laws were resolved (Whewell 1837, 3:552). Whewell emphasised throughout this part of his *History*, however, that the actual progress of geological dynamics was very limited. Thus, with regard to the aqueous causes of change, he concluded that 'probably several generations must elapse before this portion of geological dynamics can become an exact science'. Equally, he concluded that the various hypotheses on the causes of volcanoes and earthquakes could hardly be considered as sufficiently matured for calculation or derivation of the effects (Whewell 1837, 3:554, 558).

On the other hand, the doctrine of central heat, despite objections, seemed a promising part of geological dynamics. Compared to Poisson's 'arbitrary' and 'improbable' hypothesis that the earth's internal heat may be attributed to the earth having once passed through a region of space with higher temperature than at present, Fourier's doctrine of central heat 'is not only naturally suggested by the subterraneous increase of temperatures, but explains the spheroidal figure of the earth; and falls in with almost any theory which can be devised, of volcanoes, earthquakes, and great geological changes'. Furthermore, the doctrine did not necessarily imply 'the *universal* fluidity of the mass' and so objections of the kind that there must be tides in such a fluid were rendered less serious (Whewell 1837, 3:558–63). In this doctrine, as we shall see, Whewell (and especially Hopkins) had a principle of enormous generality and simplicity which could link together a great diversity of physical phenomena. Such a capability provided Whewell with his chief criterion for the truth of theories, the criterion of consilience.

In his *Philosophy*, Whewell changed the terminology from geological *dynamics* to geological *aetiology* in recognition of the fact that 'in a large portion of the subject the changes are so utterly different in their nature from any modification of motion' such that the term 'dynamics' sounded 'harsh and strange' (Whewell 1840, 2:102). In order words, this science treated not merely of subterraneous forces, but of causes of change of climate and of causes which modified the forms and habits of organic beings. Thus geological dynamics or aetiology was not limited to — though it did include — a general science of force by strict analogy with the laws of motion, but was essentially a general science of causes and the laws of those causes, having, like the laws of motion, a non-hypothetical relation to directly observable phenomena of the present.

From what has been said already, it will be evident that there were, for Whewell, three members of a fully-developed science of geology — phenomenal geology, geological aetiology or dynamics, and geological theory. As we shall

Geologists and mathematicians

see, the first two members were in effect steps towards the theory: once established, the theory becomes inclusive of, or encompasses, both the phenomena and the aetiology.

Phenomenal geology was, of course, descriptive and historical, requiring systematic classification according to natural classes such as strata distinguished either by their respective mineralogical character or by the fossil remains which they contained. Once a suitable nomenclature had been fixed, the order of strata could be applied and tested elsewhere in the world and the resulting knowledge exhibited in the form of geological maps (Whewell 1840, 2:103–4). Furthermore, phenomenal geology embraced not only classes but also laws, particularly of *geometrical* form: 'the general form of mountain chains; the relations of the direction and inclination of different chains to each other; the general features of mineral veins, faults, and fissures; the prevalent characters of slaty cleavage; — were the subjects of laws established ... by extensive observation of facts'. As we have seen, such geometrical descriptions constituted Sedgwick's distinctive style and, as we shall see in the final section of this chapter, this branch of phenomenal geology had already formed the starting point for William Hopkins' researches in physical geology. Whewell thus remarked on the importance of phenomenal geology as a first step towards theory:

> ... in all attempts to trace back the history and discover the origin of the present state of things, the portion of the science which must first be formed is that which classifies the phenomena, and discovers general laws prevailing among them. When this work is performed, and not till then, we may begin to speculate successfully concerning causes, and to make some progress in our attempts to go back to an origin. We must have a *Phenomenal* science preparatory to each *Aetiological* one. (Whewell 1840, 2:109.)

Three important additional points with regard to a phenomenal science may be noted. First, the explicit emphasis on the classification of natural objects and phenomena as the basis for inductive generalisation characterised John Herschel's 1830 *Preliminary Discourse* and much of his (and his father's) 'Herschelian survey of the heavens' (Whewell 1840, 2:104–6). Moreover, Herschel's particular reference to Kepler's laws as 'the most important and beautiful system of geometrical relations which have ever been discovered by a mere inductive process, independent of any consideration of a theoretical kind' (Herschel 1830, 268) provides an obvious contextual source for Whewell's reference to the discovery of Kepler's laws as the high point of phenomenal astronomy. Indeed, Herschel regarded the laws as comprising 'within them a compendium of the motions of all the planets' — in other words, as a geometrical classification or description of the planetary phenomena.

Secondly, the emphasis on a geometry of motion provided Whewell (and his Cambridge colleague Robert Willis) with a pure science of motion, called kinematics (following Ampère). Kinematics considered motion solely in terms of space and time, and quite independently of force. The pure doctrine of motion in turn gave rise to a science of pure mechanism, independent of force, and illustrated by the mechanical contrivances employed in manufacturing. A study of the teeth of wheels, for example, would be concerned with the transmission and modification of motion without reference to force. Such was the aim of Willis's *Principles of Mechanism* of 1841 (Whewell 1840, 1:144–7; Willis 1841, v–ix; Todhunter 1876, 2:290). This Cambridge emphasis on kinematics had far-reaching significance for the character of British mathematical physics later in the century. William Thomson, for instance, regarded kinematical description as the foundation for dynamics itself, and constructed his mathematical physics in terms of pure motion and pure mechanism without reference to force (especially Wise 1982, 185–90).

Thirdly, Whewell made geometrical conceptions fundamental to his whole philosophy of the inductive sciences (as well as to the Cambridge ideal of a liberal education through the Mathematics Tripos). Thus he enunciated two closely related rules in his *Philosophy*:

> [First] that in collecting facts which are to be made the basis of science, the facts are to be observed, as far as possible, *with reference to place, figure, number, motion*, and the like conceptions; which depending upon the ideas of space and time, are the most universal, exact, and simple of our conceptions .... [Secondly] that though these relations of time and space are highly important in almost all facts, we are not to confine ourselves to these: but are to consider the phenomena *with reference to other conceptions also*: it being always understood that these conceptions are to be made as exact and rigorous as those of geometry and number. Thus ... the science of Mechanics arose from not only observing motions as they take place in time and space, but further, referring them to *force* as their *cause*. (Whewell 1840, 2:198–9.)

Whewell regarded this 'decomposition of facts' into 'elementary facts' relating to the 'Ideas' of time, space, cause, and so on, as the beginning of all science or exact knowledge, preceding discovery of the laws of nature.

Whewell's second step towards a sound geological theory was thus to formulate a rigorous science of geological dynamics or aetiology in order to avoid foundering amid premature hypotheses. In geology, as in the mechanical sciences, the Idea of Cause was fundamental for Whewell, but the causation was historical rather than purely mechanical when we came to construct a geological *theory* — an example of what Whewell designated as a palaetiological science combining the first and second steps into a new relationship:

> ... the class of Sciences which I designate as *Palaetiological* are those in which the object is to ascend from the present state of things to a more ancient condition, from which the present is derived by intelligible causes .... Geology examines the existing appearances of the materials which form the earth, infers from them previous conditions, and speculates concerning the forces by which one condition has been made to succeed another ... [and] we may in like manner turn our thoughts towards

Geologists and mathematicians

the first condition of the solar system, and try whether we can discern any traces of an order of things antecedent to that which is now established .... (Whewell 1840, 2:95–7.)

The problem for such a science, then, was 'to determine the manner in which each term is derived from the preceding, and thus, if possible, to calculate backwards to the origin of the series'. In each palaetiological science 'we consider some particular order of phenomena now existing: from our knowledge of the causes of change among such phenomena, we endeavour to infer the causes which have made this order of things what it is: we ascend in this manner to some previous stage of such phenomena; and from that, by a similar course of inference, to a still earlier stage, and to *its* causes' (Whewell 1840, 2:100). It was here, of course, that the need for a sound knowledge of the phenomena and a sound geological dynamics was most strongly felt as vital components in any theory of the actual facts. Whewell also recognised the need for separate studies of the various kinds of causation, aetiologies which might form 'a kind of progression which we may represent to ourselves as having acted *in succession* in the hypothetical history of the earth and its inhabitants', ranging from a study of the condensation of a nebula to the progress of civilisation itself (Whewell 1840, 2:113–16). Thus the causes would form successive links of a chain of events extending 'from the beginning of things down to the present time' and together making up the sum total of past time (Whewell 1840, 2:112–13). In 'speaking hypothetically', Whewell at the same time wanted 'the true efficacy of such [natural] causes' to be made the subject of 'scientific examination' such that any *inadequacy* of these causes might be revealed, and the need for 'supernatural influences' contemplated. His view and that of Sedgwick with regard to creation and progression thus had a great deal in common. Inductive science could point to indications of an origin: inductive truths and revealed truths were ultimately consistent the one with the other (Whewell 1840, 2:116, 136–7, 157–65).

Whewell also considered the *mode* of cultivating aetiology in geological science by following his well-known method of taking 'the most successful and complete examples which we possess of such portions of science'. Limited though such examples were, Lyell in particular offered an aetiology in which causes had been studied 'by attending to their action in the phenomena of the present state of things, and by inferring from this the nature and extent of the action which they may have exercised in former times'. Whewell, however, raised the vital question of the limits of the present period. At this point, his divergence from both Lyell and Herschel becomes apparent as shown in a recent and useful study by Michael Ruse (Ruse 1976). Ruse's conclusion that 'because of their different criteria for what constitutes a *vera causa*, Herschel and Whewell made different evaluations of Lyell's work' seems to me a valid and important one, though his additional claim that 'Herschel and Whewell approached Lyell as empiricist and rationalist respectively' could be especially misleading for an understanding of the aims and methods of the author of the *History* and *Philosophy of the Inductive Sciences* (Ruse 1976, 130).

Writing in his *Preliminary Discourse*, Herschel illustrated the notion of *verae causae* — 'causes recognised as having a real existence in nature, and not being mere hypotheses or figments of the mind' — by reference to two 'actualistic' examples from cosmical physics, namely, shells found in rocks far above sea level, and changes of terrestrial climate and temperature. Of all possible causes of the phenomenon of shells, the only candidate for a *vera causa* was that supported 'by all modern geologists, with one consent': 'the life and death of real mollusca at the bottom of the sea, and a subsequent alteration of the relative level of the land and sea'. Similarly, the climatic changes had various possible causes, some of which were not candidates for *verae causae*:

> Some consider the whole globe as having gradually cooled from absolute fusion; some regard the immensely superior activity of former volcanoes, and consequent more copious communication of internal heat to the surface, in former ages, as the cause. Neither of these can be regarded as real causes in the sense here intended; for we do not *know* that the globe has so cooled from fusion, nor are we sure that such supposed greater activity of former than of present volcanoes really did exist. A cause, possessing the essential requisites of a *vera causa*, has, however, been brought forward [by Lyell] in the varying influence of the distribution of land and sea over the surface of the globe: a change of such distribution, in the lapse of ages, by the degradation of old continents, and the elevation of new, being a demonstrated fact .... (Herschel 1830, 146–7.)

Another candidate for a *vera causa* here would be the fact of the actual slow diminution of the eccentricity of the earth's orbit round the sun, but in both cases Herschel stressed the need for further examination.

When Herschel turned to geology itself in part three of the *Preliminary Discourse*, he adopted a very cautious approach towards cosmical speculation, suggesting on the one hand that 'the researches of physical astronomy are confessedly incompetent to carry us back to the origin of our system' and on the other that these researches and those of the geologist do not 'give us any ground for regarding our system, or the globe we inhabit, as of eternal duration'. Again, although the figure of the earth pointed obscurely to an origin and formation, the aims of modern geological science had to be more modest:

> ... from such indications nothing distinct can be concluded; and if we would speculate to any purpose on a former state of our globe and on the succession of events which from time to time may have changed the condition and form of its surface, we must confine our views within limits far more restricted, and to subjects much more within the reach of our capacity, than either the creation of the world or its assumption of its present figure. These, indeed, were favourite speculations with a race of geologists now extinct .... (Herschel 1830, 282.)

Herschel's cautious approach and his version of *verae causae*, then, aligned him with Lyell rather than with Sedgwick and Whewell. Like Lyell, his methodology

Geologists and mathematicians

certainly seemed to rule out a doctrine of central heat as a *vera causa* and indeed to restrict investigation to the 'present period' in Whewell's palaetiological programme of a decade later.

Whewell's question 'What are the limits of the present period?' must be seen in this context. He did not wish to restrict geological and cosmical science to the very limited region which too narrow a doctrine of *verae causae* would seem to imply. This narrow version appeared, for Whewell, in Lyell's version of uniformitarianism which implied equality of intensity as well as of kind in the agents of geological dynamics. Whewell therefore argued that such limitations were entirely arbitrary:

> ... our estimate of the efficacy of known causes will vary with the extent of the effects we ascribe to them .... Thus when the vast masses of trap rocks in the Western Isles of Scotland ... were ... identified as to their nature with the products of recent volcanoes, the amount of effect which might justifiably be ascribed to volcanic agency was materially extended. In other cases, instead of observing the current effects of our geological causes, we have to estimate the results from what we know of the causes themselves; as when, with Herschel, we calculate the alterations in the temperature of the earth which astronomical changes may possibly produce; or when, with Fourier, we try to calculate the rate of cooling of the earth's surface, on the hypothesis of an incandescent central mass. (Whewell 1840, 2:117–18.)

Our aetiology, Whewell concluded, 'is constructed partly from calculation and reasoning, partly from phenomena'. Reasoning from phenomena often led to laws of phenomena becoming well established, whereas the connection of the phenomenon with its cause was by implication a longer-term goal: 'thus the law of subterranean heat, that it increases in descending below the surface, is now well established, although the doctrine which ascribes this effect to a central heat is not universally assented to'.

If Herschel had seemed to rule out a doctrine of central heat, Whewell appeared to take an opposite tack. Whereas Herschel regarded this as a speculative hypothesis, and certainly no *vera causa*, Whewell's philosophy of theory construction opened the way for a justification of the doctrine. Having obtained the 'phenomenology' and 'aetiology' of the subject, 'we are prepared for the third member which completes the science, the *Theory* of the actual facts':

> The term Theory, when rigorously employed in such sciences as those which we here consider, ... implies a consistent and systematic view of the actual facts, combined with a true apprehension of their connexion and causes .... And if the term *Theory* be here employed we must recollect that it is to be understood, not in its narrower sense as opposed to facts, but in its wider signification, as including all known facts and differing from them only in introducing among them principles of intelligible connexion. The Theories of which we now speak are true Theories, precisely because they are identical with the total system of the Facts. (Whewell 1840, 2:121–2.)

Crosbie Smith

In order to apprehend fully Whewell's meaning here, we must realise that his epistemology consisted of the application of Ideas (space, time, cause, and so on, which did not derive from experience but which were conditions under which the mind receives impressions of sense) to Facts. Thus 'ideas are the *Form*, facts the *Material*' of our knowledge. In this sense Whewell's epistemology had a necessary, rational, *a priori* component with its immediate source in the active human mind, and a contingent, empirical, *a posteriori* component deriving from sense experience. Induction was therefore more than a mere generalisation of particular facts: 'a new mental Element is superinduced such that the facts are 'seen in a new point of view' (Whewell 1840, 1:xvii–xl; 2: 169:259). Whewell could then argue that theory was not opposed to fact, but differed only in introducing among the facts 'principles of intelligible connexion'. In other words, 'the distinction of *Fact* and *Theory* is only relative. Events and phenomena, considered as particulars which may be colligated by induction, are *Facts*; considered as generalities, already obtained by colligation of other Facts, they are *Theories*.' (Whewell 1840, 1:xli.)

As we have seen in the case of the palaetiological sciences, Whewell argued that inductive truths were of two kinds, laws of the phenomena and theories of causes. Since he did not wish to restrict knowledge to laws of the phenomena, he needed a new doctrine of *verae causae* as a key criterion for a 'true theory'. He stressed that agreement with known facts and prediction of new ones were important, but not sufficient, criteria for the truth of theories. Equally, the notion of a *vera causa* in Herschel's 'actualistic' sense was 'an injurious limitation of the field of induction. For it forbids us to look for a cause except among the causes with which we are already familiar.' To limit ourselves to everyday causes would rule out acquaintance with any new causes. To interpret the notion of *vera causa* as simply urging that 'only causes of such kinds as we have already satisfied ourselves do exist in nature' would not proscribe any hypothesis since 'some close similarity with some known kind of cause is requisite, in order that the hypothesis may have the appearance of an explanation'. But if the cause were to explain two separate classes of fact rather than one class only, we have 'the testimony of two witnesses in behalf of the hypothesis; and in proportion as these two witnesses are separate and independent, the conviction produced by their agreement is more and more complete'. Such a coincidence 'does give a reality to the cause', and as such forms Whewell's key doctrine of consilience of inductions (Whewell 1840, 2:440–8).

Elsewhere is his *Philosophy*, Whewell defined the consilience of inductions as taking place 'when an Induction obtained from one class of facts coincides with an Induction obtained from another different class'. This provided the principal 'test of the truth of the Theory in which it occurs' (Whewell 1840, 1:xxxix; 2:230–9). No example could be offered, Whewell argued, in which consilience gave testimony in favour of a hypothesis afterwards found to be false. The result of consilience was a constant convergence towards simplicity and unity,

Geologists and mathematicians

exemplified by both the theory of gravitation and the wave theory of light. Whewell's methodology, which cannot easily be categorised as rationalist in the simple sense of working *to* experience *from* the unseen (Ruse 1976, 130), may be summarised in his own words:

> It appears, then, to be required, both by the analogy of the most successful efforts of science in past times and by the irrepressible speculative power of the human mind, that we should attempt to discover both the *laws of phenomena*, and their *causes*. In every department of science ... these two great steps of investigation must succeed each other .... In both we must, as far as possible devise hypotheses which, when we ... test them, display those characters of truth of which we have spoken; — an agreement with facts such as will stand the most patient and rigid inquiry; a provision for predicting truly the results of untried cases; a consilience of induction from various classes of facts; and a progressive tendency of the scheme to simplicity and unity (Whewell 1840, 2:270).

Were the doctrine of central heat, then, to manifest such criteria, it would clearly occupy a fundamental position in the palaetiological sciences. Whewell, however, like Herschel, was cautious, almost sceptical, about the actual progress, stating that as yet no such theory as a sound palaetiological theory existed on any subject. Geological theory remained in a premature state, not having advanced beyond a few conjectures. Yet he had opened up the possibilities and loosened the restrictions of geological theorising, leaving to William Hopkins (and later William Thomson) the challenge of constructing a more mature palaetiological theory.

The physical geology of William Hopkins

William Hopkins belonged to the same generation as Sedgwick and Whewell but never reached the level of their high institutional status. Graduating as seventh wrangler in 1827 from St. Peter's College, he had entered the University comparatively late in life, and on account of marriage was ineligible for a college fellowship. As a result he became one of Cambridge's best-known mathematical coaches or private tutors with the title of 'the senior wrangler maker'. By 1849 he could claim nearly two-hundred wranglers among his pupils, of whom seventeen were senior wranglers and over forty in one of the top three places, among them Stokes and Thomson.

According to William Thomson's 1874 obituary notice of Archibald Smith, the latter joined Hopkins' reading party at Barmouth in North Wales during the summer of 1832. Sedgwick's biographers record the geology professor's excursion in North Wales after the Oxford meeting of the British Association and his failure to return to Wales in 1833. The authors of Hopkins' entries in the *Dictionary of National Biography* and the *Dictionary of Scientific Biography* note that he acquired his enthusiasm for geology from Sedgwick in 'about 1833', and the author of the latter article adds that Hopkins' intense interest followed excursions with Sedgwick near Barmouth. We may thus infer that Hopkins' introduction to practical geology took place as early as 1832. At the same time, Sedgwick and Hopkins were close personal friends in a Cambridge context (Clark and Hughes 1890, 2:153-4, 323).

Inspired by his friendship with Sedgwick, Hopkins read the first of numerous memoirs on geological subjects to the Cambridge Philosophical Society in 1835. This memoir, 'Researches in physical geology', was subsequently published in that society's *Transactions*, being supplemented by an important and extensive 'abstract', setting out more of his methodological position, in the *Philosophical Magazine* of 1836. In what follows, I shall show how closely Hopkins' investigative procedures relate to the strategy of Sedgwick and Whewell outlined in previous sections.

In the abstract, Hopkins noted that an impression had been 'too frequently created, that little hope exists of elevating the science [of geology] to any rank among the stricter physical sciences'. Such a notion, he argued, was 'most fatal to its healthy progress'. He believed that whatever might be said against Lyell's 'theoretical views', the author of the *Principles of Geology* 'must be allowed by all to have set us an example well calculated to improve in this respect the tone of geological speculation'. Hopkins therefore aimed in his own investigations to act on the same assumption — that the science *could* be elevated to some such rank among the stricter physical sciences — 'as the only one on which, if we are to speculate at all, we can speculate with safety; and if, perchance, a somewhat vague and misty sublimity which has appertained to this branch of the science should thus be diminished, ample compensation will be made if we should in return confer upon it a portion, however small, of the more naked dignity of demonstrative truth (Hopkins 1836, 365–6).

Hopkins began from a clearly defined phenomenal geology, comprehended in this case by geometrical laws. Thus 'notwithstanding the appearances of irregularity and confusion in the formation of the crust of our globe ... geologists have been able in numerous instances to detect, in the arrangement and position of its stratified masses, distinct approximations to geometrical laws'. Such general laws, Hopkins believed, were easily recognised in the phenomena of anticlinal lines, faults, fissures and mineral veins. Given the establishment of these laws as general and universal, rather than merely local, 'the phenomena alluded to are referrible not to the particular and irregular action of merely local causes, but to the more widely diffused action of some simple cause, general in its nature with respect to every part of the globe, and general in its action at least with respect to the whole of each district throughout which the phenomena are observed to approximate, without interruption, to the sàme geometrical laws' (Hopkins 1835, 1–2).

The principal aim of Hopkins' 1835 memoir, then, was 'to examine how far such relations do exist between our observed phenomena and a certain general cause to which they may be attributed'. Examination of the many phenomena

Geologists and mathematicians

and the establishment of geometrical laws pointed to 'the action of some powerful elevatory force acting beneath the superficial crust of the globe, and thus producing those elevations and dislocations which we now witness'. In particular, the law of approximate parallelism (according to which fissures within a given geological region tend to occur in two sets, each set consisting of parallel fissures but the two sets running perpendicular to one another in a cross-hatched pattern) 'which equally characterises the phenomena of anticlinal lines, faults, and mineral veins, affords, *a priori*, a strong probability that they are all assignable to the same general cause'. Again, the stratified beds originally deposited from water were now no longer horizontal but often much inclined. The inference of a powerful elevatory force seemed inescapable. Hopkins thus supposed 'this elevatory force, whatever may be its origin, to act upon the lower surface of the uplifted mass through the medium of some fluid, which may be conceived to be an elastic vapour, or in other cases a mass of matter in a state of fusion from heat'. Having thereby recognised 'certain well defined geological phenomena, distinctly approximating to geometrical laws' and having also 'a distinct mechanical cause to which geologists, with almost one consent, have agreed in considering them to be assignable', the third step for Hopkins was 'to institute an investigation, founded on mechanical and physical principles, of the necessary relations that may exist between our observed phenomena and the general cause to which we attribute them'. The nature of the investigation, Hopkins hoped, 'will be deemed a justification of my introduction of a new term into the science, that of *Physical Geology*' (Hopkins 1835, 8-11).

From our discussions of Sedgwick and Whewell in the preceding sections, two points need emphasising. First, Hopkins' employment of geometrical descriptions reflects Sedgwick's basic investigative procedure and corresponds to the first member of Whewell's palaetiological trio, phenomenal geology. Hopkins thus developed a geometry of geological motions — a kinematics — as the first stage in his physical geology. Secondly, the establishment of a simple, general cause *qua* elevatory force corresponds to the second member of Whewell's trio, geological dynamics or aetiology. The analogy with astronomy here is very striking. As in dynamics proper, the simple, general cause has a status of certainty almost equal to that of the geometrical laws. It is closely linked to observation, is non-speculative, and involves no specific reference to unobservable, hypothetical entities. It thus carries the minimum number of assumptions, the basic assumption being simply that geological phenomena have a general mechanical cause. As a result, it attracts the maximum consent from geologists. The method is, of course, *in a sense* hypothetico-deductive in the way that Newton's *Principia* is hypothetico-deductive rather than naively inductive. Yet the general, observable character of the method is very clearly demarcated from the usual hypothetico-deductive model in which hypotheses involving unobservable entities such as molecules are employed to explain a separate set of observable facts. Furthermore, the latter method appeared to

Cambridge mathematicians to be arbitrary and *ad hoc*, leading to the premature theories which had been characteristic of earlier geological science.

In conducting his investigation by the methods 'supplied by mathematical analysis', Hopkins was well aware of the objection that 'the application of these methods to geological problems may appear like an affectation of an accuracy which the nature of the subject may not be conceived to admit of'. Such an objection, however, was for Hopkins equivalent to asserting that the branch of geology 'with which we are immediately concerned presents no phaenomena characterised by general laws, or referrible to a definite and simple cause'. But 'the phaenomena do distinctly approximate to obvious geometrical laws, and there is a simple cause to which they may be referred, the effects of which it has been my object ... to investigate on mechanical principles, in order that we may compare the laws obtained from these results with those to which the observed phaenomena are found to approximate'. Thus, while the procedure required some abstraction and approximation, it nevertheless represented the best approach possible:

> The most obvious cause of deviation in our phenomena from strict geometrical laws, is irregularity in the intensity of the elevatory forces, and in the constitution of the masses on which they are supposed to act. Abstracting these sources of uncertainty, we have before us a definite problem, viz., to determine the nature of the effects produced by a general elevatory force acting at any assigned depth on extended portions of the superficial crust of the earth, and with sufficient intensity to produce in it dislocations and sensible elevations. To this simple and definite form the problem may be reduced; and at least a correctly approximate solution of it must necessarily be obtained ... before we can pronounce on the adequacy of the assigned cause to produce the observed effects .... This approximate solution is what I have now to offer .... (Hopkins 1835, 10.)

Hopkins thus challenged his potential critics to offer a solution 'equally conclusive and available by some method more adapted to the general reader, but added with confidence that the problem could in fact 'admit of no accurate solution independently of reasoning too intricate to be clearly embodied in any language but that of mathematical analysis'.

The simplest case which Hopkins considered, that of fissure formation in a thin lamina, will serve as a convenient illustration of his procedure for deriving the geometrical laws from the action of the elevatory force. In Hopkins' diagram, AB is an arbitrary line in the lamina and P any point in this line. We suppose the lamina 'acted on by external forces which place it in a state of tension, such that the direction of the tension at every point shall be parallel to a given line CD', and we seek 'the tendency of the forces of tension to separate the particles which are contiguous, but on opposite sides of the geometrical line Pp, by causing them to move parallel to CD'. The tendency is measured by pm. $F \sin \psi$ (BPQ = ψ), where F is the total tension per unit length across a line perpendicular to the tension. This result will be a maximum when AB and CD

Geologists and mathematicians

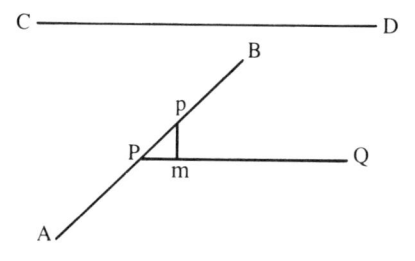

are at right angles and so the tear or fissure ought to occur along AB when it is perpendicular to the tension. In a similar manner, Hopkins worked towards more complicated systems of tension in thick strata and derived the law of parallelism and other phenomena using only analytic geometry.

Hopkins' geometrical style incorporated two other notable features. First, his assumption of 'contiguous particles' made possible his geometrical approach. Here the notion was unproblematic, since the particles were not molecules but small pieces or elements of matter containing very large numbers of molecules. Contiguous action of this kind was not only employed by Hopkins, but formed an integral part of the macroscopic approach of Stokes, Thomson and Maxwell, in which molecular considerations are eliminated (Wise 1982, 185–90). One could, therefore, with some necessary qualification, refer to a Hopkins school of mathematical physics. Although further discussion of this point would be beyond the confines of this chapter, it is clear that no adequate analysis of Cambridge mathematical physics can afford to ignore the central importance of William Hopkins.

Secondly, Hopkins derived not only statical results connecting tensions with fissures but also dynamical results for the propagation of a fissure from the point of its origin. Especially noteworthy was his claim that 'whatever may be the direction first given to the fissure by any local cause, its subsequent direction will soon become independent of that cause' (Hopkins 1835, 24). Thus the effects of propagation are the effects of contiguous action, independent of sources — an insight crucial in the development of field theory.

The *Philosophical Magazine* version of Hopkins' memoir concluded with a further important elucidation of his approach to physical geology. He emphasised again the importance of dealing with hypothetical problems which were 'chosen so as to bear the closest analogy to the corresponding ones which nature presents to us' and to which mathematical analysis was strictly applicable. Here the difference between his approach and one involving *ad hoc* hypothetical entities was evident. The closer the model came to observable nature the better. In forming 'a distinct idea of the necessary relations between any physical cause acting under complicated conditions, and its remoter consequences' or effects, we consider 'some comparatively simple but strictly analogous cause, and apply them to the actual one, with such limitations as circumstances may require'. Crucial to this procedure was the advantage of mathematical techniques; the standard case, or hypothesis, or model, to which the more complex problem was referred, was 'a definite one, from which he [the mathematician] had means of deducing his results free from that uncertainty which necessarily attends other modes of investigation'. This approach, then, had as a first step the formulation of general geometrical laws, as a second step the construction of a general, simple hypothesis as the cause of those laws and phenomena, and as a third step the exploitation of the precision of mathematical analysis to deduce accurately the phenomena. Such was Hopkins' ambition for geological theories of elevation in particular — an ambition which, if achieved, would confer upon that branch of geological science something of the 'dignity of demonstrative truth'. As such, it relates very closely to Whewell's methodology and to the ideal of physical astronomy (Hopkins 1836, 364–6).

Hopkins' 1835 memoir carefully avoided 'any speculations respecting the interior constitution of our globe' and so 'on the causes which might produce this elevatory force'. In his *Philosophical Magazine* version, however, Hopkins went beyond the mere assumption of the existence of such an elevatory force and cautiously postulated the hypothesis of a cavity existing beneath the elevated mass. Thus 'any vapour of matter in a state of fluidity from heat, forced into this cavity, or expanded there, will produce the elevatory force which I assume to have acted'. For Hopkins, this internal heat seemed the simplest conceivable mode of producing such a force. An alternative hypothesis would be to suppose the expansion by some means of a portion of solid matter of the earth at a certain depth, an expansion which would 'elevate the superincumbent mass'. This hypothesis, Hopkins believed, would prove to be insufficient to account for observed phenomena.

By comparison, the cavity hypothesis would be consistent with views of the earth's original fluidity:

> ... *if* we choose to set out from the more remote hypothesis of the earth's having been originally fluid, it might probably be shown that the formation of cavities ... would ... be the necessary consequence of that process of cooling by which we must then suppose the crust of the globe to have assumed its present solidity .... *If* we adopt the hypothesis of internal cavities, we may observe that there is no reason why we should not suppose them to exist, not only at different depths in different places, but also along the same vertical line, so that one shall be placed under another. It might, I conceive, be shown to be highly probable, *if* we should again recur to the hypothesis of the original fluidity of the globe, that the deeper cavities would in such case be more extensive. (Hopkins 1836, 230–1. My emphasis.)

In these remarks we have the first hint of Hopkins' subsequent researches in physical geology. Having investigated the geometrical laws and established the role of a general elevatory force, Hopkins now sought to trace the antecedent physical causes of the present condition of the earth's surface. If we employ

Geologists and mathematicians

Whewell's terminology of 1840, then, we can see that Hopkins was moving from phenomenal geology and general geological dynamics to palaetiology, i.e. to attempting an ascent from the present state to a more ancient condition from which the present was derived by intelligible causes and perhaps eventually to arrive at an origin. In short, Hopkins began to construct a theory of the earth according to strict methodological rules. Let us therefore look ahead to the application of these rules in his 1847 'Report on the geological theories of elevation and earthquakes' to the Oxford meeting of the British Association.

Although at first sight a survey of the recent progress of investigations in physical geology — reviewing, for example, much of his own previous work (undertaken in the period 1835–47) on elevation and on precession and nutation with reference to the earth's internal fluidity — Hopkins' Report was in fact much more than merely a documentation of past and present research. Beginning with a thorough discussion of the phenomena and current rival theories of volcanoes, Hopkins aimed to move towards his own theory of volcanic activity which could not only account for the phenomena at present observed, but which would, above all, be capable of accounting for past elevation and also be fully consistent with his recent conclusions from studies of the earth's precession and nutation that a thin crust would be incompatible with the observed phenomena (Hopkins 1840, 193–208; 1842a, 43–56). In other words, by avoiding particular and *ad hoc* hypotheses of volcanoes, and, as we shall see, by seeking *consilience* as a primary criterion of the truth of a theory, Hopkins was constructing a firm foundation for a new theory of the earth along the methodological lines advocated by Whewell in 1837–40.

Hopkins' first preliminary methodological assumption was that 'in speculating on the causes which have operated at past geological epochs, we are naturally led to refer to present causes as at least suggestive of the probable nature of those which have formerly acted, however remote may have been the period of their action' (Hopkins 1847, 33). Thus the general phenomena of elevation presented analogies with the present-day effects of volcanic action. The likeness was in kind, rather than in intensity, implying that Hopkins like Sedgwick and Whewell was keen to distance himself from Lyell's uniformitarianism. For Hopkins, as he had already made clear in the 1836 paper, the starting point in theory construction should be as close to direct observation as possible: there should be the closest possible *analogy* between hypothetical and known causes. Here the strong resemblance to Newton's second rule of reasoning (viewed as a *methodological* guide) may be noted (Newton 1972, 2:550). Sedgwick himself had in fact drawn explicit attention to the importance of Newton's second rule for geological science by explaining that 'the one safe rule in all our inquiries' was: 'Effects similar in kind to those which are produced now, must in all former times have been produced by some corresponding power of nature' (Sedgwick 1830, 211). With respect to the physical causes of phenomena of elevation, then, the existence of masses of rock of igneous origin injected into sedimentary deposits in highly disturbed regions 'presents analogies with the effects of volcanic action at the present time which cannot fail to suggest the idea that the general phaenomena of elevation are attributable to the same kind of action' (Hopkins 1847, 33). This first methodological rule was therefore a guide to the initial construction of a hypothesis to account for the phenomena of elevation.

The second preliminary methodological assumption which Hopkins laid down was that the object of physical science in general was to explain the phenomena which the physical world presents to us 'by referring them to the agency of natural causes'. Yet 'in our more extended speculations on geological causation we are necessarily driven to the consideration of the limits beyond which we are compelled to recognise the operation of those higher causes of which physical science can take no cognizance' (Hopkins 1847, 37). Hopkins, like Whewell, was thus stressing the *limitations* of natural science with respect to such questions as ultimate origins and beginnings. Because he did not wish to discuss such questions, and because he sought to examine 'theoretical views, and the hypotheses on which they are founded', he laid down the following key methodological rule for guidance on this problem of separating legitimate theorising from illegitimate speculation:

> That in treating on the natural phaenomena which geology presents to us as objects of physical investigation, we must refer them to natural causes, so far as it can be proved that the phaenomena would be the necessary consequences of such causes acting under conditions the former existence of which may be deemed admissible. But in the application of this principle we are bound to be especially careful that the dependence of the phaenomena observed on the causes assigned be established by accurate reasoning, and not by vague assumptions. It is to the gross neglect of what the rule as above enunciated dictates in this respect that we must attribute so many indeterminate speculations and objectionable opinions which have been offered to the world as geological theories. It is easy to substitute speculative notions for rigorous demonstration, but let us not forget that no one can thus build up [such speculative] geological theories without violating all the rules of inductive philosophy, and overstepping the caution and modesty which are essential for our safe guidance in these remote regions of physical science. (Hopkins 1847, 37.)

Hopkins then considered in the light of this second methodological principle the two fundamental hypotheses on which theories of volcanoes have been founded — the first of which assumed that the earth had been in a state of fluidity at some former epoch, while the second assumed the earth to have been originally constituted with an unoxidised solid nucleus. The first hypothesis, Hopkins made clear, did not assume that a fluid state had been *the* primitive state of the matter composing our planet, but 'merely one antecedent to its present state'. Thus, this hypothesis did not interfere with 'the hypothesis of a still anterior gaseous state of the earth, or of the matter composing the whole solar system'. According to this first hypothesis, furthermore:

Geologists and mathematicians

the terrestrial mass must have been more entirely free to receive all the modifications of constitution and form which physical and mechanical causes may have tended to impress upon it up to the present epoch, than if it had always existed in a state of solidity; and consequently this hypothesis (considered merely as a fundamental assumption) must I conceive be preferred to any more restrictive one by those who are disposed to adopt the [above] principle ... respecting the recognition of secondary causes as the immediate agency employed in the production of natural phaenomena, whenever it can be proved that the phaenomena would result from such causes acting under admissible conditions. (Hopkins 1847, 34.)

Hopkins emphasised, however, that he was not determining the value of a hypothesis by antecedent conditions alone, 'independently of the inductive proofs by which it may be supported'. He therefore insisted that whatever hypothesis we adopt, 'we must ever recollect that the final test of its truth must always be sought in the process of an accurate deduction of consequences resulting from our original assumptions, and a careful comparison of such calculated results with observed phaenomena'. In no science, he added, is it so essential as in geology 'to bear in mind this great single rule of inductive philosophy, because in no other science is there the same temptation to violate it. It is alone in the general accordance and harmony of our theoretical deductions with the results of observation that we must seek the ultimate evidence of the truth of our theories'.

Hopkins also carefully reviewed the second fundamental hypothesis — that of a solid nucleus — and the theories of volcanoes founded upon it. This second hypothesis could be looked at in one of two ways. If, like the first hypothesis, it was not an attempt to define the absolutely primitive state of the globe, then it might well be demanded that at least some probable explanation be given of the assumed existence of the solid unoxidised nucleus as the consequence of a previous state of the terrestrial mass. If, on the other hand, 'the state assumed by this [second] hypothesis be regarded as the primitive state of the earth, the hypothesis would seem in itself to be less simple than that which, without professing to define the primitive state of matter, supposes our planet to have once been gaseous or fluid' (Hopkins 1847, 37). Indeed, he made clear that such a view of *primitive* solidity ran counter to the concept of geological science as a study of secondary causes by excluding the possibility of antecedent physical states (Hopkins 1847, 45). On either version, moreover, chemical theories of volcanic action led to grave difficulties of a mechanical nature (Hopkins 1847, 38–9). Upon such grounds, therefore, Hopkins rejected the second of the two fundamental hypotheses as a basis for a theory of volcanic action. At the same time he rejected any theory which relied on a comparatively small thickness of the earth's crust as incompatible with the observed phenomena of volcanoes and with his earlier findings from precession and nutation (Hopkins 1847, 33–4; 39–40; 49–51).

Hopkins therefore developed his own theory of volcanoes on the hypothesis of the former fluidity of the earth. But it was a careful, step-by-step analogical procedure reflecting the Cambridge philosophy of the inductive sciences. The phenomena of existing volcanoes presented analogies with the phenomena of elevation. Past volcanic activity was thus a probable cause of elevation. Present volcanoes were in general insulated phenomena, whose most likely cause lay in the existence of cavities within the solid crust of the earth. Thus a plausible cause of both present volcanoes and past elevation was the cavity hypothesis. Finally, the general hypothesis of former fluidity seemed a plausible way of accounting for the cavities as well as other phenomena. So far, however, the procedure was one of theory construction rather than establishment, though as with Whewell, 'discovery' and 'justification' were never rigidly separated.

The *establishment* of the theory proceeded according to strict methodological rules. At this second stage, 'our attention will not be limited to the explanation of volcanic phaenomena'. Hopkins thus proceeded 'on the hypothesis of the former fluidity of the earth' to 'examine the consequences deducible from it, the degree of their accordance with the phaenomena established by observation, and the value of the evidence thus derived in favour of our fundamental hypothesis, and the theory founded upon it' (Hopkins 1847, 40).

Hopkins believed his exposition of volcanic action to be 'the first attempt to explain distinctly by the operation of simple causes the mode in which existing volcanoes may be conceived to have originated in the fluid mass which is assumed to have formerly constituted the globe'. While emphasising that the validity of his conclusions depended upon the truth of assumed conditions, he pointed out that one of his principal aims in the Report was to indicate what those conditions were, and to suggest the kinds of experiment 'by which we may hope to test their admissibility'. Thus 'it is only by a mode of investigation in which the introduction of error is to be sought in the assumed conditions, and not in the general reasoning by which we deduce our results, that we can hope to arrive at any real approximation to demonstrative truth' (Hopkins 1847, 57).

Fundamental to his procedure was consilience. The hypothesis of former fluidity seemed to offer the most general, simplest, least *ad hoc* explanation of a great range of different phenomena many of which he had himself investigated mathematically in the preceding decade — the spheroidal figure of the earth, the earth's precession and nutation, volcanic action, and the phenomena of elevation. *Ad hoc* hypotheses could be introduced to account for any one of these phenomena, but the hypothesis of former fluidity alone offered the prospect of consilience, leading to progressive simplicity and unity among very diverse phenomena and taking us backwards in time towards an origin for the present state and condition of things.

The physical geology of Mr Hopkins was carried out on as grand and massive a scale as the phenomena he sought to investigate. Although I have scarcely begun to do justice to the extent and depth of his writings, enough has been said to indicate the distinctive characteristics of geological science among the

Geologists and mathematicians

Cambridge mathematicians which developed during the period 1820–50. Hopkins spoke truly for his associates when he welcomed the diminution of 'a somewhat vague and misty sublimity' in favour of 'the more naked dignity of demonstrative truth'. Sedgwick himself, whatever his earlier reservations concerning geology as an exact science (Sedgwick 1831, 301), in his 'Three letters on the geology of the Lake District' to William Wordsworth in 1842 wrote that geology now 'claims kindred with all the offspring of exact knowledge'. Echoing Whewell, he explained:

> Exact science [i.e. pure and mixed mathematics] is the creature of the human mind — a body of necessary truths built upon mere abstractions. But when physical phenomena are well defined, and their laws made out by long and patient observations, or proved by adequate experiments: they then, by an act of thought, may be made to pass into the form of mere abstractions, and so come within the reach of exact mathematical analysis: and many new physical truths, unapproachable in any other way, and far removed from direct observation, may thus be brought to light, and fixed as firmly as are the truths of pure geometry .... We may therefore hope, that as geology advances farther towards exactness, as a science of observation, its phenomena may be brought more nearly under the government of known mechanical laws, and more closely defined by the power of exact calculation.

Few remarks could capture more succinctly the spirit of the inductive sciences characteristic of the Cambridge of Sedgwick, Whewell and Hopkins. Geology had thus acquired a special dignity such that even Wordsworth confessed to Sedgwick in 1842 that while he had once sympathised with the imaginary 'Splenetic Recluse' in his 1814 poem, he now took the view that 'geology and *mineralogy* are very different things' (Clark and Hughes 1890, 2:41).

Acknowledgements

I am greatly indebted to John Brooke, Alex Dolby, Jim Secord, and Norton Wise for their helpful insights and constructive comments during the revision of this chapter. I would especially like to thank Jim Secord for permission to refer to his forthcoming book.

= 4 =
Mathematics and mathematical physics from Cambridge, 1815–40: a survey of the achievements and of the French influences

I. GRATTAN-GUINNESS

Introduction
The revival of mathematics in England in the 1810s and the 1820s after many decades of slumber is well known to have been largely led by men teaching and/or taught at Cambridge. In this chapter I describe this renaissance during the period 1815–40 from several points of view. First, during the British hibernation, France had been enjoying a golden period in these fields, and indeed continued to exercise prominence right through the period under study; I provide a brief summary of the corpus of French study and its principal figures up to about 1825, when the momentum of Cambridge work began to increase substantially. Next I turn to Britain in general, with a survey of the efforts to *diffuse* mathematical knowledge in encyclopaedia and survey articles and in introducing treatises. The following section concentrates on Cambridge itself, with a list of the principal research papers and books brought out in the period under study. The next three sections offer some interpretative comments on these sources; they deal in turn with the calculus and related topics, astronomy and mechanics, and the wide world of 'fluids'. In all sections some note will be taken of the stimulation provided by French mathematics. A few tentative conclusions and historical questions are indicated in a closing section.

Having outlined the scope of the chapter, I must indicate its limitations. First, I restrict myself to the period 1815–40 because after that developments enter a new phase with Adams, Cayley, Thomson and Stokes (some of the achievements of the latter two are studied elsewhere in this volume) and the fad for differential operators led by Boole and D.F. Gregory (a substantial history, which is only briefly noted here). Secondly, I have taken the phrase 'Cambridge mathematician' to refer to someone resident and/or working in Cambridge, or at least operating within the traditions learnt there. (This is hardly a sharp definition; it is discussed in more detail below.) Thirdly, I have confined to a few remarks here and there the notable and rather different contemporary traditions operating outside Cambridge. Fourthly, I have concentrated largely on the

Mathematics and mathematical physics

research-level activity, and considered educational aspects only secondarily. A detailed (comparative) study of the history of mathematics education in Britain at this time, at both university and school levels, would be very desirable.[1] Fifthly, I have confined my attention to mathematics which bore closely on the mathematical physics of the time; namely, the calculus (or, later, mathematical analysis) and related topics. Sixthly, I have been unavoidably perfunctory on the prehistory of my chosen period; the historical figures themselves consciously felt that they were inaugurating new things, but features of the thought of earlier centuries were, of course, still present.

French mathematics, 1795–1825

The institutional changes in science in France after the French Revolution are well known (in general terms anyway), and so the summary here can be brief.[2] The first class of the new Institut de France, which in 1795 replaced the old Académie des Sciences, managed to build up a good deal of activity, including the presentation of papers by members and examination of submissions by outsiders. The foundation in 1794 of the Ecole polytechnique, and the gradual refurbishing of the specialist engineering schools (the *écoles d'application*), gave a special emphasis to science and engineering at various levels: not only the instruction provided but the opportunities afforded for employment in these schools (and/or the *écoles centrales* and *lycées*). The prominence of engineering gave French science a special character which was not followed anywhere else in Europe; for example, employment as engineers or engineering administrators in the various *corps* was taken by some important savants. Government bodies such as the Bureau des Longitudes also provided employment, as did the Université when it was founded in the late 1800s.

The principal scientific advances took the form of a pair of broadenings. Mechanics (celestial, planetary, terrestrial and engineering) developed substantially, with an especial interest in hydrodynamics and elasticity (in the very broad sense of 'elastic fluid' then conceived) and in understanding the prominence of the concept of work. In addition, substantial progress in optics (especially the replacement of the emissionist theory by a wave theory) and the mathematisation of heat diffusion, electricity and magnetism (especially electromagnetism and electrostatics) brought about the birth of mathematical physics. Similarly, in mathematics itself, the calculus (the theory of series and solutions to differential equations) became subsumed under mathematical analysis, based on limits. Complex-variable analysis also saw important early advances.

Interestingly, these changes took place, or at least were made public, largely in the period 1815–26, twenty years after the social and intellectual innovations just described. Further, some of the innovators were outsiders to the newly developing community, at least at the beginning of their studies; they held their posts outside Paris and/or in the engineering *corps* rather than in academic or educational life, and their efforts were often opposed by the Paris apparatchiks.

I. Grattan-Guinness

Finally, the period 1795–1815 is also the period of glory in the enrolment to the Ecole polytechnique of students who later became distinguished savants; after that time the drop in numbers of such students is rather spectacular. However, this community cannot be simply identified with the innovators, for two reasons: first, that some *polytechniciens* such as Biot and Poisson were notable opponents to the changes; and secondly, that innovators such as Ampère and Fourier were too old to have been students at the school, although they taught there.

Table I indicates the principal figures over the period 1795–1825, divided by the intellectual concerns, with indications of their status as innovators of new traditions or conservers of old ones (while, of source, still making important innovations within those traditions). Some further details of their achievements will be exhibited later in this essay in the course of exhibiting their influence on Cambridge thought.

In addition to the *content* of their theories, the *manners of working*, or *Denkweisen*, varied. Three traditions obtained: the geometrical, with the appeal to the configuration of the given (classes of) cases; the algebraic, anxious to

Table I   French savants in mathematics and mathematical physics, 1795–1825.

The division of intellectual interest is very roughly 'theoretical' versus 'practical', though more subtle ideas are needed to explicate the division (see Grattan-Guinness 1981, 109–19). For example, members of both groups worked in the calculus and mechanics, although even then their different motivations were evident. *Notations:* In, made at least one notable contribution to the introduction of a new or reworked approach to a major part of the field; Ot, spent a significant part of his career as an outsider, either professionally or intellectually, during the period 1795–1825; Al/An/Ge, preferred an algebraic/analytical/geometrical style.

Prominent	Mathematical physics/mathematical analysis				Mechanics/calculus/engineering			
By 1800	J.L. Lagrange			Al	L. Carnot			Al/Ge
	P.S. Laplace	In		Al/Ge	J.B.J. Delambre			Ge
	A.M. Legendre			Ge	G. Monge	In		Ge
					G. de Prony			Al
Between 1800 and 1815	A.M. Ampère	In	Ot	Al/An	C. Dupin	In		Ge
	J.P.M. Binet			Al/Ge	L.B. Francoeur			Ge
	J.B. Biot			Al/Ge	P.S. Girard			Ge
	A.L. Cauchy	In		An	J.N.P. Hachette			Ge
	J. Fourier	In	Ot	Ge	C.L. Navier	In		Ge
	S. Germain		Ot	Ge				
	E.L. Malus			Ge				
	L. Poinsot	In		Al/Ge				
	S.D. Poisson	In		Al				
Between 1815 and 1825	J.M.C. Duhamel			An/Ge	G.G. Coriolis	In		Ge
	A.J. Fresnel	In	Ot	Ge	J.V. Poncelet	In	Ot	Ge
	G. Lamé			Ge	L. Puissant			Ge

Mathematics and mathematical physics

avoid appeal to geometry or intuition and proceeding by the manipulations of general formulae; and the analytical (in the sense of mathematical analysis), in which *both* geometrical intuition and algebraic formalisms and classifications were superseded by deductive theories based on broadly conceived definitions drawing on the theory of limits. By and large, the geometrically inclined brought in the new ideas, especially in applied mathematics, for by definition they did not know the answers: the algebraic style could come along later and systematise the answers thus found (as in variational mechanics, for example). The savant's choice of style seems largely independent of the topics on which he worked; for example, he used his style in pure *and* applied mathematics. Socio-psychological questions arise here, on which I wish I had something useful to say.

On manifestations of mathematical interest in Britain in the early nineteenth century

> He had a great dislike of the mathematical sciences and did all in his power to suppress them; a natural consequence of which is, that mathematical knowledge has greatly depreciated in England during the last 40 years.
>
> Olinthus Gregory to Heinrich Schumacher on the recently deceased Sir Joseph Banks, 27 June 1820 (*Nachlass* Schumacher, Staatsbibliothek, East Berlin)

While all these things were taking place in France, the British muddled along as usual. But from the mid-1810s on, there were significant stirrings in the discovery, diffusion and instruction of mathematics. In this section I shall note briefly some manifestations of the second category, in the form of encyclopaedia and similar review articles, and expository books. The principal articles, whose sources I shall now describe, are listed in Table II, and are cited by description in later sections of the chapter.

A remarkable range of encyclopaedias was produced in England and Scotland in the first 30 years of the nineteenth century — far more than the French published during the same period (see Grattan-Guinness 1981, 128-9). The most substantial from our point of view is the *Encyclopaedia Metropolitana*, which appeared between the mid-1820s and 1845 in 21 volumes of text plus eight volumes of plates. Dating the publication of the articles in it is very difficult; almost all copies of the volumes now carry a title page marked '1845', although it is known that many of the articles appeared earlier; some were reprinted for the complete 1845 presentation.

Table II lists the most relevant articles with dates of preparation and publication as far as I have been able to determine them. The standard of writing is quite adequate, and often much better than that; for example, the articles by Herschel on optics and Airy on tides and waves are perhaps the best available at the time anywhere, and the massive contributions by Barlow, professor at the Royal Academy at Woolwich, are noteworthy, in particular for

I. Grattan-Guinness

Table II   Encyclopaedia and related articles published in Britain, 1820–40.

Author	Completion/ publication	Reference	Subject

Encyclopaedia Metropolitana
Some of the datings are conjectural, from internal evidence.

D. Lardner	⩾1825	1:524–632	Algebra
G.B. Airy	1824	1:672–708	Trigonometry
A. Lévy		1:769–843; 2:1–208	Differential and integral calculus
T.G. Hall		2:209–26	Calculus of variations
T.G. Hall		2:227–304	Calculus of differences
A. de Morgan	⩾1835/1836	2:305–92	Functional equations, etc.
A. de Morgan	1837	2:393–491	Probability
H. Moseley	⩾1837	2:491–594	'Definite integrals' (mostly elliptic functions)
P. Barlow	⩾1820/1829	3:1–160	Mechanics
P. Barlow	⩾1823/1829	3:161–296	Hydrodynamics (and some machines)
P. Barlow	⩾1817/1829	3:297–392	'Pneumatics' (barometers, acoustics, etc.)
P. Barlow	⩾1817/1829	3:393–484	Optics (mostly geometrical)
P. Barlow	⩾1821/1829	3:485–606	Observational astronomy and some instruments
H. Kater	⩾1825/1829	3:607–46	Nautical astronomy (tables, calculation)
J. Herschel	⩾1820/1829	3:647–729	Physical astronomy (motions, perturbations, etc.)
P. Barlow	⩾1823/1829	3:735–845	Magnetism (only) and some instruments
P. Barlow	⩾1823/1830	4:1–40	'Electro-magnetism' (theory and experiments)
F. Lunn	1820s/1830	4:41–139	Electricity
[W. Whewell]	1820s/1830	4:140–70	'Theory of electricity' (electrostatics)
F. Lunn	⩾1824/1830	4:225–340	Heat (including diffusion, gas expansion, etc.; much citation of French literature)
J. Herschel	1827/1830	4:341–586	Light (largely physical: famous survey)
J. Herschel	1830/1830	4:797–824	Sound (including wave equation, etc.)
G. Airy	1830/184?	5:165–236, 237*–240*	'Figure of the earth' (and pendulum, repeating circles, etc.)
G. Airy	1842/1845	5:241*–396*	'Tides and waves' (the major article to date)

Encyclopaedia Britannica, supplementary volumes

J. Playfair	1816	1:585–600	'Astronomy, physical' (and some mathematical)
J. Ivory	1816	1:627–44	Attraction (spheroids, etc.)
J. Leslie(?)	1817	2:116–36	Barometer
T. Young	1817	2:497–520	'Bridge' (including suspension, materials)
T. Young	1818	3:141–63	'Chromatics' (refraction, colours in plates, etc.)
T. Young	1818	3:211–22	Cohesion (especially fluids)
F.E. Bromhead	1819	3:568–72	Differential calculus (Lagrange's style)
J. Playfair	1819	4:[1–90]	Dissertation on 'natural philosophy'
J.B. Biot	1819	4:76–93	'Electricity' (mostly electrostatics)
T. Young	1819	4:283–309	'Fluents, or integrals' (mostly evaluations)
J. Ivory	1819	4:309–23	'Fluids, elevation of'
J. Ivory	1820	4:669–708	Equations
T. Young	1821	5:9–13	Hydraulics
J.B. Biot	1824	6:116–36	Pendulum
T. Young	1824	6:658–75	Tides
D.F.J. Arago	1825	6:838–63	Polarisation and double refraction

Reports of the British Association for the Advancement of Science
(The year of the meeting is given; publication took place during the year following.)

G.B. Airy	1832	125–89	Astronomy

Mathematics and mathematical physics

Author	Completion/ publication	Reference	Subject
J.W. Lubbock	1832	189–95	Tides
J.D. Forbes	1832	196–258	Meteorology
D. Brewster	1832	308–22	Optics
P. Barlow	1833	93–103	Strength of materials
S.H. Christie	1833	105–30	Earth's magnetism
J. Challis	1833	131–51	Hydrostatics and hydrodynamics (part 1)
G. Rennie	1833	153–83	Hydraulics (part 1)
G. Peacock	1833	185–352	'Certain branches of analysis' (algebras, etc.)
J. Challis	1834	253–94	Capillary attraction
H. Lloyd	1834	295–413	Physical optics (extensive survey)
G. Rennie	1834	415–512	Hydraulics (part 2)
W. Whewell	1835	1–34	Electricity, magnetism and heat
J. Challis	1836	225–52	Hydrostatics and hydrodynamics (part 2)
J.S. Russell	1837	417–96	Waves
D. Lardner	1838	197–252	Railway engineering, especially constants
B. Powell	1840	1–36	Radiant heat
J.D. Forbes	1840	37–156	Meteorology (major survey)

their historical sections. Unfortunately, many of the articles are poorly referenced, although when such information appears it shows quite good awareness of foreign (especially French) achievements. The main surprise is that the Belgian Lévy was chosen to write on the differential and integral calculus.

Another important source is the supplementary volumes of the 1810s and 1820s to the *Encyclopaedia Britannica*. The contributors (who are identified in the preface to volume 1) included Arago and Biot, whose pieces were translated by Young, himself an important contributor. Some of the most relevant longer pieces are also given in Table II. The *Edinburgh Encyclopaedia*, conducted in eighteen volumes by Brewster between 1808 and 1830, contains some substantial pieces, especially Brewster's own on optics. A few useful articles appeared in *Rees's Cyclopaedia* (thirty-nine volumes, 1819, London: unfortunately none is signed) and the *Penny Cyclopaedia* (twenty-nine volumes, 1833–46, London), published under the auspices of the Society for the Diffusion of Useful Knowledge, to which de Morgan was a prolific contributor. This society also published a four-volume series entitled *Natural Philosophy* (1829–38, London), containing introductory articles roughly on a par with volumes in Lardner's series, *The Cabinet Encyclopaedia*.

A source of a different kind is the series of reports on branches of science presented to the British Association for the Advancement of Science: although largely non-mathematical, they give useful guides (sometimes important ones). I have included them also in Table II.

Research publications by Cambridge mathematicians, 1815–40

When we turn to England, we find no single institution which offers points of comparison with the *Ecole Polytechnique* in every particular. If we consider the latter

I. Grattan-Guinness

as a school of engineers, and especially of military engineers, we can only find the Academy at Woolwich with which to compare it. But if we view the French school as an institution in which the higher mathematics are so taught as to produce teachers and investigators, we must consider it as occupying the position which in our country is filled by the University of Cambridge.

de Morgan 1835, 334–5

Cambridge mathematics began to develop with the founding of the Analytical Society in the mid-1810s (discussed below) and especially with the inauguration a few years later of the Cambridge Philosophical Society (see Hall 1969). As was mentioned earlier, the notion of 'Cambridge mathematician' is rather hard to specify. I took two principal criteria: men who were undergraduates there and showed evidence of its continued influence upon them, even if they passed some or much of their career elsewhere (examples of this are given in the final section); and men who contributed to the *Transactions* of the Cambridge Philosophical Society irrespective of their domicile. This journal was the principal source of papers: I found seventy-four in all. On this definition of 'Cambridge mathematician' I then found thirty-two papers by them in the *Philosophical Transactions* of the Royal Society and also a sextet in the *Transactions* of the Royal Society of Edinburgh; but I omitted other journals, such as the *Proceedings* of the Royal Society and *The Philosophical Magazine*, as contributions there seemed to be summary or adjunct to writings already under consideration.

The principal authors represented are Airy, Babbage, de Morgan, Green, Herschel, Murphy and Whewell, with assistance furnished mostly by Challis, Kelland, Moseley and Wallace. I have also included two Royal Society papers by the photographer Fox Talbot, as he says explicitly in one of them that his ideas began to develop during his undergraduate days at Cambridge; and two such papers by Warren, as he was based at Cambridge and his monograph on the same subject was published there.

As for content, I have covered the scope indicated earlier, from differential equations to engineering; but I have excluded papers dealing entirely in theoretical or experimental physics or using a mathematical result *solely* for numerical calculation. I also ruled out meteorology, since at this time it was barely a mathematised branch of physics; and also economics and probability as being insufficiently physical in their mathematical concerns (in the latter case, for the papers by these authors).

The information is presented in a series of tables. Table III indicates the dates of publication of the *Transactions* of the Cambridge, Royal and Edinburgh Societies. Table IV contains a list of the principal relevant branches of mathematics and mathematical physics by two-letter codes. Table V comprises an author index of the contributions to the various journals, with a coded indication of their content; the full names and dates of the authors are also given, with a two-letter code also provided to their surnames. The titles of the

Mathematics and mathematical physics

Table III Years of publication of the Cambridge *Transactions*.

Volume	1	2	3	4	5	6	7
Year of publication	1822	1827	1830	1833	1835	1838	1842
Number of pages cited	4	14	7	7	14	18	10

(For the *Philosophical Transactions* of the Royal Society, volume $100+n$ was published in year $1810+n$; for many years the Society forgot to print the volume number. The relevant volumes of the Edinburgh *Transactions* are 9(1823), 13(1836) and 14(1840).)

Table IV Abbreviation of principal subject areas.

(The division of areas, and the assignment of papers to areas, are discussed later in this chapter.)

Code	Subject area
AL	Algebra
AN	Analysis and calculus
AS	Astronomy
CR	Crystallography
DE	Differential equations
DF	Differences
EN	Engineering (machines, constructions, etc.)
EQ	Equations
FL	Fluids (aether, incompressible, etc.)
GL	Geology
GM	Geometry
HE	Heat
HY	Hydrodynamics (fluid flow, tides, etc.)
IN	Integrals (evaluation, transforms)
ME	Mechanics
MG	Magnetism
OP	Optics
PL	Planetary physics (shape of earth, attractions, etc.)
SE	Series
SO	Sound

papers are not given, but many of them are quoted, at least in part, in the context of the discussion in the later sections of this chapter. The two-letter code is then used in Table VI, which comprises a subject index of the papers as classified under Table IV. The efficacy of this classification is discussed in the final section.

In addition to papers, it is essential to note the books produced by these 'Cambridge' authors. I excluded some *very* elementary books on arithmetic and algebra, editions of Euclid, and mathematical tables, as too tangential for our concerns; but I have included not only research-level treatises and monographs but also textbooks on relevant parts of mathematics for use at university level (usually at Cambridge itself, in fact). I have also incorporated a few rather philosophical works by de Morgan and Whewell (including the first editions of his *History* and *Philosophy* of the inductive sciences), for they contained

Table V Author index of papers.

For reasons explained in Table VI, the alphabetical order of authors followed is that of their letter codes and not of their surnames. Full names and dates are given when known; they are followed on the next line by the number of papers cited, together with the number of books listed in Table VI, and the years of election to the Cambridge Philosophical and the Royal Societies (cited as 'C' and 'R' respectively), where applicable. In the references the journal titles are abbreviated as follows: C *Transactions* of the Cambridge Philosophical Society; E *Transactions* of the Edinburgh Philosophical Society; R *Philosophical Transactions* of the Royal Society.

The order of the items for each author follows the chronology of his presentations. The references are given as follows. 'C2d' cites the fourth (in order of publication) chosen paper from volume 2 of the *Transactions* of the Cambridge Philosophical Society the page numbers are then given, followed by the date of reading (day, month, year) and then an indication of contents by the codes of Table IV. The Royal Society items are cited this way, except that (say) 'R9' refers to volume 109 of the *Philosophical Transactions*.

AI Airy, George Biddell (1801–1892)
22/3
C2c	105–18			
C2e	203–17	15. 3.24	OP	
C2g	227–52	15. 3.24	AN	PL
C2h	267–71	17. 5.24	OP	
C2j	277–86	21. 2.25	OP	
C2l	379–90	2. 5.25	EN	
R16d	pt 4, 548–78	8. 5.26	AN	PL
C3c	105–28	15. 6.26	AN	PL
R17a	65–70	26.11.26	ME	
C3a	1–63	15. 2.27	AS	
R18a	23–34	14, 21. 5.27	OP	
C3e	369–72	6.12.27	AS	
C4b	79–123	14.12.29	DE	ME
C4d	279–88	21. 2.31	OP	
R22a	67–128	14.11.31	OP	
C4f	409–28	24.11.31	AS	
C5d	101–11	19. 3.32	OP	
C5j	283–91	5. 3.33	OP	
C6l	379–402	24.11.34	AN	OP
R29a	167–213	2. 5.36	IN	OP
R30b	225–43	25. 4.39	MG	
R31a	1–10	18. 6.40	OP	
		19.11.40	OP	

BA Babbage Charles (1792–1871)
9/2
	C1819		R1816	
R5a	389–423		6. 4.15	AL
R6b	179–256		14. 3.16	AL
R7a	197–216		17. 4.17	AL
E9a	337–52		1. 7.18	AL
R9a	249–82		1. 4.19	DE
C1a	63–76		1. 5.20	AL	SE
C2k	325–78		16.12.21	AL	DF
C2f	217–25		3. 5.24	SE
R16b	pt 3, 250–65		17. 1.26	AL	DF

CH Challis, John (1803–1882)
4/1
	C1826		R1848	
C3d	269–320		30. 3.29	DE	FL	SO
C3f	383–416		27. 2.30	DE	FL
C5g	173–203		3. 3.34	HY	ME
C6o	443–55		13. 2.37	HE	PL

DE De Morgan, Augustus (1806–1871)
4/3
	C1834			
C4a	71–78	15.11.30	GM	
C5c	77–94	12.11.32	GM	
C6g	185–93	16. 5.36	AL	SE
C7i	173–87	9.12.39	AL	

EA Earnshaw, Samuel (1805–1887)
3/4   C1831
C6h   203–33   21. 3.36   HY
C6n   431–42   12.12.36   OP
C7e   97–112   18. 3.39   FL

GR Green, George (1793–1841)
9/0   C1837
C5a   1–63   12.11.32   AN   FL
C5m   395–430   6. 5.33   DE   PL
E13a   54–62   11.12.33   ME
C6p   457–62   15. 5.37   DE   HY
C6m   403–15   11.12.37   SO
C7a   1–24   11.12.37   OP
C7d   87–95   18. 2.39   HY
C7f   113–20   6. 5.39   OP
C7g   121–40   20. 5.39   OP

HE Herschel, Sir John (1792–1871)
8/4   C1819   R1813
R6a   25–45   14.12.15   AL   SE
R8a   144–68   19. 2.18   AL   DF
R9b   45–100   23.12.19   OP
C1b   77–88   6. 3.20   AL   DF
R11a   222–67   22. 3.21   DF   OP
R16a   pt 2, 77–126 12. 1.26 PL
R16c   pt 3, 266–80 9, 16. 3.26 AS
       [correction in R17, 126–7]
C4g   425–40   7. 5.32   EQ

HO Holditch, Hamnett (1800?–1867)
1/0   C1822
C7c   61–86   10.12.38   EN   GM

HP Hopkins, William (1793–1866)
4/0   C1827   R1837
C5i   231–70   20. 5.33   SO
C6a   1–84   4. 5.35   GL
R29b   381–424   17. 1.39   GL   DE
R30a   193–208   7. 3.39   GL

JA Jarrett, Thomas (1805–1882)
1/0   C1828
C3b   65–104   10.11.27   AL   SE

KE Kelland, Philip (1808–1879)
9/1   C1834   R1838
C6f   153–84   22. 2.36   OP
C6i   235–88   16. 5.36   FL
C6k   323–60   1. 5.37   OP   HE   SO
C7b   25–60   13.2, 26. 3.38   FL
E14a   393–418   18. 2.39   OP
E14b   497–546   1. 1.39   HY
E14c   567–603   2.12.39   AN
E14d   604–18   20. 1.40   AN   IN   OP
C7h   153–69   30. 3.40   AN

KI King, Joshua (1798–1857)
1/0   C1819
C2b   45–6   14. 4.23   ME

MI Miller, William Hallows (1801–1880)
2/4   C1828   R1839
C5n   431–38   8.12.34   OP
C7j   209–15   21. 3.36   CR

MO Moseley, Henry (1801–1872)
2/3   C1822   R1839
C5k   293–314   9.12.33   EN
C6q   463–91   15. 5.37   EN

MR Moore, Arthur Augustus (1817–1845)
1/0
C6j 317–22 1. 1.37 AN

MU Murphy, Robert (1806–1843)
9/2 C1830 R1834
C3g 429–43 24. 5.30 AL IN
C4c 125–53 7. 3.31 IN SE
C4e 353–408 5. 3.32 AL IN
C5b 65–75 26.11.32 AL
C5e 113–48 11.11.33 AL
C5l 315–93 2. 3.35 IN IN
C6c 91–108 15.11.35 AL EQ
R27c 197–210 22.12.36 AL EQ
R27b 161–78 27. 4.37 EQ

PO Potter, Richard (1799–1886)
2/0 C1838
C6e 141–52 14.12.35 OP
C6r 555–64 30. 4.38 OP

PW Power, Joseph (1798–1868)
2/0 C1821
C2i 273–6 21. 3.25 ME
C5h 205–30 17. 3.34 HY

SM Smith, Archibald (1813–1872)
1/0 C1836
C6b 85–9 18. 5.35 OP

TA Talbot, Henry Fox (1800–1872)
2/0 R1831
R26c 177–215 10. 3.36 IN
R27a 1–18 17.11.36 IN

WA Wallace, William (1768–1843)
1/0
C6d 107–40 30.11.35 PL

WH Whewell, William (1794–1866)
13/17 C1819 R1820
C1c 179–91 18. 4.20 AS IN
C1d 331–42 25.11.21 CR GM
C2a 11–20 6. 5.22 ME
C2d 192–202 24.11.23 GM GM
R15a 87–130 25.11.24 CR GM
C2m 391–425 13.11.26 CR AL
C2n 429–39 11. 2.27 CR
                      [misdated on title page]
R23a 147–236 2. 5.33 HY
R24a 15–45 9. 1.34 HY
C5p 149–72 17. 2.34 ME
R26a 1–16 19.11.35 HY
R26b 131–48 3. 3.36 HY
R30c 255–72 18. 6.40 HY PL

WR Warren, John (1796–1852)
2/1
R19a 241–54 19. 1.29 AL
R19b 339–60 4. 6.29 AL

Mathematics and mathematical physics

extensive sections on mathematics or mathematical physics and/or bore some influence on education at Cambridge. All these works are listed in Table VII. There were also textbooks written by authors who did not write research papers and so are absent from Table V; but I have included only Peacock's textbook on algebra here, as the sole case of a significant extra work. My random sampling of texts by other such authors revealed no major deviant or alternative approaches, but they would need consideration in a more substantial study than is attempted here.

The remainder of this chapter is concerned with the content of these papers and books. Papers are cited in the manner described at the head of Table V, with the dates of publication of the volumes appended. Books which are listed in Table VII are cited by the year used in that table, enclosed in angle brackets thus: '⟨1833⟩'. This method of referencing distinguishes them from items listed in the collected bibliography of this volume.

Changes and preferences in the calculus and algebra

As a major branch of mathematics, the calculus was established in the differential form founded by Leibniz and developed by the Bernoullis and Euler. However, from the late eighteenth century on, some attention was given to Lagrange's algebraic approach based on Taylor's series. The situation became still more complicated in France in the 1820s, when Cauchy brought in a new tradition of mathematical analysis based on limits. The Lagrangian view tended to fall away, especially with Cauchy's demonstration in 1822 of counter-examples such as e^{-1/x^2} to Lagrange's belief that a function always took a Taylor expansion; but the differential tradition was still maintained, for its geometrical ground was most suitable for applying to the configuration of a given case (or class of cases) in geometry and applied mathematics.

In other words, between 1815 and 1825 France tended to change its basic formulation of the calculus from a mixture of Euler and Lagrange (with a little interest in limits) to a mixture of Euler and Cauchy (with some effort made to conserve Lagrangism) (see Grattan-Guinness 1980, Chapter 3). But at Cambridge the story was quite different. The Newtonian fluxional approach died down (I believe that Dealtry (1810) was the last freshly written volume to be produced within that tradition), and was replaced by advocacy of 'the Continental notation' by the Analytical Society from the early 1810s (see Enros 1983). However, the summary above shows that this phrase, which unfortunately historians often repeat, is too unclear. In fact, Babbage, Herschel and Peacock adopted the *algebraic* line of Lagrange; when they translated Lacroix's *Traité Élémentaire* of the calculus in 1816, they reproached their author for substituting 'the method of limits of d'Alembert, in the place of the most correct and natural method of Lagrange, which was adopted in the former' edition of the book to the one which they were translating (Lacroix ⟨1816⟩, iii–iv)!

So Cambridge moved in a direction *opposed* to Paris, in that the method of

I. Grattan-Guinness

Table VI Subject index of papers.

This is divided into two main areas: pure and applied mathematics. These areas are split up in turn into the topics listed in Table IV, and the two-letter codes used there are recalled. Several topics are further divided, so that the maximal length of a column is around ten entries. The layout of the columns under main topics is only partly guided by intellectual concerns; some of the adjacencies are caused by needs of layout.

In each entry is the chronological order of presentation, as determined by Table V. The author is indicated by the two-letter code used there (which is why authors were listed there by alphabetical order of these codes). The papers are cited only by journal and volume reference; the page numbers are not repeated.

Analysis (AN)				Integral (IN)			
foundations, etc.		Legendre polynomials, etc.		transforms		evaluation	
C6j	MR	C2e	AI	C3g	MU	C1c	WH
C7h	KE	C2l	AI	C4c	MU	C5d	AI
E19c	KE	R16d	AI	C4e	MU	R26c	TA
E14d	KE	C5a	GR	C5e	MU	R27a	TA
		C5j	AI	C5l	MU	C6l	AI
						C7h	KE
Differential equations (DE)		Differences (DF)		Series (SE)		Geometry (GM)	
C3d	CH	R8a	HE	R6a	HE	C1d	WH
C3e	AI	R9a	BA	R9a	BA	C2d	WH
C3f	CH	C1a	BA	C2e	BA	R15a	WH
C5m	GR	C1b	HE	C3b	JA	C2m	WH
C6p	GR	R11a	HE	C4c	MU	C4a	DE
R29b	HP	R16b	BA	C6g	DE	C5c	DE
		C6b	MU			C7c	HO
		R27c	MU				
		Algebra (AL)				Equations (EQ)	
foundations, notation		operators		of functions			
R16b	BA	R6a	HE	R5a	BA	C4g	HE
C2n	WH	C2k	BA	R6b	BA	C5b	MU
C3b	JA	C3g	MU	R7a	BA	C6c	MU
R19a	WR	C5b	MU	R8a	HE	R27b	MU
R19b	WR	C5e	MU	E9a	BA	R27c	MU
C7i	DE	C6c	MU	C1a	BA		
		R27c	MU	C1b	HE		
				C4e	MU		
				C6g	DE		
				C7i	DE		

Mathematics and mathematical physics

Mechanics (ME)				Planetary (PL)				Geology (GL)	
dynamics		statics		attractions, etc.		geodesy, etc.			
C2a	WH	C2b	KI	C2e	AI	R16a	HE	C6a	HP
C3c	AI	C2i	PW	C2l	AI	R16d	AI	R29b	HP
C3e	AI	C6a	HP	R16d	AI	R26b	WH	R30a	HP
C5f	WH			C5m	GR	C6d	WA		
C5g	CH			R29a	AI	C6o	CH		
E13a	GR			R29b	HP				

Hydrodynamics (HY)				Fluids (FL)				Astronomy (AS)	
waves, etc.		tides		aether, etc.		'elastic', etc.			
C5g	CH	R23a	WH	C6i	KE	C3d	CH	C1c	WH
C5h	PW	R24a	WH	C6k	KE	C3f	CH	R16c	HE
C6h	EA	R26a	WH	C7b	KE	C5a	GR	R17a	AI
C6p	GR	R26b	WH	C7e	EA			R18a	AI
E14b	KE	R30c	WH					R22a	AI
C7d	GR								

				Optics (OP)					
diffraction		polarisation		reflection/refraction		geometrical		rainbow	
C5d	AI	R9b	HE	C5n	MI	R11a	HE	C6e	PO
C5j	AI	C4b	AI	C6b	SM	C2c	AI	C6l	AI
C6n	EA	C4d	AI	C6f	KE	C2g	AI		
C7h	KE	C4f	AI	C6k	KE	C2h	AI		
R30b	AI	C7g	GR	C6l	AI	C3a	AI		
R31a	AI			C6r	PO	C6r	PO		
				C7a	GR				
				E14a	KE				
				C7e	EA				
				C7f	GR				

Magnetism (MG)		Crystallography (CR)		Sound (SO)		Heat (HE)		Engineering (EN)	
R29a	AI	C1d	WH	C3d	CH	C6i	KE	C2j	AI
		R15a	WH	C5i	HP	C6o	CH	C5k	MO
		C2m	WH	C6i	KE			C6q	MO
		C2n	WH	C6m	GR			C7c	HO
		C7j	MI						

I. Grattan-Guinness

Table VII   List of principal monographs and textbooks.

This list gives the (short titles of the) principal relevant book publications as specified in the text; new editions are not listed when they are merely reprints. The level of the text is indicated as follows: 'TE' refers to a textbook for use at Cambridge University (or some comparable institution); 'TR' indicates a (fairly) substantial and/or comprehensive treatise, written for an audience already past the textbook level; and 'MO' denotes a monograph on some *specific* topic(s) or aspect(s) of the field, with its apparent level indicated by the appropriate letters of the two levels just defined written in *lower* case. (I do not always follow the authors' *own* uses of words like 'treatise' in their titles!) Cambridge and London, as places of publication, are marked by 'C' and 'L', respectively. (The Cambridge house was usually Deighton, who would be worth a study.) Subscripts indicate editions. The order of the 46 items within a year is alphabetical, of the authors.

1813	[Babbage, Herschel], *Memoirs of the Analytical Society*, C. MO tr
1816	S.F. Lacroix (trans. Babbage, Herschel and Peacock), *An elementary treatise on the differential and integral calculus*, C. TE
1819	Whewell, *An elementary treatise on mechanics*$_1$, C. TE
1820	Babbage, *Examples of the solution of functional equations*, [C.] MO te
1820	Herschel, *A collection of examples of the application of the calculus of finite differences*, C. MO te
1823	Whewell, *A treatise on dynamics*, C. TE
1824	Whewell, *Mechanics* $\langle 1819 \rangle_2$, C. TE
1826	Airy, *Mathematical tracts on physical astronomy, the figure of the earth, precession and nutation, and the calculus of variations*$_1$, C. MO te
1828	Warren, *A treatise on the geometrical representation of the square roots of negative quantities*, C. MO te
1828	Whewell, *Mechanics* $\langle 1819 \rangle_3$, C. TE
1830	Herschel, *A preliminary discourse on the study of natural philosophy*, L. MO tr
1830	Moseley, *A treatise on hydrostatics and hydrodynamics*, C. TE
1830	Peacock, *A treatise on algebra*, C. TR
1831	Airy, *Tracts* $\langle 1826 \rangle_2$, C. MO te
1831	Miller, *The elements of hydrostatics and hydrodynamics*$_1$, C. TE
1832*a*	Earnshaw, *On the notation of the differential calculus*, C. TE
1832*b*	Earnshaw, *Dynamics*$_1$, C. TE
1832*a*	Whewell, *An introduction to dynamics*, C. TE
1832*b*	Whewell, *On the free motion of points*, pt. 1, C. TE
1832*c*	Whewell, *The first principles of mechanics*, C. MO te
1833	Herschel, *A treatise on astronomy*, L. TE
1833	Miller, *An elementary treatise on the differential calculus*$_1$, C. TE
1833	Murphy, *Elementary principles of the theories of electricity, heat, and molecular actions*, Part 1 [and only], *Electricity*, C. TE
1833*a*	Whewell, *Mechanics* $\langle 1819 \rangle_4$, C. TE
1833*b*	Whewell, *Analytical statics*, C. TE [Supplement to 1833*a*.]
1833*c*	Whewell, *Astronomy and general physics considered with reference to natural theology*, C. MO te [A Bridgewater treatise.]
1834	Airy, *Gravitation*, L. MO te [First published as an article in the *Penny cyclopaedia*.]
1834	Earnshaw, *The theory of statics*, C. TE
1834	Moseley, *A treatise on mechanics*$_1$, L. TE
1834	Whewell, *Points* $\langle 1832b \rangle$, pt. 2, C. TE
1835	De Morgan, *The elements of algebra*, L. TE
1835	Miller, *Hydrostatics* $\langle 1831 \rangle_2$, C. TE
1835	Whewell, *Thoughts on the study of mathematics as part of a liberal education*, C. MO te
1836	De Morgan, *Mathematics*, L. MO te [Somewhat philosophical.]
1836–42	De Morgan, *The differential and integral calculus*, L. TE
1837	Kelland, *Theory of heat*, C. TE/TR
1837	Moseley, *Mechanics* $\langle 1834 \rangle_2$, L. TR

Mathematics and mathematical physics

1837a	Whewell, *The mechanical Euclid*, C. TE [On mechanics and hydrostatics.]
1837b	Whewell, *The history of the inductive sciences*, 3 vols, C. MO tr
1838	Challis, *Syllabus . . . on the equilibrium and motion of fluids, and on optics*, C and L. TE [Lecture course outline and proofs.]
1838a	Whewell, *The doctrine of limits*, C. TE
1838b	Whewell, *On the principles of English university education*, L. and C. MO te [Some comments on mathematics education.]
1839	Earnshaw, *Dynamics* ⟨1832b⟩$_2$, C. TE
1839	Miller, *A treatise on crystallography*, C and L. TR
1839	Murphy, *A treatise on the theory of algebraic equations*, L. TE
1840	Whewell, *The philosophy of the inductive sciences founded upon their history*$_1$, 2 vols, C. MO tr

limits, which Lacroix was supporting to some extent and which Cauchy was soon to elevate to prime status, lay in *harmony* with the fluxional calculus which Cambridge was abandoning. The analogy between Cauchy and Newton should not be exaggerated, however; Cauchy offered a much more explicit and developed theory of limits than Newton, and differed entirely from him on the important related question of the convergence of infinite series. Of this latter point Babbage quickly became aware; a paper on summing infinite series (Rqa (1819), 281–2) he reported a conversation on the dangers of divergent series with Poisson (who, however, remained regrettably adept at bringing them in, especially when using limiting values of variables).

The preference for Lagrange is only one sign of the infatuation which Cambridge exhibited in algebras of various kinds. The papers by Babbage and Herschel on functional equations and operators, written mostly in the decade from 1815, was an extension of an interest pursued by various minor French mathematicians following in Lagrange's footsteps (see Grattan-Guinness 1979, in criticism of Dubbey 1978). Babbage carried his interests into methods of calculation by machine, with his famous difference and analytical engines; and his view was taken up by Herschel, who discussed in a paper C4g (1833) on equations the possibility of constructing a machine for resolving transcendental equations.

The emphasis on algebraic manipulations continued in the 1830s with Murphy's methods of finding integral transforms and solving equations (see Burkhardt 1908, 1109–30); and in C5b (1835) 'On eliminating between an indefinite number of unknown quantities' he furthered the study of infinite matrices, which Fourier had pioneered (see Bernkopf 1968). When the *Cambridge Mathematical Journal* began in 1837, its early volumes contained many short papers on operator solutions to differential equations, in some cases with applications to certain branches of mathematical physics. The principal studies, by Boole and Gregory, appeared in this journal and elsewhere after 1840 (see Koppelman 1972). During our period differential equations as such were not deeply studied, although their general theory or solution methods often occurred in papers on mathematical physics. The entry under 'Differential equations' in Table VI includes only items of particular interest in this regard: the principal one is Green's pioneering contribution in his paper C6p (1838) on waves to the 'WKB' (Wentzel–Kramers–Brillouin) theory of approximative solutions (see Schlissel 1977, 313–14). The 'new' topic of elliptic functions, fostered by Abel and Jacobi in the 1820s, also inspired little interest.

Another line of algebraic research sought for the foundations of algebra (as then understood). This interest went back at least to Woodhouse (see Becher 1980*b*) and informed the concerns of Babbage and Herschel; and it received a powerful stimulus with Peacock's treatise ⟨1830⟩ on algebra and his long British Association report of 1833 (see Richards 1980 and Pycior 1981). It was taken up by de Morgan in various writings, including his textbook ⟨1835⟩ on algebra and the paper C7i (1842) on the 'foundations of algebra' (see Pycior 1983). These researches influenced to a notable extent his later studies in syllogistic logic (in contrast to Boole's contemporary Boolean algebra, where analogies from differential operators were dominant). He also wrote the first extensive study on functional equations, for the *Encyclopaedia Metropolitana*. Although varying, even uncertain, philosophical positions were adopted, there was a notable tendency to advocate contentual aspects of algebra (as opposed to the modern approach, where abstraction is adopted *ab initio* and interpretations are made afterwards). The proposals even included the use of divergent series, a view which goes back to Woodhouse.

Not all tendencies in pure mathematics led to algebras. One exception concerned complex numbers, where there was another revival of the geometrical interpretation by Warren in his two papers R19a and R19b (1829) and the contemporary monograph ⟨1828⟩. (The 'Argand diagram' as such belongs to French developments of the 1800s.) However, apart from touches such as Murphy C4e (1833) on solving algebraic equations, there was little Cambridge interest in Cauchy's current development of complex-variable calculus (on which see Burkhardt 1908, 671–704).

The other tendency away from algebra belongs, interestingly enough, to de Morgan; for in his massive textbook, The *Differential and Integral Calculus*, published in parts between 1836 and 1842 by the Society for the Diffusion of Useful Knowledge, he took Cauchy's theory of limits as a ground. For him the derivative was the basic notions for the differential calculus but he took 'dy/dx' to be a whole symbol (⟨1836–42⟩, 50) rather than the ratio '$dy \div dx$' that Cauchy had adopted (using his own new definition of a differential). De Morgan also defined continuity in the 'here is an ε, find a δ' form, rather than Cauchy's sequential approach (⟨1835⟩, 154–5).

By then, of course, de Morgan was London based; at Cambridge, Moore C6j (1838) was arguing that Cauchy's counter-example e^{-1/x^2} to Lagrange's calculus, which W.R. Hamilton had also found and upon which Moore was commenting, could be explained (away) via some (vague) definitions of orders of in-

Mathematics and mathematical physics

finitesimal. However, the new doctrine of limits gradually came in to Cambridge (see Becher 1980a, 17-23). Whewell's text ⟨1838a⟩ drew more patriotically on Newtonian grounds when advocating 'the doctrine of limits', but his approach was rather *passé* by then.

An interesting variant of these approaches was taken by Airy, who used the Eulerian term 'differential coefficient' and yet defined it in Cauchy's style as the limiting value of the difference quotient! He also proposed the notation '$\int_\theta f(\theta)$' for the intergral, in order to avoid using the differential notation 'dθ' (⟨1826⟩, 1). The notation gained a few followers, including Whewell in some writings (⟨1832b⟩ and ⟨1834⟩, for instance) and Murphy's textbook ⟨1833⟩ on electrostatics.

Contributions to astronomy and mechanics

Airy proposed this notation in his first book ⟨1826⟩. Entitled *Mathematical Tracts* ..., it covered three topics in astronomy (lunar theory, the shape of the earth, and precession and nutation) together with a short presentation on the calculus of variations, since the account in Lacroix's elementary tractise on the calculus (1814) was held to be 'confused and unintelligible'. 'Designed for the use of students in the university' as a (partial) replacement for the current textbooks in astronomy used at Cambridge (Vince 1816 and Woodhouse 1821-23), the tracts followed fairly closely the French traditions of Lagrange and Laplace without entering into the excruciating lengths (including lengths of expressions!) which Laplace and his successors were then employing. Herschel's article on physical astronomy for the *Encyclopaedia Metropolitana* gave a similar coverage in somewhat more detail.

In astronomical research Airy took up the preparation of solar tables, which had long been a concern of both British and French astronomers. In the papers R17a (1827) and R18a (1828) he discussed the corrections needed in perturbation theory for Delambre's tables (1806) (see Wilson 1980, 283-300). He recalled in his autobiography that he 'slipped ... under Whewell's door' his final calculations (Airy 1896, 78); for Whewell himself was already active in astronomy. Indeed, in Whewell's first contribution to the Cambridge Philosophical Society, C1c (1827) 'On the position of the apsides of orbits of great eccentricity', he made use of various approximative means. The reasoning is hardly rigorous even for its time (see Becher 1980b, 16-18), but it is noteworthy in its search for such methods, in contrast to the French penchant for 'exact' solutions. In this respect he showed sympathy with *German* theoretical astronomy — quite suitable, perhaps, for an enthusiast for Kant's philosophy.

Whewell also wrote extensively on dynamics, both in several monographs (Table VII) and some papers (especially C2a (1827) 'On the rotation of bodies' and C5p (1835) 'On the nature of the truths of the laws of motion'). Taking many of his scientific leads from Newton (his view of limits was noted above), he gave new prominence to Newton's laws of motion as a foundation for dynamics, while basing statics on the lever principle. I believe that his *Elementary Treatise on Mechanics* ⟨1819⟩ gave something of a fresh treatment to Newton's approach. French mechanics had long granted roughly equal status to other principles (see Bailhache 1975), and at this time certain lines in engineering mechanics in France were granting preference to kinetic energy/work conversion (see Grattan-Guinness 1984). While Whewell criticised the French, especially in 'On the principles of dynamics, as stated by French writers' (Whewell 1828), for conflating the laws of momentum and of impact, and put some considerations of work and friction into certain of his textbooks (see, for example, ⟨1833c⟩, 229–68), the full importance of this approach was not recognised at Cambridge until the 1840s.[3]

Whewell was also responsible for pioneering the mathematisation of crystallography. Already developed as a science on the physical side, he tried to make more explicit the geometrical thinking involved in the specification of faces and axes. His interest in the topic seemed to decline during the 1830s; but it was picked up by Miller, who followed broadly the same methods, presenting them in his treatise ⟨1839⟩. Their system of notation became widely used.

In planetary mathematics Airy and Whewell again stand out. Airy C2e (1827) on the shapes of rotating fluid bodies invoked methods alternative to Laplace's for the analysis of spheroids (see Todhunter 1886, 495–9), and in papers R16d (1827) and C21 (1827) on the attraction of spheroids, he defended Laplace's treatment of points near to the point of attraction against criticism by Ivory but offered a new proof of the expansion of a function in a series of Legendre polynomials (as we now call them).[4]

In his *Encyclopaedia Metropolitana* article on tides and waves, Airy referred in places to Whewell's studies, which were contained in a long series of papers presented to the Royal Society. In Table VI I have cited (under 'tides') only those with some mathematical content; this constituted the calculation of high-water levels and the effect of solar and diurnal inequalities.

Airy also described the theory and design of various instruments in planetary science. For example, in a paper C3c (1830) on clockwork escapements, he discussed the modification to simple harmonic motion caused by a general function, and in C3e (1830), 'On certain conditions under which perpetual motion is possible', he modified the equation for simple harmonic motion with a general delay term. Either wittingly or not, he extended or built on some of the studies of the simple pendulum carried out by Laplace and Poisson in the 1810s in connection with its use in Borda's sophisticated design of this pendulum for geodetic studies (see Wolf 1889). He also wrote on pendula and related devices in his large *Encyclopaedia Metropolitana* article on the figure of the earth, dated as of 1830. Then, in a paper (Airy 1840) presented to the Astronomical Society, he pioneered the study of the dynamics of servomechanisms with an analysis of 'the regulation of the clock-work for effecting uniform motion of equatoreals' (see Fuller 1976).

Mathematics and mathematical physics

Among other geodetical papers, we may note Herschel R16a (1826) on the longitude of the Paris and London observatories, and Wallace C6d (1838) on the use of spherical trigonometry in geodesy, following Delambre (1806–10). Both topics had aroused much interest in Paris already, both for Delambre and for some of his colleagues. However, there was no Cambridge interest in cartography.

An unusual, indeed pioneering, contribution to planetary mathematical physics was made by Hopkins' studies of physical geology. Beginning with a paper C6a (1838) of 'Researches' in this area, he drew on basic notions of the composition of forces to try to determine the maximal direction for the production of fissures in a system of strata. He followed up with a pair of papers, R29b (1839) and R30a (1840), 'On precession and notation' of the earth, assuming its interior to be fluid; his methods, drawn from the presentation in Airy's *Mathematical Tracts* ⟨1826⟩, involved some routine but fairly intricate stuff on four simultaneous ordinary differential equations in the first paper (where the densities of both fluid and outer shell were taken to be equal and uniform) and some theory of attractions in the second (where these densities were allowed to vary). Both treatments display a heavily Laplacian character, mediated no doubt through Airy and perhaps directly; indeed, these studies could be regarded as a certain extension of Laplace's and Fourier's studies around 1820 of the cooling (and thereby shrinking) earth. For a more detailed discussion of Hopkins' ideas, both in these and in later papers, see Smith's chapter in this volume.

Most of the citations made above are of the papers listed in Tables V and VI, but Table VII shows that textbooks and treatises in mechanics were very popular among Cambridge authors; counting in new editions, nearly a score of such works appeared, especially in the basic parts of dynamics and statics. This situation is partly explained by Whewell's prolificity on these topics, but several other authors contributed also. By contrast, far fewer books were produced for other areas of mathematical physics, although in some of them research papers were plentiful. To these areas we now turn.

Contributions to the mathematical physics of 'fluids'

When Airy published the second edition of his *Mathematical Tracts* in 1831, he added two new chapters. One of them, 'Planetary theory', covered various aspects of perturbation theory and the expressions for the astronomical variables (⟨1831⟩, 61–126). The other addition was a more substantial piece: a lengthy essay 'On the undulatory theory of optics' (pages 248–409), presented for 'the students of the university' because the theory had 'the same claims to his attention as the Theory of Gravitation: namely that it is certainly true, and that by mathematical operations of great elegance it leads to results of great interest' (pages iii–iv). After describing simple harmonic notion and the principle of interference, he outlined the undulatory explanations of reflection, refraction,

I. Grattan-Guinness

Newton's rings, diffraction, double refraction and polarisation (including the various surfaces of Fresnel's theory, and reflection and refraction of polarised light).

Airy's own researches in optics had begun in the 1820s with various studies in geometrical optics (that is, topics such as tracing the paths of rays *without* considering the mode of generation, which study belonged to physical optics); the French had been rather silent on this area for some decades. He contributed especially to the study of the design of lenses, and in an interesting submission C2h (1827), 'On a peculiar defect in the eye, and a mode of correcting it', he described the astigmatism in his own eyesight.

Airy's 1831 tract, together with Herschel's long article of 1830 on optics for the *Encyclopaedia Metropolitana*, did much to increase interest in England in Fresnel's contributions to physical optics (see Burkhardt 1908, 1130–65; Cantor 1975), and many papers were written in it (see Table VI under 'Optics'). Among Airy's contributions, in his C4b (1833) on double refraction quartz he posed various 'conjectures' on the elliptic polarisation of doubly refracted light, and C4d and C4f (1833) vindicated Fresnel's opinion on the formation of Newton's rings with polarised light. In C61 (1838) 'On the intensity of light in the neighbourhood of a caustic', he came across the so-called 'Airy functions' in the form $\int_0^\infty \cos(\frac{1}{2}\pi(w^2 - mw))\,dw$ and compiled a table of values for various values of m, like the table he gave for Fresnel's integrals $\int_0^m \frac{\sin}{\cos} w^2 dw$ at the end of ⟨1831⟩. Fresnel himself had so acted in 1816 in connection with diffraction, a branch of optics taken up in Kelland's study C7h (1842) of 'the quantity of light intercepted by a grating placed before a lens'. This paper contained some unusual mathematical details, especially the use of fractional derivatives, which were not then widely studied; he took them up in a pair of companion papers E14c and E14d (1840) 'On general differentiation' (see Ross 1977).

It was then customary in physical optics to invoke the ether as the medium for the passage of light, and to study its alleged properties and its interaction with matter, and from there to ponder on the applicability of such reasoning to other branches of mathematical physics.[5] Green is a good example of this view: in his C5a (1833) he studied the 'equilibrium of fluids analogous to the electric fluid', and he boldly assumed an nth-power force law and developed, both there and and in a sequel paper C5m (1831) on 'the exterior and interior attractions of ellipsoids of variable densities', the corresponding potential theory and Legendre polynomials (see Büttner 1900). His readiness to assume laws other than the inverse square was a controversial matter at that time; Cauchy had done so in connection with dispersion, especially in his monograph (Cauchy 1836), to the encouragement of Earnshaw C7e (1842) on 'the constitution of the luminiferous

Mathematics and mathematical physics

aether', but Kelland's study C6f (1838) expressed grave doubts (see Buchwald 1980, 257–9). Kelland seems to have wanted to restrict these laws to the preferred inverse square.

Kelland bears some comparison with another minor figure of the period: Challis, Airy's successor at Cambridge as professor of astronomy and Director of the Observatory. Challis's famous contribution to that subject was to fail to identify Neptone in 1846, but his principal activity lay in advocating (so-called) Newtonian views for 'all' physical phenomena, especially the motion of 'fluids'. Sometimes these were of the tactile kind, as when he contributed two sections to Moseley's textbook ⟨1830⟩ on hydrostatics and hydrodynamics (the first such work of its kind in our period written for use at Cambridge University). In his research in these areas, Challis continued along lines of thought pursued by Poisson; for example, his treatment C3d (1830) of 'the theory of the small vibratory motions of elastic fluids' drew on various functional and trigonometric-series solutions to the wave and related equations. Although his British Association reports reveal his wide knowledge of contemporary and recent literature, he rarely added significantly to knowledge, and over the years he came more and more under attack from others, especially Stokes (see Burkhardt 1908, 1056–107). Airy's 1842 article on 'Tides and waves' for the *Encyclopaedia Metropolitana* gave a much more substantial treatment.

Perhaps the most interesting papers in hydrodynamics are by Green. In C6p (1838) 'On the motion of waves in a variable canal of small depth and width' and in the supplementary 'Note' C7d (1842), he went a little beyond the general theory produced by Lagrange, Cauchy and Poisson (see Burkhardt 1908, 429–47) to find analytical solutions to the motion of waves in a shallow and narrow canal; his E13a (1836) 'On the vibration of pendulums in fluid media' (where this time the fluid would be a gas or vapour) contained a new closed-formula solution of the differential equations (see Stokes's report of 1847, in 1880–1905, 1:157–87). His readiness to range across the world of fluids is evident also in the fact that his C7a (1838) 'On the laws of the reflexion and refraction of light at the common surface of two non-crystalised media' (in which he also cited Cauchy's monograph (1836) on dispersion) was presented to the Cambridge Philosophical Society on the same day as his C6m (1838) 'On the reflexion and refraction of sound'.

Green's papers in fluids often drew upon the elastic properties of the medium, including the 21-parameter equation and the reduction in the number of parameters by assumption of special properties (especially in C7g (1842) 'On the propagation of light in crystallised media': see Saint-Venant 1864, 708–32). These equations had been introduced in the 1820s by Cauchy, Navier and Poisson in an atmosphere polemical even by Paris standards (see Burkhardt 1908, 526–41).

I conclude with a few remarks on areas of mathematical physics in which the interest was limited, beginning with electricity and magnetism. The lack of

I. Grattan-Guinness

attention given to Green's monograph (1828), published in Nottingham by a private person who was not yet in the Cambridge circle, is not very astonishing. However, I am surprised that he did not submit a summary paper at the time to the Cambridge Philosophical or Royal Societies, or write such an account for one of the commercial scientific journals. He gave three copies of his book to Hopkins, who was to hand one to Thomson; but otherwise he granted it little publicity. His own later papers, surveyed above, rarely cite it. His monograph, Whewell's article of 1830 for the *Encyclopaedia Metropolitana* on the 'theory of electricity',[6] and Murphy's textbook ⟨1833⟩ on electricity for use at the University, dealt largely with electrostatics, building on the various contributions of Poisson; on the mathematics side, they were heavily involved with potential functions and Legendre polynomials. Otherwise, there is no paper in Table VII centrally concerned with electromagnetism or electrostatics — a notable silence.

The most noteworthy contributions to magnetism are Airy's efforts R29a (1839) at 'discovering correction for the deviation of the compass produced by the iron of the ships' in which it was mounted. He rejected Poisson's theory of magnetism (presumably that given in Poisson 1838) for its 'great, perhaps insuperable' difficulties, and tried instead a simpler theory based on (Greenian?) inverse nth-power force laws, which gave 'comparative though not the same absolute results' (page 177). He applied the results to the adjustment of the compass on the *Ironside*, the first iron sailing vessel built in Britain (see Cotter 1976).

The other area of comparative neglect was heat diffusion — something of a relief after Paris, where the Bright Young Things (such as Duhamel, Lamé, Liouville and Sturm) were pursuing studies in Fourier's style to rather absurd lengths, and Poisson was continuing his own Laplacian line of research, especially with his *Théorie de la chaleur* of 1835 (see Bachelard 1928). Two papers stand out: Kelland C6i (1838), in which he compared and contrasted heat diffusion and sound as caused by 'the motion of a system of particles' (an example of his unifying approach mentioned earlier); and Challis C6o (1838) 'On the decrement of atmospheric temperature depending on the height above the earth's surface'. In addition, Kelland published at Cambridge a textbook/treatise ⟨1837⟩ on heat diffusion and radiation, the first such volume to be written in this area in English after the ideas of Fourier became well known in the 1820s (although Lunn provided a very useful survey of heat theory in his *Encyclopaedia Metropolitana* article of 1830). For physical theory Kelland appealed to Dalton's view that 'the particles of matter are surrounded by particles of caloric' with a 'mutual repulsion between particles of the same medium and a mutual attraction between those of an opposite one' ⟨1837⟩, v). He rejected Poisson's methods of modelling as 'devoid of simplicity' (page x), and for mathematical methods he applied Fourier series, a technique conspicuously absent from Cambridge mathematics and here given a rather poor

Mathematics and mathematical physics

treatment, as the young William Thomson was soon to find out (see Burkhardt 1908, 1174–81). But that, like so many developments of the 1840s, is another story.

Concluding remarks

As is usual in discussions of nineteenth-century Cambridge mathematics, the contributions to this volume tend to concentrate on Thomson and Maxwell. It is quite legitimate to regard these men as reaching the peaks of their subject during the heyday of Victorian physics. But they constitute members of the second and third generations of Cambridge figures; the first generation belongs to the reigns of George IV and William IV, when Babbage, Herschel, Whewell, Airy, de Morgan and Green set the pace in laying down some of the traditions on which their followers were to prosper. So far, studies of this generation have been modest in scale: perhaps Babbage and Herschel have been the most popular, but even then often in connections other than the mathematics; only Green has received an edition of his papers (Green 1871); and little biographical material is yet in print.[7] The fates of historical fashion have fallen just as favourably upon their contemporaries based elsewhere, who contributed with such significance to the encyclopaedias and book series described and listed here. It should be added that many of these contemporaries also published research works of various kinds which the choice of geographical centre for this book has dictated exclusion from appraisal here.

This chapter can be no more than a prolegomena, a draft outline for somebody's Ph.D. one day, perhaps; in this book it serves only as a curtain-raiser for later and greater things described in the pages following. I conclude by drawing attention to the following features, and posing a variety of historical questions.

Table II and my surrounding description reveals a mass of expository writing on almost all aspects of mathematics and mathematical physics, especially in the 1820s and 1830s. By and large the accounts are competent, and often reveal considerable knowledge of sources, both home and foreign. What are the intellectual and educational backgrounds of these authors, many not based at Cambridge and several poorly remembered today? As a related question, was there consonance or dissonance of intention at Cambridge during this period between the education for the Tripos and the possible advance to a research career? Certainly 'the training of the mind' provided there was intended to produce absorbers rather than creators of knowledge; and yet, for those of this latter intention, some of the textbooks listed in Table VII and encyclopaedia articles of Table II would have provided a pretty good grounding in skills and techniques, although not much of an eye for open questions. The Smith's Prize questions gave some stimulus to the creatively inclined.

Mention of activities outside Cambridge raises the question of the status of the city itself in our story. It was the Alma Mater of all our figures, and its

Philosophical Society published the majority of their papers; many of their textbooks and treatises also appeared there. Some of our major figures were permanently resident there throughout much of their creative lives, especially Green (though only after the publication of his Nottingham monograph (1828) on electricity and magnetism) and Whewell; among lesser lights, Challis and Hopkins were based there. But the other of our principals conducted much, in some cases nearly all, of their creative work elsewhere: in chronological order of departure, Herschel around 1816, after graduation in his 25th year; Babbage a few years later; the 22-year-old de Morgan, London-bound in 1828 with little so far published; Murphy also to London, in 1832; Airy to follow them in 1835, aged 34, though with a distinguished Cambridge career, including two chairs, behind him; and the 29-year-old Kelland, travelling north to an Edinburgh professorship in 1837 at the beginning of his publishing career.

These details rightly temper the status of Cambridgehood in this story. I see Cambridge playing its role most effectively as an educational centre and as a forum for research activity and publication, the centre of gravity for a partly scattered collection of mass points: geographical departure did not entail intellectual emigration, as the regular submission of papers to the Cambridge Philosophical Society shows. One should also not forget the university and college libraries, among the few comprehensive sources of literature in Britain at that time.

I turn now to the classification of the work produced by 'Cambridge' as presented in Table VI, the subject index. This table is an exercise in bivalent logic: a paper does or does not gain citation under a given topic. But such judgements are artificial, for we are in the world of degrees rather than yes-or-no; at which point, for example, does a paper on optics give sufficient attention to differential equations to gain a mention there also? Several of my decisions have been subjective, borderline and tentative, and might have come out the other way on another occasion. Unfortunately, although there are non-classical logics available (especially fuzzy set theory), I have not yet found an efficacious means of applying them. As a first approximation these tables may be a source of initial insights; when the Ph. D. aforementioned has been written, they may be safely forgotten.

Let us take up some features of the work which, however imperfectly, Table VI (and its successor on books) suggest. I discussed earlier the strong predeliction for algebras of various kinds in the 'pure' mathematics of the time, a tendency which England was to show all century long. (Indeed, it was evident before 1815 in Woodhouse, with roots back to the late 18th century.) What was the motivation for it? In France, where much of its parentage was found, there were explicit lines of influence from the neo-semiotic philosophies of Condorcet, Condillac and Degérando (see Auroux 1981); I do not see the same imperative in Cambridge mathematicians. Enros (1983) persuasively argues that they associated 'analysis' with *discovery* in mathematics, as opposed to the mere

Mathematics and mathematical physics

training of the mind furnished by synthetical geometrical reasoning. However, they might have realised that the French had shown that the associations of analysis with algebra and synthesis with geometry were breaking down at that time — as witness, for example, the establishment of 'analytic geometry' by Biot, Lacroix and others (see Boyer 1956).

I also indicated that Cambridge mathematicians between them made a somewhat uneven contribution to mathematical physics. The coverage of optics, 'fluids' and planetary physics was quite extensive, in contrast to the modest attention paid to topics *á la mode* such as heat theory. In addition, Cambridge uninterest in electricity and magnetism reminds one of the rather patchy Paris performance.[8] How did these preferences emerge? Is there any general rationale for the *choice* of problems upon which the Cambridge men attacked, or the chronology of that choice? In several cases a 'local' motivation is clear; some particular case noticed not to have been explored, a need for educational material at Cambridge University, and so on. However, I have not spotted a general strategy — and I would be prepared to believe that none exists. After all, the French (especially) had produced a mass of major work during the decades of Fenland torpor, so providing almost an *embarras du choix* for the first generation to choose from, as the fancy of choice and flash of inspiration may have taken them. I was slightly surprised, however, to see so little correlation between the chosen topics of research and the articles commissioned by encyclopaedias.

An interesting contrast of *Denkweisen* arises. As I have just emphasised, algebras dominated pure mathematics; but the applied mathematics is quite notably geometrical in character. Herschel is particularly interesting here, as he shows both aspects. The contrasts apply also to other parts of Britain; Scottish work was strongly geometrical (in line with commonsense philosophy), while the Irish had the Irish answer, with a penchant for rendering geometrically oriented mathematics in algebraic form. (Hamilton is a particularly good example.) Thus *Denkweisen* have some association with national characteristics, and seem to have influenced the nature of mathematics adopted from France (where all three modes were used). Crosland and Smith (1978) provide much useful information on 'The transmission of physics from France to Britain', during our period, although they do not distinguish the component countries.

My final remark concerns an individual. Among the Cambridge mathematicians, Airy stands out for the scope and depth of his researches. His election to the Royal Society in his 35th year is astonishingly late for that time. Best remembered today as an astronomer and geodeter, a notable mathematician remains to be discovered, the brightest light in an overly neglected community of late Georgian mathematicians.

Acknowledgements

I am grateful to Miss E.A. Stow, of Cambridge University Library, for

I. Grattan-Guinness

information on membership of the Cambridge Philosophical Society; and to the *Staatsbibliothek*, Berlin (DDR), for permission to quote from the Schumacher correspondence.

Notes

1. Some indications are provided in Ball (1889) and Garland (1980), but neither is satisfactory. Challis (1878) contains some interesting reminiscences of Cambridge education, although his own prejudices interfere in places. Howson's recent history (1982) of mathematical education has a little discussion on pages 76–9 in its chapter on de Morgan. Wilson's chapter in this volume contains further information.
2. Crosland (1967), based on the Société d'Arcueil, in fact contains a great deal of information on other aspects of French science. Grattan-Guinness (1981) surveys pure and applied mathematics (including engineering) for the period to 1840, and includes on pages 125–32 data on developments outside France, and at the end (pages 135–8) an extensive bibliography of secondary literature. This section and also parts of later ones draw on this paper.
3. See Whewell (1841) (written for Willis, the Professor of Natural and Experimental Philosophy at Cambridge from 1837 and himself much concerned with Kinematics), and Moseley (1843) (based on his lectures at King's College, London). Ferguson (1962) briefly surveys French and English research in these areas; Cambridge scientists seem to have been less interested than their London and Royal Society oriented colleagues in them (compare Hilken 1967, 45–57). Note also that the translation of Poisson's *Traité de mécanique* into English was made on a Dublin initiative, rather than on one from Cambridge (Poisson 1842).
4. See Sturm (1828) and Todhunter (1886, 258–71); the Sturm piece was published in the *Bulletin universel des sciences et de l'industrie*, a remarkably comprehensive reviewing and abstracting journal in which several of the books and papers produced by our Cambridge figures were noticed.
    The name 'Legendre polynomials' (or functions) became popular in the later 19th century; at this time Whewell introduced 'Laplace's functions' in his *Encyclopaedia Britannica* article of 1830 on 'the theory of electricity' (on which see note 6 and text).
5. See Whittaker (1951–53, 1: chapters 6 and 7); Buchwald (1980, 1981); and Siegel (1981). A lot more study is required of the uses made at the time of the words 'fluid', 'solid' and 'elastic', in order to unravel the intricacies of the theories and the disagreements of the protagonists.
6. Whewell's article is a good example of the bibliographical pitfalls surrounding this encyclopaedia, which were mentioned earlier. It appears, unsigned, as the second part of an article on electricity written by his Cambridge colleague Lunn; Whewell identified his authorship in one of his textbooks on mechanics (⟨1832b⟩, xviii). Becher (1980a, 18) points out the authorship problem, but awards Whewell the whole article (which Whewell himself did not claim); and he dates it as of 1845 instead of 1830.
    On the developments described in the text here, see Burkhardt (1908, 1111–19; Buchwald (1977); and Cross's chapter in this volume.
7. The autobiographical material is disappointing, too; Babbage (1864) is entertaining but superficial, and Airy (1896) tedious. Both are useful on some details, however; for example, Babbage recalled that while an undergraduate he bought (the first edition of) Lacroix's large *Traité* on the calculus (1864, 26–27: he and his colleagues were to translate the elementary version).
    Another dimension of this story awaiting exploration is manuscript sources: the massive collections left by Airy, Babbage, de Morgan, Herschel and Whewell have as yet not been much raided for their mathematical contents. Historical Manuscripts Commission (1982) is a useful compact guide to the location of manuscripts of British scientists.
8. The establishment of electricity and magnetism, with electrodynamics and electromagnetism, as major disciplines within mathematical physics, owes much to the inspiration of Thomson (compare Knudsen's chapter in this volume), and of Weber and F. Neumann in Germany during the 1840s. The Parisian patchiness of interest becomes all the more puzzling; for to a considerable extent these *étrangers* applied methods and techniques of French mathematical physics to their problems.

Mathematics and mathematical physics

physics to their problems. As evidence of the Cambridge apathy, note that the first initiative to study electrodynamics came from the professor of *chemistry*, James Cumming, who published in 1827 a translation of J.F. Monferrand's *Manuel d'électricité dynamique* (1823), together with his own additions and updatings. His edition made little impact.

9. It is recognised that Scotland and Ireland provide such British centres. I suspect that London may be worth examining as a centre in which various figures lived and worked and/or were related to the Royal Society (where, of course, Babbage and others promoted a famous ruction in 1830).

= 5 =

Integral theorems in Cambridge mathematical physics, 1830-55

J.J. CROSS

Introduction

The aim of this chapter[1] is to sketch the development of the relations between certain types of integral in Cambridge during the period 1830-55. The date 1830 is chosen from Murphy's first paper, and 1855 is taken from the start of Maxwell's work. One does not need to add 'in mathematical physics' because all papers written on this topic dealt with the contemporary physics in a mathematical way.

What are integral theorems? Why are they important for the progress of physical mathematics and physics in general? What forces them on us?

Potential theory[2] is rooted in the theory of mechanics, in the motion of particles, the flow of fluid, the shape of the earth. Early work by Huygens on clocks led Leibniz and then the Bernoullis to the conservation of kinetic energy (*vis viva*) of motion due to the fall of a body as the potential energy (*vis viva potentialis*) of its height as it rises and falls during the motion. Daniel Bernoulli and Leonhard Euler built on these foundations in 1738; their theory of the minimisation of energy integrals was the cornerstone. Euler's simple (i.e. one independent variable) integrals expressed the potential energy of a particle or a body, and its minimisation was regarded by Daniel Bernoulli as the program to bring all of mechanics under one principle. In general particle mechanics it is significant that all velocities in one-dimensional motions were expressed by Euler in terms of equivalent heights risen or fallen under constant gravity.

Alexis-Claude Clairaut derived a potential for the forces in a rotating mass of self-gravitating fluid by integrating along and around curves. From this integral he deduced a formula for the shape of the earth under his equilibration of the forces and motions. His integral is represented as a function rather than a definite integral and his ideas were subjected to severe challenge by Jean-le-Rond d'Alembert. D'Alembert applied the calculus to fluid flow in 1750 and derived a complex function as a solution to the so-called Cauchy–Riemann equations; this function of a complex variable acted as a potential for the velocity of the fluid. His integration of these equations for the fluid velocity in two dimensions followed the pattern of his solution of the equation of motion for the vibrating string given several years before, and does not concern us here.

The first actual integral transformation recorded in the literature seems to be

Integral theorems

that of Lagrange.[3] In 1760–61 he transformed a volume integral into a surface integral by using integration by parts. He and Euler had been working on the theory of sound propagation in an elastic medium, and they had arrived at the following set of partial differential equations:

$$\frac{T^2}{2h}\frac{d^2x}{dt^2} = \frac{E}{D}\left(\frac{d^2x}{dX^2} + \frac{d^2y}{dXdY} + \frac{d^2z}{dXdZ}\right)$$

$$\frac{T^2}{2h}\frac{d^2y}{dt^2} = \frac{E}{D}\left(\frac{d^2y}{dY^2} + \frac{d^2x}{dYdX} + \frac{d^2z}{dYdZ}\right) \quad (1)$$

$$\frac{T^2}{2h}\frac{d^2z}{dt^2} = \frac{E}{D}\left(\frac{d^2z}{dZ^2} + \frac{d^2x}{dZdX} + \frac{d^2z}{dZdY}\right)$$

where x, y, z are the coordinates of a moving elastic particle, t is the time, X, Y, Z are its original coordinates at time $t = 0$, and h, E, D, T are constants associated with the elasticity of and the stress in the medium. Briefly, what Lagrange does is multiply these three equations respectively by functions L, M, N of the coordinates X, Y, Z, add the results, and then integrate over the volume occupied by the elastic system. After integration by parts with respect to each of X, Y, Z where appropriate, he obtains nine equations of which a typical one is

$$\int \frac{d^2z}{dXdZ} L \, dX dY dZ = \int z \frac{d^2L}{dXdZ} dX dY dZ + \int \frac{dz}{dZ} L \, dY dZ - \int z \frac{dL}{dX} dX dY \quad (2)$$

Lagrange realised that the last two integrals in each transformation were over a surface 'slice' of the fluid and could be made zero by suitably choosing the region the medium occupies, and values L, M, N, x, y and z take on the surface bounding the region. He stated that one can do this for surfaces of arbitrary shape but his calculations were for a rectangular box centred at the origin with sides parallel to the coordinate planes. This reduced his differential equations to a very simple one,

$$\frac{d^2s}{dt^2} = kcs \quad (3)$$

where $s = \int (xL + yM + zN) dX dY dZ$, k is an arbitrary constant and $c = 2Eh/DT^2$.

Our interest is not the Fourier integral transform which follows but the transformation expressed by (2). This expresses a relation between an integral over a volume region in space and integrals over the surface bounding that region. Although the surfaces are somewhat primitively expressed without explicit terminals, the idea is clear.

Euler certainly knew of Lagrange's transform but he did not apply it in his fluid mechanics at the appropriate point.[4] Hence his fluid mechanics ends at this step: the theory of integration over curved surfaces had not then been developed. Fifty years were to pass before Gauss wrote his paper 'Theoria attractionis corporum sphaeroidicorum ellipticorum' in 1813 (Gauss 1863–1933, 5:6) which helped rekindle interest in general integrals.

This important paper does not seem to have been translated into any language other than German, something unusual for a paper of such weight but understandable because of its timing. The paper treats bodies formed by one or more arbitrary continuous closed finite surfaces which together constitute the surface of the body. The infinitesimal element of area of this surface is denoted by ds, with P a point in this area, PQ the exterior normal to the surface, and $\lambda = \widehat{QPX}$ the angle between the normal and the positive x-axis.

To show that $\int ds \cos \lambda = 0$ when the integral is extended over the whole surface of the body, Gauss proceeds as follows: he takes a plane perpendicular to the x-axis far enough back so that is does not cut the surface, i.e. the plane $x = \alpha$ for α less than the minimum value of x on the surface. Then he chooses an element dΣ of area in this plane, where dΣ lies in the shadow or projection of the surface on to the plane. At a point Π in dΣ he erects a perpendicular to the plane which cuts the surface at P′, P″, P‴, ... entering the body at P′, exiting at P″, and so on. Then about this line he constructs a cylinder whose generators are the lines perpendicular to the plane $x = \alpha$ at the boundary points of dΣ. This cylinder defines infinitesimal areas ds', ds'', ds''', ... at the corresponding points P′, P″, P‴, ... It is easy to see that

$$d\Sigma = -ds' \cos \lambda' = +ds'' \cos \lambda'' = -ds''' \cos \lambda''' = \ldots \quad (4)$$

because the angles are alternately obtuse and acute at entry and exit. Hence the result:

$$\int ds \cos \lambda = 0 \quad (5)$$

Similarly, $\int ds \cos \mu = 0 = \int ds \cos \nu$ where μ and ν are the angles between the exterior normal and the y- and z-axes, respectively.

Further, if $T = T(y, z)$, $U = U(z, x)$ and $V = V(x, y)$, then a simple modification of the argument gives

$$\int (T \cos \lambda + U \cos \mu + V \cos \nu) ds = 0 \quad (6)$$

These two equations (5) and (6) represent Gauss's first divergence theorems.

Neither in the derivation of these equations in §3 of the 'Theoria attractionis' nor in the expression of the attraction as a surface integral in the later §§8 and 9 is there any hint of *transformation* of integrals by integration by parts. Certainly this paper does not show the relation between general surface and volume integrals, it does not approach Lagrange's idea of integral transformation, and it does not do the Poisson–Ostrogradsky versions of the divergence theorem nor Green's relations for integrals. To be specific, the functions T, U and V in (6) are not differentiated at any stage.

The contributions Gauss's paper makes are twofold: first, the idea of the

Integral theorems

surface of a body and its general local infinitesimal area ds, and secondly, the process of lines entering and exiting from the surface. These two concepts remained long after the rest of the paper had been forgotten.

So now we can answer the second and third of our original questions. Integral theorems are important in mathematical physics because they help solve partial differential equations, because they seem to form an essential step in the transformation of such solutions, because they occur in many different areas of physical mathematics, and because the giants among our mathematical forefathers invented them and paid great attention to them. Let us now see how these transforms developed in the period after Napoleon.

Integral Theorems 1820–30

... about 1812, ... M. Poisson presented to the French Institute two memoirs of singular elegance, relative to the distribution of electricity on the surface of conducting spheres ... It would be quite impossible to give any idea of them here: to be duly appreciated they must be read. It will therefore only be remarked that they are in fact founded on the consideration of what have, in this Essay, been termed potential functions .... (Green, *Essay* 1828, Preface.)

The major problems in physical mathematics in this decade lay in elasticity, electricity and magnetism, and in heat propagation. It was in these three areas and in the associated field of complex analysis that integrals in volumes, over surfaces, and along curves were used. Further, the partial differential equations derived from the physical processes held for regions whereas the external conditions governing the processes were imposed on the boundary of the region: for example, the Laplace–Poisson equation

$$\frac{\partial^2 V}{\partial x^2} + \frac{\partial^2 V}{\partial y^2} + \frac{\partial^2 V}{\partial z^2} = 4\pi\rho \tag{7}$$

holds in a spatial region R in which there is a charge density ρ, and V is the electric potential. On the boundary surface of R it is usual to prescribe the values $\bar{V}$ of V; for example, the boundary may be earthed so as to make $\bar{V} = 0$.

This tension between the equation governing a function's values in a region and the values of this same function specified on the boundary of the region led to attempts to match the two sets of values in some way. Lagrange had already done so for the wave equation in a particular case; and the volume integrals for the gravitational forces and their potential from (7) were well known. Green was to complete the pairing in the potential theory case in the *Essay*, but his work is based on that of Poisson who is the major external contributor to the development of integral transformations in Cambridge during the next three decades. Poisson strongly influenced Green and certainly affected Stokes and Thomson; he was a key figure in Paris,[5] and it is the Paris school which affects both Green[6] and Gauss[7] as well as Dirichlet.[8] Therefore let us begin with an

J.J. Cross

examination of that work[9] of the Paris school which affected, directly or indirectly, the work of the Cambridge school using integral transformations in physical mathematics; directly, if the work is read and quoted by the Cambridge men themselves, or indirectly, if the effect comes through others as, for example, Ostrogradsky's ideas may have been passed on by Poisson. We begin with Ampère.

Between 1820 and 1825 Ampère constructed an extensive mathematical analysis of electrodynamic phenomena. These researches were published as a whole in a long paper, 'La théorie mathématique des phénomènes électrodynamiques' (Ampère 1827). The major mathematical problem involved in Ampère's work was the transformation of integrals around circuits or closed curves in the plane into integrals over the planar region the circuit encloses, and vice versa. For example, Ampère takes the integral around a small planar circuit enclosing a current-carrying wire.

$$C = \int \frac{x\mathrm{d}y - y\mathrm{d}x}{r^{n+1}} \tag{8}$$

in cartesian coordinates x, y and re-expresses it as an integral in plane polar coordinates u, ϕ using $x = u \cos \phi$, $y = u \sin \phi$:

$$C = \int \frac{u^2 \mathrm{d}\phi}{r^{n+1}} \tag{9}$$

where r is the distance from the wire to the plane circuit. He then proceeds to express this line integral as an integral over the plane region enclosed by the circuit:

$$C = \int \left(\frac{n-1}{r^{n+1}} - \frac{(n+1)qz}{r^{n+3} \cos \xi} \right) u \delta u \delta \phi \tag{10}$$

in which $u\delta u\delta\phi$ is the element of area in polar coordinates (Ampère 1827, 225–7). He then extends his integral from small regions to finite ones by dividing the plane into congruent parallelograms and integrating around these; this has become a classic[10] idea in complex function theory for proving Cauchy's theorem.

Cauchy wrote on complex integrals along paths in the complex plane between any two complex numbers z_0 and z_1 in his *Mémoire sur les intégrales définies* early in 1825:

$$\int_{z_0}^{z_1} f(z)\mathrm{d}z \tag{11}$$

The limits were explicit in the integral, and so too was the curve between the two points, $x = \phi(t)$, $y = \chi(t)$, t real; this was the curve in the z-plane along which the integral was taken, $z = x + iy$. He did not here explicitly use the idea of an integral around a circuit, but introduced the idea of homotopy, the continuous

Integral theorems

deformation of one curve to another nearby. From this, the integral along the (ϕ, χ)-curve was shown to be equal to that along the (ϕ + εu, χ + εv)-curve if ε was small, f was bounded and continuous, and both curves have the same endpoints; he did this by showing that the first variation of the integral was zero.

There are two further comments to be made about this *Mémoire sur les intégrales définies*. First, there is a comment that if we switch to the notation of Lagrange, and put $u = \delta x$, $v = \delta y$, and apply the ideas of the calculus of variations, then the statement that the first variation $\delta \int f(z) dz = 0$ means that 'the function [basically f] under the integral sign reduces to an exact differential' (Cauchy 1825, 45). This is an idea later used by Riemann in a similar context,[11] the context where Dirichlet's principle arises. Also in Cauchy's *Mémoire* is the idea of the Cauchy principal value; this concept was introduced to deal with the case where a singularity of the function f(z) occurred on the path of integration. It is this which prevented Ostrogradsky from submitting for publication[12] his manuscript on the same subject, read at the Paris Academy meeting of 7 August 1824.

The contribution of Ostrogradsky to theorems on integrals is as immense as its influence was small: Ostrogradsky lacked the effective propagandist that Gauss found in Dirichlet (number theory), Dirichlet found in Riemann (physics), Weierstrass found in Cantor, Heine and Paul du Bois-Reymond (analysis), and Green posthumously found in Thomson (the *Essay*). There is no adequate discussion in English of Ostrogradsky's work though there are many in Russian (Ostrogradsky 1959; Antropova 1965; Youschkevitch 1965 and 1967 [in French]; Sologub 1975).

Briefly, Ostrogradsky and Dirichlet were in Paris from 1822 to 1827. The former was a protégé of Cauchy, taught by Poinsot, Binet, and Lamé, and a friend of Sturm, Poisson and Fourier, as well as the Russian number theorist Buniakowsky.[13] Dirichlet became a protégé of Fourier and took a dislike to Poisson; traces of this can be seen in his letters to Gauss, in his influence on Riemann,[14] and in his writings on Fourier series.

The original French texts of Ostrogradsky's papers on heat conduction, read to the Paris Academy on 13 February 1826 and on 14 May and 6 August 1827, have not yet been published; a Russian translation has appeared (Ostrogradsky 1965a,b) with some notes by Antropova and Youschkevitch. On his return to St Petersburg in 1828, he read these Paris papers to the Academy as two abbreviated Notes (Ostrogradsky 1831a,b) on 5 November 1828 and 28 July 1829, and these were published in September 1831 and reprinted in Russian translation in his *Collected Works* (Ostrogradsky 1959–61). It is recorded in the *Procès Verbaux* of the Paris Academy[15] that Poinsot and Legendre were to report on the paper of 1826. The report exists but it does not seem complete and was not printed; this report has not yet been fully studied.

However, the published version of his theory in the two St Petersburg 'Notes' (Ostrogradsky 1831b, 130–1; 1959, I:62–3) make his contribution clear: his divergence theorem explicitly transforms a triple integral over the volume enclosed by a surface:

$$\int \left(\frac{dp}{dx} + \frac{dq}{dy} + \frac{dr}{dz}\right)\omega = \int \left(\frac{dp}{dx} + \frac{dq}{dy} + \frac{dr}{dz}\right) dz dy dx \qquad (12)$$

via integration by parts; he gives no reference. The triple integral (12) is taken over the volume of a region bounded by an arbitrary surface; he supposes that p, q, r are finite functions of x, y, z. His language is reminiscent of Lagrange and Lamé, and his style recalls that of Lagrange and Gauss. He takes

$$\int \frac{dr}{dz}\omega = \int \frac{dr}{dz} dz dy dx = \int dx dy \int \frac{dr}{dz} dz \qquad (13)$$

and thinks of a rectangular box perpendicular to the xy-plane (at some point) and with cross-section $dxdy$. This long box cuts the boundary surface at $2n$ points, whose z-values are, in order, $z_1, z_2, \ldots z_{2n}$. Then r takes values $R_1, R_2, \ldots R_{2n}$ at these points, so that

$$\int dy dx \int \frac{dr}{dz} dz = \int (R_2 + R_4 + \ldots + R_{2n}) dy dx$$
$$- \int (R_1 + R_3 + \ldots + R_{2n-1}) dy dx \qquad (14)$$

where the even (exit) terms are positive and the odd (entry) terms are negative. If ν is the angle between the positive z-axis and the exterior normal at the point on the surface and s the element of area formed there by the prism, then

$$\int \frac{dr}{dz}\omega = \int (R \cos \nu) s \qquad (15)$$

where the surface integral is taken over the whole boundary surface. Hence

$$\int \left(\frac{dp}{dx} + \frac{dq}{dy} + \frac{dr}{dz}\right)\omega = \int (P \cos \lambda + Q \cos \mu + R \cos \nu) s \qquad (16)$$

where the left side is a triple integral through the volume, the right side is over the boundary surface so that P, Q, R are the values of p, q, r on the surface, λ, μ, ν are the angles the exterior normal to the surface makes with the positive x-, y-, z-axes, respectively, and ω, s are the differential elements of volume and area. Ostrogradsky then applies this integral theorem to partial differential operators and to the heat conduction equation.

We have purposely portrayed the work of Ampère, Cauchy and Ostrogradsky before that of Poisson, even though Poisson's work appeared before and after theirs in the period 1812–30. Seven of Poisson's memoirs in this period had a decisive influence on our Cambridge quartet. In particular, his 'Mémoires sur la distribution de l'électricité' (Poisson 1812) were very widely read and used by Green. Stokes paid particular attention to his 'Mémoire sur … corps élastiques' (Poisson 1829) and his 'Mémoire sur … fluides' (Poisson 1830).

Integral theorems

Further, his 'Mémoires sur le magnétisme' (Poisson 1826a,b, 1827) were certainly read and used by Green and Thomson. What attracted Cambridge's attention to these papers?

His 'Mémoires sur l'électricité' (Poisson 1812) are basic: they influenced Green and Murphy through the potential for the electrostatic forces (Poisson 1812, 16–17) and its expansion in terms of the Laplace coefficients (Poisson 1812, 18–22). The integral

$$\int \frac{\rho(x, y, z)\mathrm{d}x\mathrm{d}y\mathrm{d}z}{\sqrt{\{(x - a)^2 + (y - b)^2 + (z - c)^2\}}} \tag{17}$$

over a specified volume region gives a function of a, b, c corresponding to a distribution of mass or charge at points (x, y, z) with density $\rho(x, y, z)$; this integral dates back to Lagrange and Euler, and was well known by 1812. The derivatives of this function with respect to a, b and c give the components of the force exerted at (a, b, c) by the charge inside the volume. It was not easy to calculate the integral (17) even in simple cases and certainly not for spheroids. So it was usual to approximate true ellipsoids as surfaces differing little from spheres, and to expand the various integrals as series of functions depending on the coordinates and the geometry of the sphere. Thus the Laplace coefficients, widely disseminated through the *Mécanique céleste* though due to Legendre, were intensively studied and their orthogonality conditions were well known. Hence his 'Seconde mémoire sur l'electricité' contains a long section on the Poisson integral and its use to calculate various definite integrals and to sum series involving the Laplace coefficients or Legendre polynomials (Poisson 1812, 212–30). Green adopted the idea of the potential and its association with the Laplace equation. Murphy generalised the whole approach to families of functions, their orthogonality relations, and the use of such expansions to express forces and potentials, and he did this by inverting integral operators, i.e. by solving integral equations.

Poisson's set of papers on magnetism in 1824 are not dominated by the sphere as his papers on static electricity were in 1812. These 'Mémoires sur magnétisme' (Poisson 1826a,b, 1827) contain ideas used by Green, Stokes, Thomson and Maxwell. First, the divergence theorem is given in a paper read to the Paris Academy on 2 February 1824 but published in 1826 (Poisson 1826a, 294–6): as a means of 'simplifying formulae', he transforms the volume integral (over the whole extent of a volume region)

$$\iiint \left(\frac{\mathrm{d} \cdot \frac{\alpha' k'}{\rho}}{\mathrm{d}x'} + \frac{\mathrm{d} \cdot \frac{\beta' k'}{\rho}}{\mathrm{d}y'} + \frac{\mathrm{d} \cdot \frac{\gamma' k'}{\rho}}{\mathrm{d}z'} \right) \mathrm{d}x'\mathrm{d}y'\mathrm{d}z' \tag{18}$$

by integration by parts into the surface or double integral

$$\int (\alpha' \cos l' + \beta' \cos m' + \gamma' \cos n') \frac{k'}{\rho} \mathrm{d}\omega' \tag{19}$$

J.J. Cross

in which $d\omega'$ is the differential element of area, and l', m' and n' are the angles between the exterior normal to the surface and the positive x-, y- and z-axes, respectively, and the integral extends over the whole surface bounding the volume. The simplification is made in a manner similar to that of Lagrange and Ostrogradsky, an integration by parts along the z-axis with the boundary surface split into an upper and a lower part, and the expression of the difference of these two integrals over the xy-plane as a single integral over the surface.

These equations (18) involve the components of the magnetic force along the coordinate axes expressed as the derivative of a potential, and these components are presumed invariant under rotations of the axes, i.e. they are cartesian tensors.

The third 'Mémoire sur magnétisme' (Poisson 1827) treats the divergence theorem as a known theorem, and is read after Ostrogradsky's 'Proof of a theorem in the integral calculus' (Ostrogradsky 1965a) read on 13 February 1826. One does not know how much, *if at all*, Poisson rewrote these three papers after they were originally read. The surface expressions form an integral part of the first paper, dated 1826 at the Imprimerie Royale. The dates of these three papers should be noted: the first (above) was read on 2 February 1824, the second on 27 December 1824, and the third on 10 July 1826.

So far we have dealt with Poisson's papers on electrostatics and magnetism. He used these same ideas in a slightly extended and combined form in his 'Mémoires sur ... corps élastiques et fluides' (Poisson 1829 and 1830). He again transforms volume integrals of the internal forces into surface integrals involving the surface stress tensor, in the manner described above: integration by parts with respect to each coordinate separately and then a combination of the two (not $2n$ as in Ostrogradsky) integrals to form a single integral over the surface bounding the volume:

$$\int \left(\frac{dP_1}{dz} + \frac{dP_2}{dy} + \frac{dP_3}{dx} \right) \rho \, dxdydz = \int (\gamma P_1 + \beta P_2 + \alpha P_3) ds$$

$$\int \left(\frac{dQ_1}{dz} + \frac{dQ_2}{dy} + \frac{dQ_3}{dx} \right) \rho \, dxdydz = \int (\gamma Q_1 + \beta Q_2 + \alpha Q_3) ds \quad (20)$$

$$\int \left(\frac{dR_1}{dz} + \frac{dR_2}{dy} + \frac{dR_3}{dx} \right) \rho \, dxdydz = \int (\gamma R_1 + \beta R_2 + \alpha R_3) ds$$

where ds is the differential surface area element and α, β, γ are the cosines of the angles which the exterior normal makes with the x-, y-, z-axes, respectively. The Ps, Qs and Rs are the stresses and the nine form the stress tensor. Their combination with the direction cosines α, β, γ form the stress along the x-, y-, z-axes on the right side of the equation.

The particular form of these equations is based on a molecular model of the forces in elastic solids, fluids and liquids which was peculiar to the Laplacian style of the French school. Under the action of rotations and reflexions in three

Integral theorems

dimensions the expression reduced to one arbitrary constant for longitudinal extension under tension and a fixed ratio (Poisson's ratio of $\frac{1}{2}$) relating the transverse extension to the longitudinal extension. Despite experimental evidence and mathematical argument to the contrary, this error persisted in the French engineering schools to the end of the Second Empire.[16]

Reviewing developments to the end of the third decade of the nineteenth century we see that the transformation of integrals from one type (volume/surface) to another (surface/line) is a prominent feature of the physical mathematics of the time. Lagrange had used it for the partial differential equations of sound propagation, Gauss had been on the fringes of it for gravitational attraction, Ampère had used it in electrodynamics, Cauchy had adapted it for homotopy in complex function theory, Ostrogradsky had applied it in heat conduction, and Poisson had actively employed it in magnetism and elasticity. Coupled with it were the expressions of forces as the derivatives of potentials and the expansion of these in terms of series involving Laplace coefficients for almost spherical objects in gravitational attraction and electrostatics.

Robert Murphy, 1830–38

There are peculiar difficulties in judging Murphy and his work. Murphy was plucked from a poor family in Ireland and, after graduating as third wrangler in 1829, was placed in a position of some responsibility after being made a Fellow of Gonville and Caius. He left Cambridge under a cloud in December 1832 and did not return. Academically he had five productive years and then the quality of his work evaporated as he struggled with debt and then illness. He died in 1843.[17] Despite the excellence of his early major papers, there is a large disparity between his promise and his performance.

Murphy and Green were not in college together because Green entered Caius late in 1833. However, through Whewell and Bromhead they exchanged papers, including Green's *Essay* and Murphy's 'Definite integrals'; Murphy acted as referee for Green's 'On the laws of equilibrium of fluids' late in 1832 (Bowley 1976, 72–3). Despite this, Murphy's *Electricity* (1833c) has several places where the word 'potential' could have been used and the concept clearly identified, but Murphy did not do so. The only mention of Green occurs in the 'First memoir on inverse methods' (Murphy 1833b, 357) and it is this reference which later stirred Thomson's interest.

Murphy wrote on three related topics: algebraic equations, integral equations and operator calculus. His work was almost totally neglected at the time, and modern evaluations[18] seem to have overlooked the content and purpose of his 'Definite integrals' (Murphy 1830), his long memoir on 'Inverse methods' (Murphy 1833b, 1835a,b), and his *Electricity* text (Murphy 1833c). These, along with his 'Resolution of algebraical equations' (Murphy 1833a), form a corpus of 400 pages intended to show the central role in physical mathematics of definite integrals and certain families of functions associated with both these integrals and Laplace's equation. Whereas his *Electricity* was written as a text for students and was suited to the Cambridge syllabus of the time with its emphasis on the Laplace coefficients, the other five papers listed above are written at a very high level of mathematical sophistication; further, they betray a high degree of planning and organisation, they are tightly written for the time, and they contain developments of substance far beyond those of his contemporaries. What Murphy did was develop a general and systematic method for solving integral transform equations, both directly and by series; part of the solution involved special and general sets of orthogonal functions and a systematic treatment of delta functions.[19] This general theory was far in advance of current British mathematics and ahead of Continental practice.

Let us now look at Murphy's work in some detail. As early as May 1830, Murphy had conceived the plan of taking phenomena observed in a physical system and deriving from them the laws governing the elementary actions of the particles forming the system; to him this meant, for example, taking the observed gravitational attraction of a body on a particle and either deriving the inverse square law of attraction between the given particle and the particles of the body or deriving the mass distribution required to give the observed attraction. Since gravitational and electrical attractions were expressed as simple, double or triple integrals, he realised that 'considerable advantages ... might be expected to result from the study of *general* properties of definite integrals', especially in the theories of heat conduction and electric charge distribution; for in these cases 'the form of the function under the sign of definite integration is unknown' (Murphy 1830, 429). This first paper on 'Definite integrals' is preoccupied with the expression of definite integrals as certain derivatives of a function, or as power series, or as series of special types.

His second paper on 'Algebraical equations' (Murphy 1831) shows how to find the roots and certain functions of the roots of polynomials. Most methods involve power series, and one uses the Cauchy residue theorem. In particular, if $\phi(x)$ is any power series in integer powers of x, finite or infinite, then the coefficient of $1/x$ in $-\log\{\phi(x)/x\}$ gives the least root of the equation $\phi(x) = 0$, where 'least' means least in absolute value. These first two papers involve the changing of the terminals of definite integrals in the complex plane, but in a rather formal way; the change in the path of integration involved is not adverted to.

We now come to the central and massive part of Murphy's work: 170 quarto pages on the 'Inverse method of definite integrals' (Murphy 1833b; 1835a,b) in the *Cambridge Philosophical Society Transactions*. Here he presents the systematic solution of integral equations, and related matters. The treatment is Eulerian, based on many special examples but giving a general theory. It deserves a somewhat detailed treatment. Before we do so we comment that Murphy claims acquaintance with the work of Cauchy, Fourier, Poisson, Gauss

Integral theorems

and Green as well as Euler, Lagrange, Laplace, Servois, Français, Lacroix, Navier and Jacobi. He does not mention Dirichlet and this is not surprising since Dirichlet had not yet published on definite integrals.

The essence of these papers is: given the definite integral of a function, find the function. In terms of a formula, given a function $\phi(x)$ which is the definite integral of function $f(t)$ with respect to some kernel function, initially t^x, find the function f:

$$\phi(x) = \int_0^1 f(t) t^x dt \qquad (21)$$

The terminals of the integral may be chosen arbitrarily, but 0 and 1 were adopted by Gauss and taken over by Murphy. To invert the transformation from f to ϕ in (21), he restricted x to be an integer parameter, and ϕ to be a polynomial, a rational function, a non-rational power, or a logarithmic function, either of x or $x + a$. An example of a fairly complicated type is provided by $\phi(x) = 1/(x + m)^n$ for all positive real numbers n, where the inversion gives

$$f(t) = t^{m-1}(-\log t)^{n-1}/\Gamma(n) \qquad (22)$$

How was the inversion accomplished? In the beginning, for the kernel t^x as in (21) and rational functions $\phi(x)$ which tend to zero as x goes to infinity, he gives an algebraic algorithm: multiply $\phi(x)$ by t^{-x}, expand this product in powers of $1/x$ for x large, take the coefficient T of $1/x$ and divide it by t. This algorithm works for any number of variables, and can be extended to irrational functions also. The algorithm is established by a power series argument completed by saying that since

$$\int_0^1 \frac{T}{t} t^x dt = \phi(x) = \int_0^1 f(t) t^x dt \qquad (23)$$

it follows that $f(t) = T/t$, as given in his 'First memoir' (Murphy 1833b, 363).

It was well known in 1832 that the gravitational attraction of spheroids (Gauss 'Theoria attractionis') and the electrostatic attraction of charged spheres (Poisson 1812a) underwent discontinuities at the surface of the spheroid or sphere. These attractions were represented as integrals over these surfaces. The problem was to devise a single formula, '*an analytical expression, which without altering its form, may continue to represent the true value of the definite integral, or discontinuous function, under all circumstances*' (Murphy 1833b, 374; emphasis in the original). Using the techniques he had devised in his paper on 'Algebraical equations' (Murphy 1833a) he developed a series which had values 0 or $1/\alpha^n$ *as* α was less than or greater than some value β, and from this he could obtain attractions derived from a potential. For example, the attraction of a spherical shell of radius β is given as

J.J. Cross

$$A(\alpha) = \begin{cases} 0, \text{ if } \alpha < \beta \\ 1/\alpha^2, \text{ if } \alpha > \beta \end{cases} \qquad (24)$$

where naturally α and β are positive numbers. A single formula for $A(\alpha)$ can be derived as follows: we put $\phi(x) = (x + \alpha)(x + \beta)$ and look for the coefficient of $1/x$ in $-\log\{\phi(x)/x\}$ or in $-\log\{\phi(x)/x(\alpha + \beta)\}$ since the addition of the constant does not affect this coefficient. We expand

$$-\log\left\{\frac{(x+\alpha)(x+\beta)}{x(\alpha+\beta)}\right\} = -\log\left\{1 + \frac{\alpha\beta + x^2}{x(\alpha+\beta)}\right\} \qquad (25)$$

and take the coefficient of $1/x$ in each term to get:

$$\min(\alpha, \beta) = \frac{1}{2}\frac{\alpha\beta}{\left(\frac{\alpha+\beta}{2}\right)} + \frac{1\cdot 1}{2\cdot 4}\frac{(\alpha\beta)^2}{\left(\frac{\alpha+\beta}{2}\right)^3} + \frac{1\cdot 1\cdot 3}{2\cdot 4\cdot 6}\frac{(\alpha\beta)^3}{\left(\frac{\alpha+\beta}{2}\right)^5} + \ldots \qquad (26)$$

If we replace α, β by their reciprocals $1/\alpha$, $1/\beta$ respectively, then:

$$S(\alpha, \beta) = \min\left(\frac{1}{\alpha}, \frac{1}{\beta}\right)$$

$$= \frac{1}{2}\frac{1}{\left(\frac{\alpha+\beta}{2}\right)} + \frac{1\cdot 1}{2\cdot 4}\frac{(\alpha\beta)}{\left(\frac{\alpha+\beta}{2}\right)^3} + \frac{1\cdot 1\cdot 3}{2\cdot 4\cdot 6}\frac{(\alpha\beta)^2}{\left(\frac{\alpha+\beta}{2}\right)^5} + \ldots \qquad (27)$$

and a very quick calculation shows that:

$$\frac{d^{n-1}S}{(-1)^{n-1}\cdot 1\cdot 2\ldots(n-1)d\alpha^{n-1}} = \begin{cases} \frac{1}{\alpha^n} \text{ when } \alpha > \beta \\ 0 \text{ when } \alpha < \beta \end{cases} \qquad (28)$$

Hence, for $n = 2$, we get

$$A(\alpha) = -\frac{dS}{d\alpha} \qquad (29)$$

Murphy gives several more mathematical examples, one involving the generating function for the Laplace coefficients, but the examples of interest are those devoted to electricity where this same generating function is prominent. For example, he uses a spherical shell of radius α to which a rod is attached radially at the surface. Using the x-axis along the rod and the origin at the point diametrically opposite the point of attachment and β the distance *from the centre* of the sphere to any point of the rod, Murphy writes down the usual formula for the potential (he calls it 'tension' but he has already identified the two words in the famous footnote on page 357 of this 'First memoir', when referring to Green's *Essay*) at any point on the rod as

Integral theorems

$$V(\beta) = \frac{m}{\beta} = \int_0^{2\alpha} \frac{2\pi\alpha A dx}{\{\alpha^2 + \beta^2 + 2\beta(\alpha - x)\}^{1/2}} \tag{30}$$

and the general formula for the potential inside or outside the shell along the line of the rod as

$$V(\beta) = \begin{cases} m/\alpha & \text{for } 0 \leq \beta < \alpha \\ m/\beta & \text{for } \beta > \alpha \end{cases} \tag{31}$$

where m is the total charge on the spherical shell and $A = A(x)$ is now the distribution of charge on the surface. He points out that V is observed and/or deduced from experiment but A is an unknown function. He puts $x = 2\alpha t$ to change the limits as required by his method:

$$V(\beta) = \int_0^1 \frac{A dt}{\{\alpha^2 + \beta^2 + 2\alpha\beta(1 - 2t)\}^{1/2}} \tag{32}$$

and notes that the requirement that m be the total charge implies that

$$m = \int_0^1 A dt \tag{33}$$

Hence the problem has been reduced to that of finding $A = A(t)$ such that the integral

$$\int_0^1 \frac{A(t) dt}{\{\alpha^2 + \beta^2 + 2\alpha\beta(1 - 2t)\}^{1/2}} = \min\left(\frac{m}{\alpha}, \frac{m}{\beta}\right) \tag{34}$$

By expanding the denominator on the left in (34) as

$$\frac{1}{\alpha + \beta} \left\{ 1 - \frac{(\alpha\beta)t}{\left(\frac{\alpha + \beta}{2}\right)^2} \right\}^{-1/2} \tag{35}$$

and using the series $S(\alpha, \beta)$ from (27) in the right side, he shows that

$$\int_0^1 A(t) \cdot t^x \cdot \frac{1 \cdot 3 \cdot \ldots (2x - 1)}{2 \cdot 4 \cdot \ldots (2x)} 2^{2x} dt = m \frac{1 \cdot 1 \cdot 3 \cdot \ldots (2x - 1)}{2 \cdot 4 \cdot 6 \cdot \ldots (2x + 2)} 2^{2x+1}$$

or

$$\int_0^1 A(t) t^x dt = \frac{m}{x + 1} \tag{36}$$

from which function $\phi(x) = m/(x + 1)$ it is easy to show that the function $f(t) = A(t) = m$. Hence he has derived the standard result that the potential (31) is due to a uniform distribution of charge.

He proceeds with further examples which involve special distributions of charge and the Laplace coefficients in particular, but we have given enough detail to portray an outline of his methods. One comment must be made: the identification of two integrands in (23) or (36) because they give the same definite integral is unjustified; Murphy himself will point this up in next two papers we examine, in equations (39) and (44) below, but the error here is in line with one Gauss will make in 1839. There are infinitely many distributions on a two-dimensional surface which will give the same potential in three-dimensional space: Chasles conjectured this in 1838 and Poincaré proved the conjecture correct in 1895.[20]

Let us follow Murphy into his 'Second and third memoirs on the inverse method' (Murphy 1835a,b). In the 'Second memoir' he developed the idea of general families of orthogonal functions; this memoir should be read in conjunction with his book on *Electricity* (Murphy 1833c) where only the Laplace coefficients are developed. He treated the question of finding a family of functions F(t, x) required to solve the integral equations for f(t) when the function $\phi(x)$ is given: either

$$\int_0^1 f(t)F(t, x)dt = 0 \qquad (37)$$

when $\phi(x) \equiv 0$, or

$$\int_0^1 f(t)F(t, x)dt = \phi(x) \qquad (38)$$

when $\phi(x)$ is not identically zero. Since the Legendre polynomials $P_n(t)$ satisfying $\int P_n P_m = 0$ for $m \neq n$ and $\int P_n P_n = 1$ were known,[21] it was easy to see that, when F(t, x) = t^x wherein x is a non-negative integer, these $P_n(t)$s could be a family of such f(t)s. Further he found such families for other kernels F(t, x), and a general theory of such functions parametrised by the integer n and other (integer) parameters was developed in this 'Second memoir'. Given an integer n, the general solution of (37) was found for the case F(t, x) = t^x where (37) holds for $x = 0, 1, 2, \ldots, n - 1$. Then

$$f(t) = \frac{d^n}{dt^n}(t^n(1 - t)^n V(t)) \qquad (39)$$

where V is any arbitrary (analytic) function of t.

Having found these families and general solutions for (37) he redevelops them for (38) as follows: for suitable functions $\phi(x)$ he considers the n equations obtained from (38) and the function

$$f(t) = A_0 + A_1 t + A_2 t^2 + \ldots + A_{n-1} t^{n-1} \equiv T_{n-1}(t) \qquad (40)$$

when the values $0, 1, 2, \ldots, n - 1$ are substituted for x. For these $\phi(x)$ we assume that

$$\int_0^1 T_{n-1}(t)t^x dt = \phi(x) \qquad (41)$$

Then the general function f(t) satisfying (38) is obtained from

Integral theorems

$$\int_0^1 T_{n-1}(t)t^x dt = \phi(x) = \int_0^1 f(t)t^x dt \tag{42}$$

or

$$\int_0^1 \{f(t) - T_{n-1}(t)\}t^x dt = 0 \tag{43}$$

i.e. f(t) is composed of the particular solution $T_{n-1}(t)$ and the general solution (39):

$$f(t) = T_{n-1} + \frac{d^n}{dt^n}(t^n(1-t)^n V(t)) \tag{44}$$

In the remainder of the paper this idea is developed at length for several kernels with one or more (integer) parameters.

In his 'Third memoir on the inverse method of definite integrals' (Murphy 1835b) he continues his development of these families of reciprocal functions, deriving the Legendre functions of both types and of various orders. He then goes on (Murphy 1835b, 347): 'Let $\phi(h, t)$ be such that when h has a particular value assigned, the whole function vanishes whatever may be the value of t, except in one case; $\phi(h, t)$ under these circumstances, is a transient function having only a momentary existence.' For example, with $t \neq 0$, we have the transient function

$$\phi(h, t) = \frac{(1-h)(1+h)}{\{1 - 2h(1-2t) + h^2\}^{3/2}} \tag{45}$$

equal to zero for $h = 1$, while for $t = 0$ the function ϕ tends to infinity as h tends to 1. These transient functions developed the idea of reciprocal functions but their properties are quite spectacular. These families of functions satisfy the usual integral relations for the reciprocal functions but they are zero everywhere except at one point. That is, the transient family reciprocal to t^n is a family $V_m(t)$ such that

$$\int V_m t^n = 0 \text{ for } m \neq n \text{ and } \int V_n t^n \neq 0 \tag{46}$$

and each V_m is zero except at $t = 0$ where it is infinite. For the kernel t^n the family is generated from the Legendre generating function $R = \{1 - 2h(1-2t) + h^2\}^{-1/2}$ which Murphy has used so often; he derives the general transient function $V_n(t)$ to be

$$V_n(t) = \lim_{h \to 1} 2h^{-(2n-1)/2} \frac{d}{dh}\left(h^{(2n+1)/2} \frac{d^{2n} h^n \{1 - 2h(1-2t) + h^2\}^{-1/2}}{1 \cdot 2 \cdot 3 \cdot \ldots (2n+1) \cdot dh^{2n}}\right) \tag{47}$$

In particular, the transient function $\phi(h, t)$ in (45) is the function $V_0(t)$, where

$$V_0(t) = \lim_{h \to 1}\left(R + 2h\frac{dR}{dh}\right) \tag{48}$$

J.J. Cross

He derives the properties of this particular family and other such families in detail, including, as before, series expansions of functions in terms of these transients. He concludes this 'Third memoir' with thirteen examples of various types, most of them mathematical. For example, to find $\phi(t)$ such that

$$\int_0^\pi \phi(t) \cos(at) dt = 1 \qquad (49)$$

he adopts a family of reciprocal functions $\cos(nt)$ where n is any non-negative integer, so that with

$$\phi(t) = \sum_{n=0}^{\infty} c_n \cos nt \qquad (50)$$

and

$$\cos(at) = \frac{\sin(a\pi)}{a\pi} + \sum_{n=1}^{\infty} (-1)^n \frac{2a \sin a\pi}{\pi(a^2 - n^2)} \cos nt \qquad (51)$$

he obtains from (49) the equation

$$1 = \sin a\pi \left\{ \frac{c_0}{a} - \frac{a \cdot c_1}{a^2 - 1} + \frac{a \cdot c_2}{a^2 - 2^2} - \frac{a \cdot c_3}{a^2 - 3^2} + \ldots \right\} \qquad (52)$$

If $a = 0, 1, 2, 3, \ldots$ are successively substituted into (52), we get (correcting a typographical error of Murphy's)

$$c_0 = \frac{1}{\pi}, \ c_1 = \frac{2}{\pi}, \ c_2 = \frac{2}{\pi}, \ \ldots, \qquad (53)$$

from which

$$\pi\phi(t) = 1 + 2\cos t + 2\cos 2t + 2\cos 3t + \ldots, \qquad (54)$$

$$= \lim_{h \to 1} \frac{(1-h)(1+h)}{1 - 2h\cos t + h^2} \qquad (55)$$

i.e. $\phi(t)$ is a transient function (Murphy, 1835b, 379).

It is to be emphasised that Murphy's approach to these definite integrals has a deep physical motivation, in gravity, electrostatics and heat. He was on the verge of productive work in physical applications as the last example with Fourier series shows, but his life was disturbed at this stage and he was diverted into an algebraic cul-de-sac. For example, he could have approached the classic triple integral for the attraction of an ellipsoid,

$$V(x', y', z') = \int_{-a}^{+a} \int_{-y_0}^{+y_0} \int_{-z_0}^{+z_0} \frac{\rho \, dz \, dy \, dx}{\sqrt{\{(x-x')^2 + (y-y')^2 + (z-z')^2\}}} \qquad (56)$$

where $z_0 = c\sqrt{\{1 - x^2/a^2 - y^2/b^2\}}$, $y_0 = \sqrt{\{1 - x^2/a^2\}}$, and $\rho(x, y, z)$ is an unknown function to be found when V is given. He did not;[22]

Integral theorems

... what he might have been if the promise of his boyhood had not been destroyed by the unfortunate circumstances we have described, it is difficult to say: for he had a true genius for mathematical invention. Before however he had more than commenced his career, his departure from Cambridge, and the necessity of struggling for a livelihood, made it impossible for him to give his undivided attention to researches which, above all others, demand both peace of mind and undisturbed leisure!

His *methods* extended as far as to handle (56) but he never did make the *application*; the methods were revolutionary, but not redeveloped nor applied for a century.

George Green, 1833–41

... Mr Green's memoirs are very remarkable, both for the elegance and rigour of the analysis, and for the ease with which he arrives at most important results. (Stokes 1880–1905, 1:178.)

George Green wrote *An Essay on the Application of Mathematical Analysis to the Theories of Electricity and Magnetism* (1828) in the period up to late 1827, i.e. during Ostrogradsky's stay in Paris. Green's reading for the *Essay* was not all that wide[23] — the classics of Laplace and Lacroix, Poisson on electricity and magnetism, Biot's *Traité de physique* (1816), and probably Fourier, Lagrange, and Coulomb — but his preparation was first class. We see that he had the integration by parts method through Poisson, the potential function through Laplace, Lagrange and Poisson, and the divergence theorem through Poisson. From his reading he could see the importance of the 'new' calculus as treated by Lacroix and currently (1819–30) so popular in Cambridge, along with the central problems in gravity, electricity and magnetism: and he saw that these problems were similar, to be treated in the same way, as he states in the *Essay* (Green 1871, 9).

What ideas did Green contribute to mathematics and mechanics? We shall concentrate on his papers on electricity, magnetism, gravity and elasticity; his theories involving integrals are in these areas.

The *Essay* contains five major advances presented in an explicit, 'modern' way, in a fashion most Continental readers would have been able to understand immediately along with new Cambridge graduates. Green is pellucid, with writing aimed directly at the point and not obscurely wrapped around it: one must put this down to his French reading. Green's notation is clear and consistent, not always chopping and changing as in Murphy and Lagrange.

First, there are the formulae relating surface integrals to volume integrals, developed with and without singularities of the functions involved, and with a nice sense of orders of magnitude. These formulae are not the divergence theorems of Poisson and Ostrogradsky, because they do not have the direction cosines but rather the derivative with respect to the coordinate normal to the surface. What Green has done is use integration by parts in two ways on an

J.J. Cross

'energy' integral to get the surface and volume terms. His triple integral is similar to those involving products of orthogonal functions in contemporary Laplace coefficient work in Duhamel and Lamé but nevertheless essentially different;[24] the integrals in Gauss's 'Allgemeine Lehrsätze' (1839) are similar but nevertheless different because Green's integrals involve two continuous functions U and V rather than just one, V, as in Gauss.[25] From the following triple integral, in which the derivatives do not become infinite,

$$\int dxdydz \left\{ \left(\frac{dV}{dx}\right)\left(\frac{dU}{dx}\right) + \left(\frac{dV}{dy}\right)\left(\frac{dU}{dy}\right) + \left(\frac{dV}{dz}\right)\left(\frac{dU}{dz}\right) \right\} \quad (57)$$

he proceeds to integrate by parts up to the surface of the body (say, with respect to x):

$$\int dxdydz \left(\frac{dV}{dx}\right)\left(\frac{dU}{dx}\right) = \int dydz \left(\frac{dV}{dx}\right)\left(\frac{dU}{dx}\right) dx \quad (58)$$

$$= -\int dydzdx \, V\frac{d^2U}{dx^2} + \int dydz \int \frac{d}{dx}\left(V\frac{dU}{dx}\right) dx$$

The integral inside the last integrand in (58) he expressed as $\left(V\frac{dU}{dx}\right)$ evaluated at the surface, i.e.

$$\int \frac{d}{dx}\left(V\frac{dU}{dx}\right) dx = V''\frac{dU''}{dx} - V'\frac{dU'}{dx} \quad (59)$$

where V'' and dU''/dx correspond to the greater values of x, and V' and dU'/dx correspond to the smaller. He then notes that if dw is an infinitely small line element perpendicular to the surface and directed towards the interior, then the element of surface area corresponding to $dydz$ is

$$dydz = -\frac{dx}{dw}d\sigma'' \quad (60)$$

at the greater values of x, and

$$dydz = -\frac{dx}{dw}d\sigma' \quad (61)$$

for the smaller values of x. Consequently the last integral in (58) becomes

$$\int dydz \, V''\frac{dU''}{dx} - \int dydz \, V'\frac{dU'}{dx} = -\int \frac{dx}{dw}d\sigma'' \, V''\frac{dU''}{dx} - \int \frac{dx}{dw}d\sigma' \, V'\frac{dU'}{dx}$$

$$= -\int d\sigma \, V\frac{dU}{dx}\frac{dx}{dw} \quad (62)$$

where the combination of greater and smaller values of x gives an integral over the whole surface. Similarly, for the other two parts of the energy integral, we

Integral theorems

get a volume and a surface integral, and these combine so that (57) becomes equal to

$$-\int d\sigma\ V\left(\frac{dV}{dx}\frac{dx}{dw} + \frac{dV}{dy}\frac{dy}{dw} + \frac{dV}{dz}\frac{dz}{dw}\right) - \int dxdydz\ V\left(\frac{d^2U}{dx^2} + \frac{d^2U}{dy^2} + \frac{d^2U}{dz^2}\right) \tag{63}$$

If the factors are interchanged in (58), we get (63) with U and V interchanged, i.e. the energy integral (57) is the negative of the two equal expressions

$$\int d\sigma\ V\frac{dU}{dw} + \int dxdydz\ V\delta U = \int d\sigma\ U\frac{dV}{dw} + \int dxdydz\ U\delta V \tag{64}$$

where δ is the Laplace operator $d^2/dy^2 + d^2/dz^2 + d^2/dz^2$. This symmetric formula (64) is Green's formula (Green 1871, §3).

The next step forward, taken also by Liouville and Sturm several years later, was to replace distributions over volumes by values over surfaces, particularly for cases where Laplace's or Poisson's equation applies. To do this, Green first investigates what happens when U and its derivatives may become infinite at some point inside the body. He fixes a point P', say (x', y', z'), and puts $r = \sqrt{\{(x-z')^2 + (y-y')^2 + (z-z')^2\}}$: he assumes that U is essentially equal to $1/r$ in the neighbourhood of P' and he encloses P' inside a sphere of radius a which lies within the body; and so he modifies the integrals in (64), which have to be taken over the body with this sphere excised. Since $\delta U = \delta(1/r) = 0$, the integral containing this term can be extended over the whole body including the sphere enclosing P'. Further, the other volume integral has $U = 1/r$ and $dxdydz$ of order $r^2 dr$ inside the sphere, so that the total error is of order

$$\int_0^a \frac{1}{r} \cdot r^2 dr \text{ or } a^2$$

Hence (as a goes to zero) this integral too can be extended over the whole body including the sphere enclosing P'. The original surface integrals must be supplemented by an integral over the surface enclosing P'. Then

$$\int d\sigma\ U\frac{dV}{dw} \text{ is of order } a^2 \times \frac{1}{a} = a \tag{65}$$

and so may be neglected while

$$\int d\sigma\ V\frac{dU}{dw} = \int_0^{2\pi}\int_0^{\pi} a^2 \sin\theta\ d\theta\ d\phi\ V \cdot \left(\frac{-1}{a^2}\right) \tag{66}$$

and this tends to $-4\pi V(P')$ as a tends to zero. Therefore (64) becomes

$$\int dxdydz\ U\delta V + \int d\sigma\ U\frac{dV}{dw} = \int dxdydz\ V\delta U + \int d\sigma\ V\frac{dU}{dw} - 4\pi V(P') \tag{67}$$

where the integrals are now over the volume and surface of the body with the sphere included. Finally, suppose we wish to solve

$$\delta V = 0 \tag{68}$$

when its values $\bar{V}$ are given on a closed surface and V is not singular within the surface; then we use a function U such that $\delta U = 0$, while U has only one singularity at a point P' within the body enclosed by the surface at which point it is 'sensibly equal to $1/r$' (Green 1871, §4). Equation (67) loses the two volume integrals by Laplace's equation and will lose the surface integral on the left if $U = 0$ on the surface. Then

$$4\pi V(P') = \int d\sigma \, \bar{V} \overline{\frac{dU}{dw}} \tag{69}$$

where the bar over V and dU/dw in (69) indicates values taken on the surface. Such a function U exists, says Green in his *Essay* (Green 1871, §5), because of the following physical situation: we think of the closed surface as an earthed perfect conductor; we put a unit positive electric charge at P' and then U is the total potential of the charge at P' and the charge it induces on the surface. For U is approximately $1/r$ near P' and obviously has no other singularities inside the surface; and on the surface U must be zero since the surface is earthed. This Green's function U marvellously relates the physical singularity to the geometry of the surface so intimately that it solves *all* problems associated with Laplace's equation and this particular geometry. Note that if V satisfies Poisson's equation $\delta V = 4\pi\rho$ in the body, then the Green's function will still solve the problem given the boundary values $\bar{V}$ on the surface, with

$$4\pi V(P') = \int d\sigma \, \bar{V} \overline{\frac{dU}{dw}} - \int dxdydz \, 4\pi\rho U, \tag{70}$$

and Green derives an equivalent formula in the next paragraph of his *Essay*. So Green has stated and 'solved' the so-called Dirichlet problem, the problem of finding a potential satisfying Laplace's equation when its values are given on a surface, and this before Dirichlet returned to Germany from Paris in 1828 and began to lecture on the equations of mathematical physics in Berlin.

Green later applied his techniques to deriving the equations of motion (of light) in an elastic medium and the boundary conditions which apply in certain cases. The light was looked on as a motion of the medium itself, i.e. the two motions (of light and of the medium) were identified. His method was put forward in two papers, 'On the laws of reflexion and refraction of light' (1837) and 'On the propagation of light in crystalline media' (1839) and we combine the two here for convenience (Green 1871, 245–69, 293–311). He takes the medium to be a system of particles acting on each other. A particle at x, y, z in some state of the medium will be at x', y', z' at the end of a time t, so that

Integral theorems

$$x' = x'(x, y, z, t) = x + u$$
$$y' = y'(x, y, z, t) = y + v \qquad (71)$$
$$z' = z'(x, y, z, t) = z + w$$

where u, v, w are the displacements of the particle originally at x, y, z. A particle originally in the form of a rectangular box with sides dx, dy, dz becomes a general parallelepiped under the motion, and he takes as a measure of the deformation of the medium the lengths of the sides of the parallelepiped and the cosine of the angles between these sides:

$$\left[\left(\frac{dx'}{dx}\right)^2 + \left(\frac{dy'}{dx}\right)^2 + \left(\frac{dz'}{dx}\right)^2\right] dx^2 = a^2 dx^2 \qquad (72)$$

and similarly for $b^2 dy^2$, $c^2 dz^2$, and

$$\alpha = \left[\frac{dx' dx'}{dy\, dz} + \frac{dy' dy'}{dy\, dz} + \frac{dz' dz'}{dy\, dz}\right] \bigg/ bc \qquad (73)$$

and similarly for β, γ. After substituting u, v, w, he obtains

$$a^2 = 1 + 2\left\{\frac{du}{dx}\right\} + \left(\frac{du}{dx}\right)^2 + \left(\frac{dv}{dx}\right)^2 + \left(\frac{dw}{dx}\right)^2$$

$$b^2 = 1 + 2\left\{\frac{dv}{dy}\right\} + \left(\frac{du}{dy}\right)^2 + \left(\frac{dv}{dy}\right)^2 + \left(\frac{dw}{dy}\right)^2$$

$$c^2 = 1 + 2\left\{\frac{dw}{dz}\right\} + \left(\frac{du}{dz}\right)^2 + \left(\frac{dv}{dz}\right)^2 + \left(\frac{dw}{dz}\right)^2 \qquad (74)$$

$$bc\alpha = \alpha' = \left\{\frac{dv}{dz} + \frac{dw}{dy}\right\} + \frac{du\, du}{dy\, dz} + \frac{dv\, dv}{dy\, dz} + \frac{dw\, dw}{dy\, dz}$$

$$ac\beta = \beta' = \left\{\frac{du}{dz} + \frac{dw}{dx}\right\} + \frac{du\, du}{dx\, dz} + \frac{dv\, dv}{dx\, dz} + \frac{dw\, dw}{dx\, dz}$$

$$ab\gamma = \gamma' = \left\{\frac{du}{dy} + \frac{dv}{dx}\right\} + \frac{du\, du}{dx\, dy} + \frac{dv\, dv}{dx\, dy} + \frac{dw\, dw}{dx\, dy}$$

These equations (74) represent the Cauchy–Green tensor (Truesdell 1966, 18) whether the displacements u, v, w are large or small; in the 'Laws' (1837), Green restricted his attention to the terms in the braces { } in all six components, but in the 'Propagation' (1839) paper, he gives the full form.[26]

The motion of the medium is then derived from an energy balance under a virtual displacement $\delta u, \delta v, \delta w$, the corresponding variation $\delta \phi$ of a function ϕ representing the actions of the particles on each other (ρ is the density and δ here is the Lagrange variational notation):

$$\iiint \rho \, dx dy dz \left(\frac{d^2 u}{dt^2} \delta u + \frac{d^2 v}{dt^2} \delta v + \frac{d^2 w}{dt^2} \delta w\right) = \iiint dx dy dz \, \delta \phi \qquad (75)$$

Further progress depends on the nature of function ϕ. Green argues that ϕ is a function of the deformation so that

$$\phi = f(a^2, b^2, c^2, \alpha', \beta', \gamma') \tag{76}$$

and he then applies various principles and arguments to simplify ϕ. These are

(1) u, v, w and their derivatives are small;
(2) ϕ can be expanded as a power series $\phi = \phi_0 + \phi_1 + \phi_2 + \ldots$ (77)
where each ϕ_i is a homogeneous polynomial in the six variables of total degree i;
(3) ϕ is invariant under reflexions in the coordinate planes;
(4) ϕ is invariant under (small) rotations about the coordinate axes.

Let us now concentrate on the 'Laws' (1837) paper in which we use (1) immediately to reduce the six variables:

$$a^2 = 1 + 2\frac{du}{dx}, \text{ with } s_1 = \frac{du}{dx}$$

$$b^2 = 1 + 2\frac{dv}{dy}, \text{ with } s_2 = \frac{dv}{dy}$$

$$c^2 = 1 + 2\frac{dw}{dz}, \text{ with } s_3 = \frac{dw}{dz} \tag{78}$$

$$\alpha' = \frac{dv}{dz} + \frac{dw}{dy}$$

$$\beta' = \frac{dw}{dx} + \frac{du}{dz}$$

$$\gamma' = \frac{du}{dy} + \frac{dv}{dx}$$

and ϕ reduces to a function

$$\phi = \phi(s_1, s_2, s_3, \alpha', \beta', \gamma') \tag{79}$$

The homogeneous expansion (77) now implies several useful results under the assumption that $u = v = w = 0$ gives equilibrium for the medium. Obviously ϕ_0 is a constant and $\delta\phi_0 = 0$ makes it irrelevant in (75). Then, in this equilibrium, (75) reduces to

$$0 = \iiint dxdydz \, \delta\phi_1$$

neglecting $\phi_2, \phi_3, \phi_4, \ldots$, so that ϕ_1 may be discarded. Finally, $\phi_3, \phi_4, \ldots$, may be discarded since they are of higher degree than ϕ_2, and hence (75) is reduced to

$$\iiint \rho \, dxdydz \left\{ \frac{d^2u}{dt^2}\delta u + \frac{d^2v}{dt^2}\delta v + \frac{d^2w}{dt^2}\delta w \right\} = \iiint dxdydz \, \delta\phi_2 \tag{80}$$

Integral theorems

Now ϕ_2 is the most general polynomial of degree two in the six variables, and when terms such as $s_1\alpha'$, $\alpha's_1$ are combined, there are $36 - 15 = 21$ arbitrary coefficients. Next, when (3) is applied, the 21 coefficients are reduced to 9, viz.,

$$\phi_2 = Gs_1^2 + Hs_2^2 + Is_3^2 + L\alpha'^2 + L\alpha'^2 + M\beta'^2 + N\gamma'^2$$
$$+ 2P\frac{dv}{dy}\cdot\frac{dw}{dz} + 2Q\frac{du}{dx}\cdot\frac{dw}{dz} + 2R\frac{du}{dx}\cdot\frac{dv}{dy} \tag{81}$$

in which terms such as $(du/dx)(du/dz)$ have been eliminated. If (4) is now applied to (81) in the form of small rotations about each of the three coordinate axes, the energy ϕ_2 is reduced to

$$\phi_2 = G\left\{\left(\frac{du}{dx}\right)^2 + \left(\frac{dv}{dy}\right)^2 + \left(\frac{dw}{dz}\right)^2\right\} + L(\alpha'^2 + \beta'^2 + \gamma'^2)$$
$$+ (2G - 4L)\left\{\frac{dv}{dy}\cdot\frac{dw}{dz} + \frac{du}{dx}\cdot\frac{dw}{dz} + \frac{du}{dx}\cdot\frac{dv}{dy}\right\} \tag{82}$$

or

$$= G\left(\frac{du}{dx} + \frac{dv}{dy} + \frac{dw}{dz}\right)^2$$
$$+ L\left\{\left(\frac{du}{dy} + \frac{dv}{dx}\right)^2 + \left(\frac{du}{dz} + \frac{dw}{dx}\right)^2 + \left(\frac{dv}{dz} + \frac{dw}{dy}\right)^2\right.$$
$$\left. - 4\left(\frac{dv}{dy}\frac{dw}{dz} + \frac{du}{dx}\frac{dw}{dz} + \frac{du}{dx}\frac{dv}{dy}\right)\right\} \tag{83}$$

In this form we can now proceed to manipulate $\delta\phi_2$ in (75). If we use Lagrange's calculus of variations technique and then integrate by parts in the result, we get that the triple integral reduces to the sum of the surface and volume integrals whose particular form need not be quoted here. With the help of these, Green obtains the equations of motion:

$$\rho\frac{d^2u}{dt^2} = -2G\frac{d}{dx}\left(\frac{du}{dx} + \frac{dv}{dy} + \frac{dw}{dz}\right) - 2L\left\{\frac{d^2u}{dy^2} + \frac{d^2u}{dz^2} - \frac{d}{dx}\left(\frac{dv}{dy} + \frac{dw}{dz}\right)\right\} \tag{84}$$

and two similar equations for the variables v, w. The surface integrals give the conditions at the boundary which are to be imposed on the displacements u, v, w and their derivatives.

What has Green contributed here? First, he acknowledges that the methods are due in part to Cauchy.[27] Secondly, he has eliminated the molecules from his derivation of the stress–strain relation through $\delta\phi_2$. Thirdly, he has shown the key to crystalline materials via the various symmetries which he has emphasised, symmetries arbitrarily imposed rather than derived from intermolecular force laws.[28] Fourthly, he was read by Stokes, and it is his influence there to which we now turn.

J.J. Cross

George Gabriel Stokes, 1835–55

> Indeed, both Green and Stokes may be regarded as followers of the French school of mathematicians. Rayleigh (in Stokes 1880–1905, 5:xii)

Before covering Stokes' work on integral transformations leading up to Thomson, we present a brief biography of Stokes' life in this period, 1835–55. Stokes seems to have been strongly influenced by Francis Newman, at Bristol College in 1836 and 1837 (Stokes 1907, 1:6). Newman had reacted strongly against his brother John's stand[29] during the Tractarian movement in Oxford, although not himself an orthodox Anglican; he also wrote some elementary mathematical texts, and a book on elliptic functions and integrals in 1888–89. Stokes was taught mathematics by Newman and showed his ability; later in 1837 he entered Pembroke College, Cambridge, aged 18. Stokes' brother William resided in Gonville and Caius around this time along with George Green whom William certainly knew (Bowley 1976, 87–8).

During 1838 George Stokes began working with the famous tutor William Hopkins. In 1841 Stokes was Senior Wrangler and First Smith's Prizeman, and was elected to a fellowship in Pembroke; Green died in Nottingham that year. The taking of a degree in Cambridge required one to take an oath on graduation to be a practising member of the Church of England. Just as Stokes completed his exams in 1841, John Newman's Tract No. 90 appeared in Oxford; it created a sensation, as it interpreted the Thirty-nine Articles basic to the English church in a way harmonious with the doctrines of the Roman church. In those stirring times one of John Newman's colleagues, William Ward, was stripped of his Oxford degrees in 1845 while Thomson was taking his exams in Cambridge, and a similar but veiled threat against John Newman was halted by the Proctors (Church 1891, 308–11). Hence we see that Stokes's sensitivity to matters of religious principle was developed both in his cradle by his clergyman father and in the fervid atmosphere of the early Victorian years.

Stokes in 1841–42 looked around for something to work on while a Fellow and tutor. Hopkins suggested hydrodynamics, then a field whose theory was well established due to the work of Euler but nevertheless a field in which there was a scarcity of solutions to particular problems compared with the abundance in particle mechanics. Stokes's first papers were on particular cases of fluid flow, but his critical faculty gave him an ability to strip arguments and descriptions down to their very basic elements and assumptions; it is difficult to say how he developed this, but it may have come from his education with Francis Newman and William Hopkins, his wide reading, particularly of Green and the French, and his friendship with William Hallows Miller,[30] professor of mineralogy in Cambridge throughout this period. It was this ability to penetrate to the core of a physical situation which was to be of such great service to Thomson.

In the period 1842–52, Stokes wrote nearly two-thirds of his mathematical

Integral theorems

and physical papers; these papers were concentrated in two areas: continuum mechanics and optics. Most of his papers have an experimental or practical object in view, and at the end of this period (1850–52) he wrote a long paper on the apparent changes in the wavelength (refrangibility) of light, a paper which gained him the Copley medal. This paper was researched and written in the style of Faraday, with notebooks detailing the day-by-day testing of hundreds of chemical solutions. He was admitted to the Royal Society in 1850–51 and took part in its administration from 1852: first on its Council, then on its various committees, finally as Secretary starting in 1854. He became Lucasian Professor in Cambridge in 1848 after negotiating with Thomson over the chair in Glasgow in 1847; he refused to stand for the Glasgow chair as it meant taking an oath to be a practising member of the Church of Scotland, a change of principle for him, especially in the context of the times. He seldom missed any of his university lectures and almost always attended the Royal Society on Thursday; the minutes of the Council and his lecture notebooks attest to his regularity, year by year, from 1848 to 1898.

So we have a picture of the man, a man of integrity and singular wholeness; he goes from age 21 to age 31 from a dedicated, patient, thorough researcher intent on fundamental work in hydrodynamics and optics, capable of extraordinary mathematical invention and deep physical insight, to a teacher–administrator, a man starting to become somewhat obsessed with the running of the Royal Society, its Council, its publications, and its Committees. Let us therefore return, with some regret for our loss, to his work on hydrodynamics and theoretical optics.

His first major paper 'On the theories of the internal friction of fluids in motion, and of the equilibrium and motion of elastic solids' was read in 1845 but appeared much later (Stokes 1880–1905; 1:75–129). This paper repeats work done by Navier, Poisson and Green but from quite a different point of view. The equations he obtains are those of Cauchy and Green but there are some differences in detail. There is no molecular model behind Stokes's arguments, so that the derivation of the relation between the stress and the rate of strain depends on his two axioms and the use of principal axes and rules for changes of axes. Instead of Green's energy potential and the molecular models of Poisson's 'Mémoire sur les équations générales ... des corps solides élastiques et des fluides' (Poisson 1829) and Navier's 'Mémoire sur les lois du mouvement des fluides' (Navier, 1827: 393, 400–14) and the works of Cauchy,[27] Stokes uses the principles:

> That the difference between the pressure on a plane in a given direction passing through any point P of a fluid in motion and the pressure which would exist in all directions about P if the fluid in its neighbourhood were in a state of equilibrium depends only on the relative motion of the fluid immediately about P; and that the relative motion due to any motion of rotation may be eliminated without affecting the differences of the pressures above mentioned. (Stokes 1880–1905, 1:80.)

J.J. Cross

This principle means that the relative motion reduces to a rate of extension and the relation between the force or stress and this rate of extension is invariant under rotations, just as Cauchy had shown in a molecular model.[31] This means, in modern terminology, that the stress tensor is invariant under the special orthogonal group and is thus an isotropic tensor whose special form allows only two arbitrary constants dependent on the material. The Laplacian molecular models used by Cauchy, Poisson and Navier as quoted above reduced these two material constants to one, because an extra symmetry was imposed from the assumed form of the force law between the molecules.

This idea of invariance under certain actions (rotations of axes or objects) and symmetries was quite widespread and certainly not new: it was over a century old, dating back to Lagrange and Euler at least. In the present context of Stokes's application of it to the stress (tensor), as well as previous British applications of it to the scalar energy by Green and to certain optical vectors by MacCullagh, it served to emphasise a point of view: that certain equations describing physical situations, such as the motion of fluids, elastic bodies, or light, were independent of the detailed physical models used to derive them. In other words, these equations of mathematical physics could be derived from mathematical principles rather than physical ones.

One other paper of Stokes in this period deserves detailed attention, the paper 'On the dynamical theory of diffraction' (Stokes 1880–1905, 2:243–328). It contains an application of the divergence theorem to represent the Laplace operator as the limit of an integral over a surface, and presents the solution of the Poisson equation as an integral involving a 'Green's function'. But its essence is as follows: it is a treatise on the two types of wave possible in isotropic elastic media, the dilation waves for which the curl of the velocity (u, v, w) is zero:

$$\frac{dw}{dy} - \frac{dv}{dz} = \frac{du}{dz} - \frac{dw}{dx} = \frac{dv}{dx} - \frac{du}{dy} = 0 \tag{85}$$

and the distortion waves for which the divergence of the velocity (u, v, w) is zero:

$$\frac{du}{dx} + \frac{dv}{dy} + \frac{dw}{dz} = 0 \tag{86}$$

These two types of motion, with and without rotation, had been known for some time. Stokes knew of potential flows and of rotational ones as well. Hence the vorticity vector or the curl of the velocity,

$$\left(\frac{dw}{dy} - \frac{dv}{dz}, \frac{du}{dz} - \frac{dw}{dx}, \frac{dv}{dx} - \frac{du}{dy}\right) \tag{87}$$

was well established in mechanics when he came to write this paper. Further, James MacCullagh had used a vector such as (87) in optics, where he had shown it to be invariant under rotations of the coordinate axes (MacCullagh 1880,

Integral theorems

150–1); hence the vector was also prominent in optics, where light was refracted, reflected and diffracted as a wave moving through a medium, or in their terms, the motion of the light was the motion of an elastic medium, and hence arose this distortion wave with non-zero vorticity vector (87).

We conclude this discussion of Stokes and his work by drawing several of his ideas together and making some comparisons. The 'Diffraction' paper (Stokes 1880–1905, 2:243–328) represents a change in approach to the equations of motion and their solution: he adopts Green's style of using integrals rather than the Eulerian style of small particles and rectangular prisms; he uses the divergence theorem and the accompanying transformation between surface and volume integrals. This is a departure from his 'Internal friction' paper (Stokes 1880–1905, 1:75–129), but there is no break: the same equations of motion are used in both, with two material constants to provide the two types of wave. Below we shall indicate how Thomson came to read both these papers in November 1850, and how the ideas which led to the so-called Stokes's theorem on transformation of integrals came together in his mind.

William Thomson, 1842–52

> Both had great power and great insight, but while Stokes was uniformly calm, reflective and judicial, Thomson's enthusiasm was more outspokenly fervid, and he was apt to be at times vehement and impetuous in his eagerness to push on an investigation; .... (Gray 1908, 81.)

It is not easy to trace Thomson's thoughts over this period. The two main printed sources are his *Reprint of Papers on Electrostatics and Magnetism* (Thomson 1872a) and his *Mathematical and Physical Papers* (Thomson 1882–1911). There are severe disadvantages in using these: they are incomplete, they are not chronological, they have notes added at various dates, they are quite disorganised by topic; in other words, they reflect Thomson. Thomson's manuscript *Notebooks* kept in the Cambridge University Library are useful, but suffer from the same five disadvantages. However, a search of these sources shows that during his eight years as an undergraduate at Glasgow and Cambridge (1835–45) he read widely and well: Laplace, Lagrange, Poisson and Fourier before Cambridge, and then a selection of the very best from contemporary authors: Gauss, Liouville, Jacobi, Chasles, as well as well as Bertrand, Dupin, Kelland, Lamé and Murphy. All these authors and several more were quoted in his papers to 1850, but Dirichlet was not, nor were Fresnel and Sturm. His early work and reading were in two areas — on the figure of the earth, with gravity, the potential, and second-order surfaces or quadrics; and on heat conduction, with integrals and expansions in series.

During a visit to Paris in 1845 he met Liouville who turned him towards more general questions to do with partial differential equations. Thomson had just read Green's *Essay*, and had recently been introduced to Michael Faraday

J.J. Cross

whose experimental results on electromagnetism cried out for mathematical explanation. It is this period which unites in his mind the theories of heat, gravity, current flow and lines of force, and he produced several important papers. First let us look at his approach to the Dirichlet problem and the method of proof for the existence of its solution, the Dirichlet principle.[32]

From 1843 to 1847 Thomson wrote many papers on gravitational attraction; these revolve around the ideas of Gauss's 'Allgemeine Lehrsätze' (1839) and Green's *Essay* (Green 1828) and Green's paper 'On the determination of the ... attractions of ellipsoids ...' (Green 1871, 187–222). The major paper is Thomson's 'Note sur une équation aux différences partielles ... de physique mathématique' (Thomson 1872a, 139–43; see also Thomson 1882–1911, 1:93–6). The equations mentioned combine the ideas of Fourier's heat conductivity and of Poisson's charge density to form the partial differential equation

$$\frac{d}{dx}\left(\alpha^2 \frac{dV}{dx}\right) + \frac{d}{dy}\left(\alpha^2 \frac{dV}{dy}\right) + \frac{d}{dz}\left(\alpha^2 \frac{dV}{dz}\right) = 4\pi\zeta \tag{88}$$

for a function V of x, y, z to be found satisfying this equation when α is a real-valued continuous or discontinuous function of x, y, z; ζ is a real valued function which vanishes outside some closed surface; and V is to vanish as x, y, z become infinite. This is the Dirichlet problem for the infinite region.

He forms the potential

$$U = \iiint \frac{\zeta' dx' dy' dz'}{\sqrt{\{(x - x')^2 + (y - y')^2 + (z - z')^2\}}} \tag{89}$$

and the energy integral

$$Q = \int_{-\infty}^{\infty} \int_{-\infty}^{\infty} \int_{-\infty}^{\infty} \left\{ \left(\alpha \frac{dV}{dx} - \frac{1}{\alpha}\frac{dU}{dx}\right)^2 + \left(\alpha \frac{dV}{dy} - \frac{1}{\alpha}\frac{dU}{dy}\right)^2 \right. $$
$$\left. + \left(\alpha \frac{dV}{dz} - \frac{1}{\alpha}\frac{dU}{dz}\right)^2 \right\} dx dy dz \tag{90}$$

He observes that Q is always positive and has no upper bound as V varies. Since Q has a lower bound, he says that as we vary V there must be a particular function V such that Q is made a minimum. Hence, he says, if we apply the calculus of variations and vary V by δV, we obtain the variation δQ as follows when we integrate by parts and discard the resulting surface integrals since V vanishes as the variables x, y, z go to infinity:

$$-\frac{1}{2}\delta Q = \iiint \delta V \left\{ \frac{d}{dx}\left(\alpha^2 \frac{dV}{dx} - \frac{dU}{dx}\right) + \frac{d}{dy}\left(\alpha^2 \frac{dV}{dy} - \frac{dU}{dy}\right) \right.$$
$$\left. + \frac{d}{dz}\left(\alpha^2 \frac{dV}{dz} - \frac{dU}{dz}\right) \right\} dx dy dz \tag{91}$$

Integral theorems

But the definition of U in (89) gives the Poisson equation

$$\frac{d^2U}{dx^2} + \frac{d^2U}{dy^2} + \frac{d^2U}{dz^2} = 4\pi\zeta \qquad (92)$$

and hence the variation δQ becomes

$$-\frac{1}{2}\delta Q = \iiint \delta V \left\{ \frac{d}{dx}\left(\alpha^2 \frac{dV}{dx}\right) + \frac{d}{dy}\left(\alpha^2 \frac{dV}{dy}\right) + \frac{d}{dz}\left(\alpha^2 \frac{dV}{dz}\right) - 4\pi\zeta \right\} dxdydz \qquad (93)$$

Hence, since δV is arbitrary and since $\delta Q = 0$ for the minimum, the integrand inside the braces { } vanishes and we get the Fourier–Poisson equation (88) for V. Hence Thomson has 'solved' the Dirichlet problem for the infinite region by a method which is similar to that which Riemann will soon learn[33] from Dirichlet and use in his *Habilitationschrift* (Riemann 1892, 1). Note that Thomson's method covers both Laplace's and Poisson's equations, and he also states and solves the Neumann problem where not the function but its normal derivative is given on the surface or at infinity.

This paper appeared when Paris was in uproar during the revolution of 1848; also at this time Riemann was attending Dirichlet's lectures on the differential equations of mathematical physics in a Berlin also disturbed by revolution. Liouville moved into number theory and corresponded with Dirichlet while Thomson stayed with mathematical physics and his long correspondence with Stokes began. During 1848 Thomson and Stokes produced a series of notes on hydrodynamics, establishing for students and for the hard-to-persuade Professor Challis the equations of motion and several types of flow, including waves and flows governed by potentials. In these notes there occur various standard conditions that the flow be in or perpendicular to certain surfaces; these ideas Thomson borrows from Stokes rather than their older sources such as Euler, Clairaut and Fontaine,[34] and uses in the papers of 1849–50 discussed below.

In the 1846 paper 'On a mechanical representation of electric, magnetic, and galvanic forces' (Thomson 1882–1911, 1:76–80) the operators divergence, curl and gradient all appear,[35] and Thomson's interest in these operators continues through 1847–49 as he cajoles Stokes into writing the notes on hydrodynamics. Thomson is driven, as entries in his notebook for 1848–50 show,[36] by the problem of replacing a surface distribution of charge by a galvanic current running around its edge. These ideas, and that of replacing a volume distribution of charge or mass by a surface distribution, are present in his mind when he is writing the grand papers of 1849–50, on 'A mathematical theory of magnetism' (Thomson 1872a, 340–431). The sections of particular interest to us are those which detail solenoidal and lamellar distributions of magnetism. A distribution of magnetism is a vector (α, β, γ) whose components are functions of the spatial coordinates x, y, z. Such a distribution is solenoidal, if the magnet corresponding to it can be divided into infinitely thin rings whose direction of magnetisation is uniformly around the ring, i.e. (α, β, γ) is tangential to the ring and oriented round it tangentially in the one sense. Hence, by his identification of magnetism and fluid flow through Laplace's equation, the analogous fluid flow is a rotation and it is not surprising that the corresponding 'vorticity' vector (α, β, γ) must be the curl of some vector and its divergence must be zero; the corresponding concept in Stokes 'Diffraction' paper (Stokes 1882–1905, 2:255, 260) is the distortion wave. Thomson reinforces this by giving the result (Thomson 1872, 384):

A distribution of magnetism expressed by $\{(\alpha, \beta, \gamma) \text{ at } (x, y, z)\}$ is solenoidal if, and is not solenoidal unless,

$$\frac{d\alpha}{dx} + \frac{d\beta}{dy} + \frac{d\gamma}{dz} = 0 \tag{94}$$

Further, a distribution of magnetism is lamellar, i.e. forms an infinitely thin sheet which is bounded on each side by a closed surface and is magnetised uniformly in the direction of the (say, exterior) normal, if the vector (α, β, γ) is the gradient of some function of x, y, z (Thomson 1872, 385) and so its curl is zero (Thomson 1872, 386):

$$\frac{d\beta}{dz} - \frac{d\gamma}{dy} = 0, \frac{d\gamma}{dx} - \frac{d\alpha}{dz} = 0, \frac{d\alpha}{dy} - \frac{d\beta}{dx} = 0$$

This corresponds to the dilation wave in Stokes's 'Diffraction' paper (Stokes 1882–1905, 2:254, 259).

Then Thomson derives the following result. The distribution (α, β, γ) is lamellar if the corresponding magnet can be divided into infinitely many magnetic shells either closed or with their edges in the surface bounding the magnet. The distribution is complex lamellar if the intensity of the field may vary, e.g. it may be along the exterior normal in one place, the interior normal in another. In that case (α, β, γ) is some factor times the gradient of some function, and the well-known condition for this from the time of Euler and Clairaut[34] is

$$\alpha\left(\frac{d\beta}{dz} - \frac{d\gamma}{dy}\right) + \beta\left(\frac{d\gamma}{dx} - \frac{d\alpha}{dz}\right) + \gamma\left(\frac{d\alpha}{dy} - \frac{d\beta}{dx}\right) = 0 \tag{95}$$

We recall (α, β, γ) is always along the normal and hence its curl, whose components are in the parentheses, must be tangential to the surface.

So by July 1850 when Thomson read Stokes' paper on 'Diffraction' on his way to London, he had in his mind the following ideas: (1) the direction cosines of tangents and normals; (2) transformations of 'material' lines, surfaces and volumes; (3) transformations of integrals on surfaces and volumes from Stokes's 'Diffraction' paper; (4) transformation of integrals on lines and surfaces, as evidenced from his citation of Ampère's result given above, in his paper 'On the potential of a closed galvanic circuit of any form' (Thomson

Integral theorems

1872a, 426); (5) the calculus of variations; (6) integration by parts; (7) the concept of complete differentials of various kinds; (8) scalar and vector potentials; and (9) the divergence, curl and gradient operators. If these did not occur in his paper on the 'Mathematical theory of magnetism' then they were in Stokes's 'Diffraction' paper.

When he arrived in London, Thomson sent the following letter to Stokes:[37]

> 9 Barton Street,
> Westminster,
> July 2, 1850

My dear Stokes,

As I have not a copy of your paper on the Equil[ibrium] & Motion of Elastic solids, nor any other work of reference for the purpose, by me, I shall be much obliged by your sending me the equation of equil[ibriu]m of a non crystalline elastic solid under the action of any forces, and the formulae for the mutual actions betw[een] any two contiguous portions of the body. I have been trying but as yet without success, to make out something about the interpretation of the equations for the case of a solid of any form, with each point of its surface displaced to a given extent & in a given direction from its natural position. I think I see how it can be done when the solid is a rectangular parallelepiped, but not in a very inviting way. It was reading your paper on diffraction on my way from Cambridge[38] that made me take up the subject again.

Do you know that the condition that $\alpha dx + \beta dy + \gamma dz$ may be the diff[erentia]l of a function of two independent variables for all points of a surface is

$$l\left(\frac{d\beta}{dz} - \frac{d\gamma}{dy}\right) + m\left(\frac{d\gamma}{dx} - \frac{d\alpha}{dz}\right) + n\left(\frac{d\alpha}{dy} - \frac{d\beta}{dx}\right) = 0? \qquad (96)$$

I made this out some weeks ago with ref[eren]ce to electromagnetism.[36] With ref[eren]ce to an elastic solid, the cond[itio]n may be expressed thus — the resultant axis of rotation at any point must be perp[endicula]r to the normal.

> Yours very truly
> William Thomson

P.S. The following is also interesting & is of importance with reference to both physical subjects.

$$\int (\alpha dx + \beta dy + \gamma dz) = \pm \iint \left\{ l\left(\frac{d\beta}{dz} - \frac{d\gamma}{dy}\right) + m\left(\frac{d\gamma}{dx} - \frac{d\alpha}{dz}\right) \right.$$
$$\left. + n\left(\frac{d\alpha}{dy} - \frac{d\beta}{dx}\right) \right\} dS \qquad (97)$$

where l, m, n denote the dir[ectio]n cosines of a normal through any el[emen]t dS of a surface; & the integ[ratio]n in the sec[on]d member is performed over a portion of this surface bounded by a curve round wh[ich] the int[egratio]n in the 1st member is performed.

J.J. Cross

The postal service functioned quite quickly in those days, even with its mixture of trains and coaches. Here is part of Stokes's reply:[39]

Pembroke College, Cambridge
July 4th 1850

Dear Thomson,

The equations of equilibrium are ...

The theorems which you communicated are very elegant and are new to me. I have demonstrated them for myself, the first by the calculus of variations, the second (which includes the first as a particular case) by simple considerations like what I have employed at the beginning of my paper. I suppose from what you say that you arrived at these theorems by working with magnetic ideas.

...

Yours very truly
G.G. Stokes

And so the Stokes theorem (97) was born, but not yet in print. A careful search of Stokes's lecture diaries for 1850–55 and Maxwell's record of Stokes's lectures for 1853 show no trace of this theorem.[40] Its first appearance in print is as an examination question for Maxwell and Routh (Stokes 1880–1905, 5:320):

Smith's Prize Examination Paper
February, 1854

8. If X, Y, Z be functions of the rectangular co-ordinates x, y, z, dS an element of any limited surface, l, m, n, the cosines of the inclinations of the normal at dS to the axes, ds an element of the bounding line, show that

$$\iint \left\{ l\left(\frac{dZ}{dy} - \frac{dY}{dz}\right) + m\left(\frac{dX}{dz} - \frac{dZ}{dx}\right) + n\left(\frac{dY}{dx} - \frac{dX}{dy}\right) \right\} dS$$
$$= \int \left(X\frac{dx}{ds} + Y\frac{dy}{ds} + Z\frac{dz}{ds} \right) ds \qquad (98)$$

the differential coefficients of X, Y, Z being partial, and the single integral being taken all round the perimeter of the surface.

Its next appearance in print was in a prize essay of Hermann Hankel's, written for Bernhard Riemann (Hankel 1861, 36). It is intriguing to note that, just as no British author quoted Dirichlet, the German Hankel quoted the Germans Dirichlet, Riemann and Helmholtz but no British author at all. Later in the decade when Thomson (or T) and Tait (or T') came to write their *Treatise on Natural Philosophy* (1867) commonly referred to as T & T', they naturally included a paragraph on the Stokes theorem (Thomson & Tait 1867, Appendix A to §204). When Maxwell came to write his *Treatise on Electricity and Magnetism* (1873), he corresponded on various intriguing points with Thom-

Integral theorems

son, Tait and Stokes. Maxwell became interested in the source of the theorem and wrote asking Stokes on 11 January 1871 (Stokes 1907, 2:31); no reply to this letter seems to have been preserved. But the following correspondence on postcards between Maxwell (pseudonym dp/dt, from the thermodynamic formula dp/dt = JCM, Maxwell's initials) and P.G. Tait (pseudonym T') in part concerning Herman Helmholtz (pseudonym H^2):

Dr dp/dt   Edinburgh 31/3/71

H^2's address is now Koniginn Augusta Str. 45 Berlin. Did you get the proofs of your book and a letter[41] I sent to the Athenaeum?

Yours. T'.

Dr T' Proofs be come to hand. Corrections thankfully received and honour given to whom honour is due. But the history of

$$\iint \left\{ l\left(\frac{dZ}{dy} - \frac{dY}{dz}\right) + m\left(\frac{dX}{dz} - \frac{dZ}{dx}\right) + n\left(\frac{dY}{dx} - \frac{dX}{dy}\right) \right\} dS$$
$$= \int \left(X\frac{dx}{ds} + Y\frac{dy}{ds} + Z\frac{dz}{ds} \right) ds$$

ascends (at least) to Stokes Smith Prize paper 1854 and it was then not altogether new to yours truly. Do you know its previous history? Poisson? On light??? ...

[Dated 4/4/71]   Yours dp/dt

Edin. 5/4/71

Dr dp/dt ... As[42] to $r \iint S \cdot Uv \nabla\sigma \, ds = \int S\sigma \, d\rho$, I really thought it due to T, and first published by the Archiepiscopal pair. ... T'

Conclusion

What is the importance of mathematics in Cambridge mathematical physics in the period 1830–55? In particular, of what importance were these theorems involving the transformation of various types of integral? Would physics have progressed without these results or would there have been gaps, disruption, or even error?

The partial differential equations of Laplace and Poisson occurred in heat conduction, electricity, magnetism, elasticity, fluid flow and gravitational attraction. The pervasiveness of these equations led to physical ideas being passed from one context of applications to another, with both heat and electricity being treated as 'fluids' at various stages. Therefore any advance in one field of application was quickly transferred to the others. Our examples show this: the Dirichlet principle begins with applications in gravity and ends as a broad tool in heat and electricity; the Green's function starts life in electricity and magnetism and has invaded all of contemporary physics and engineering.

The integral transformations were essential for the correct formulation of

J.J. Cross

boundary value problems corresponding to partial differential equations, where the equation for a function holds sway in a region and the function has its own values or those of its normal derivative specified on a surface. For electricity and magnetism, the presence of electric currents along conducting wires (lines) makes the Stokes theorem, which transforms line integrals into surface integrals as Ampère began to do, a key element in a physicist's repertoire. The solution of these boundary value problems today involves the Green's function to a degree surpassing any other single idea or method in physics or mathematics; no field of physics has been left untouched by its influence.

The progress of that branch of physics which, until the World War of 1939–45, had been least touched by mathematics shows what neglect of advances in physical mathematics entails: one would rather the glorious confusion of Thomson's mathematical physics than the chaos of contemporary thermodynamics.

Notes

1. It is not our intention here to present a complete picture of the work of Green, Stokes, Thomson and Murphy, nor to give a detailed mathematical description of the integral theorems. This will appear in my 'History of Green's Theorems and Associated Ideas, 1730–1930', which will include a history of potential theory.
2. See Cross 1983 and the entry on the 'History of potential theory' in the forthcoming *Bibliography of the History of Mathematics*, ed. J. Dauben.
3. See Lagrange 1760–61. See also Deakin 1981–82.
4. See Euler 1954–55 (vols. XII and XIII of second series of *Opera Omnia*). At page LXXIV of vol. XII Green's theorem could be applied, and at pages XLV–LIV of vol. XIII the divergence theorem is sorely missed.
5. There are several recent studies of Parisian science by Fox, Arnold, Home and Frankel. See Grattan-Guinness 1981 and 1984.
6. See H.G. Green 1947, where specific references to, among others, Lacroix, Lagrange, Cauchy and Poisson are detailed.
7. The correspondence of Gauss gives some details. For example, Gauss read Ampère and defended him against Wilhelm Weber twenty years later; see the letters Weber to Gauss, 1 February 1845, and the reply Gauss to Weber, 19 March 1845, held in the Niedersächsische Staats und Landsbibliothek, Göttingen.
8. The letters between C.F. Gauss and P.G.L. Dirichlet appear in Gauss 1863–1933, 12:309–11, and in Dirichlet 1889–97, 2:373–87. These letters detail Gauss's negative reaction to Poisson and Poinsot. Further to this, Paris had a strong influence on Bernhard Riemann through Dirichlet.
9. Certain Paris figures such as Fresnel and Sturm are not mentioned, principally because they are not quoted by our Cambridge men. Ostrogradsky is included even though he was first quoted in Cambridge by Maxwell in the 1870s; he is included because his work influenced both Cauchy and Poisson.
10. Ampère 1827, fig. 27, plate 1. See Goursat 1884 and 1900.
11. G.F.B. Riemann 1892, in §§7–11, 16–18 of his Inaugural dissertation of 1851, 'Grundlagen für eine allgemeine Theorie der Functionen einer veränderlichen complexen Grösse', pages 1–48.
12. The manuscript of this memoir of Ostrogradsky's is in his dossier in the Archives de l'Académie des Sciences in Paris; cf. Youschkevitch 1967, 17–19.
13. See Youschkevitch 1967, 12–14.
14. See Riemann's 'Ueber die Darstellbarkeit einer Function durch eine trigonometrische Reihe' (Riemann 1892, 227–91), especially page 233 where the rivalry between Poisson and Fourier is resolved in favour of the latter. This paper was written at Dirichlet's instigation; cf. Riemann 1892, 546–7.

Integral theorems

15. Institut de France, Academie des Sciences, *Procès Verbaux*, VIII, 1824–27, page 349, right-hand column: 'M. Michel Ostrogradsky remet un manuscrit contenant la "Démonstration d'une proposition de calcul intégral qu'il applique a des équations aux différences partielles du second ordre et a la détermination des fonctions arbitraires", le 13 fev. 1826, MM. Legendre et Poinsot sont nommés commissaires pour l'examen de ce Mémoire.' Further, on pages 525 and 574, Poisson and Fourier are named as the referees for the later two papers of 1827; no reports exist.
16. See C.L.M.H. Navier, in B. de Saint-Venant, ed., 1864.
17. See G. Smith, *Robert Murphy*, Monash University Preprint, 1984, for a life of Murphy, particularly the reference to de Morgan 1856.
18. See Wise 1981.
19. He calls them 'transient functions', and he calls the orthogonal functions 'reciprocal functions' from the properties of the P_n listed under equation (38).
20. See the article on 'History of potential theory' in note 2.
21. Hence his name 'reciprocal functions' for all families of this type. Of course the Legendre polynomials are an example of a family of orthogonal functions usually not normalised as stated here.
22. The next quote is from de Morgan 1856, 338.
23. See H.G. Green 1947, 573–4.
24. Duhamel 1833, 68–9, and Lamé 1833, 203–4. These present standard calculations of coefficients in the Legendre polynomial or Fourier series expansions for an arbitrary function in terms of an orthogonal (complete) basis for the function space.
25. Gauss already assumes that V is an harmonic function.
26. See Green 1871, 249 and 295–6. There are similar expressions in Cauchy, in particular in his paper 'Sur la condensation et la dilatation des corps solides' (Cauchy 1882–, Ser. 2, vol. 7, 82–93). This paper was first published in 1827.
27. See three papers of Cauchy's in Cauchy 1882–, Ser. 2, vol. 8, 195–277. In particular, pages 239–44 give a reduction process for the constants in an elastic medium whose stress is governed by a function of the distance between pairs of particles in the body; and pages 203–9 give a form of derivation of the Green–Stokes equations with two constants, and this derivation dates back, in part, to 1822. What Cauchy did was to use a mixture of *ad hoc* assumptions about possible motions and about molecular forces to obtain equations similar to those of Poisson. These papers were published in 1828. Green's development is systematic, free from the molecular hypothesis and focused on the material rather than on the motions to which the material is subjected.
28. The derivation by Cauchy on page 238 of the reference in note 27 is clearly tied to his molecular model, but he does make explicit the arbitrary choice of the coordinate axes on pages 239–41 — which is the invariance under rotations, and the final effect of the molecular model to reduce the constants to one is on page 243. The difference between the derivations of Cauchy and Green lies in part in the influence of the molecular model used by Cauchy and in part in the Lagrangian, *a priori* approach used by Green. Certainly Cauchy's argument can be freed from its molecular casing and this Green has done and more.
29. See his entry in the *Dictionary of National Biography, Supplement*, III, 221–3.
30. Miller and Stokes were members of the Royal Society and often acted jointly as referees of papers submitted for its *Philosophical Transactions*.
31. The difference between Stokes and Cauchy lies in the abandonment of the molecular model by Stokes, following Green, and in the more vigorous and thoroughgoing application of mathematical arguments and methods which were certainly present in Cauchy, but not always to the same extent as in Stokes: invariance under rotations or changes of cartesian axes, the use of principal stresses and strains, and a reduced reliance on purely physical considerations. Stokes was more a development of Cauchy than a break in the tradition.
32. See Cross, 'History of potential theory' as in note 2.
33. Riemann spent several years in Berlin, 1848–50, and his friendship with Dirichlet lasted till the latter's death in 1859.
34. See, for example, Engelsman 1982 and Greenberg 1982.
35. Naturally they are not so named there!
36. Notebook 34, for 1845–56, as in Wilson 1976, page 350 of the Kelvin section. The dates of the

J.J. Cross

entries (and the corresponding pages in the notebook) are 20 November 1848 (page 108); 3 February 1849 (pages 112–13); 12 July 1849, which is a letter to Liouville on the subject of Thomson's 'Note sur une équation' (Thomson 1872a, 139–43) (pages 124–31); 7 and 14 November 1849 (pages 134–7 & 138–41); 20 November 1849 (pages 142–3); 1 December 1849 (pages 144–6); and finally 7 May 1850 (pages 146–53, at the address 9 Barton Street, Westminster).

37. Letter K39 as in Wilson 1976, page 232 of the Stokes section. Hopefully the whole of the Stokes–Kelvin correspondence edited by Wilson will soon appear. The square brackets enclose those parts omitted by Thomson when he abbreviated words.
38. Thomson visited Cambridge each summer from 1846 to 1852. The 'Diffraction' paper is Stokes 1882–1905, 2:243–328.
39. Letter S356 as in Wilson 1976, page 258 of the Kelvin section.
40. See Notebooks 2 to 6 and 10 (definitely 1853 by internal evidence) as listed in Wilson 1976, page 584 of the Stokes section. For Maxwell, see MS Add. 7655, Vm7 and 8 in the Cambridge University Library.
41. These three items are from MS Add. 7655, I, in the Cambridge University Library. The letter Tait mentions does not seem to have been preserved.
42. See Maxwell 1873a, 2:211–38. The capital 'S' denotes the 'scalar part' q_1 of a quaternion $q_1 + iq_2 + jq_3 + kq_4$, in which the i, j, k naturally are Hamilton's quaternions, with $i^2 = -1$, etc., and $ij = k = -ji$, etc. The vector v is the normal $li + mj + nk$, and the vector $\sigma = iX + jY + kZ$, and ρ the vector $xi + yj + zk$, with ∇ the usual gradient operator but expressed in terms of Hamilton's i, j, k instead of Gibb's geometric vectors.

= 6 =

Mathematics and physical reality in William Thomson's electromagnetic theory

OLE KNUDSEN

Introduction

William Thomson's work in the field of electricity and magnetism ranged from profound achievements on the foundation of electromagnetic theory to important theoretical and practical contributions to the development of electrical technology. As the title suggests, this chapter will deal almost entirely with Thomson the natural philosopher, and Thomson the electrical engineer will make only a very brief appearance.

Thomson worked on fundamental aspects of electromagnetic theory during two periods of his life, separated by a gap of almost 15 years. The first period began with his first paper on electrostatics in 1841 and ended in about 1856 when he became completely absorbed in the Atlantic cable enterprise. During this early period he not only reached a number of important new results in electromagnetic theory, but he also originated an entirely new trend in mathematical physics, based on Faraday's field-theoretical ideas. The first sections of my chapter deal with this period. In them I describe Thomson's invention and use of the method of analogies, and I try to delineate Thomson's views on the function of analogical reasoning in the construction of physical theory and to show how these views were shaped by his study of Fourier's work, as well as by his own successes.

The second period lasted from about 1870, when Thomson started working on the *Reprint of Papers in Electrostatics and Magnetism* which appeared in 1872, to the end of his life. In the intermittent period many exciting things had happened, the most important being the development of Maxwell's electromagnetic theory of light, in which Thomson had not been actively involved; and Thomson's later work is to some extent marked by his having been left behind. He was still able to produce valuable results, not least in the additional sections he wrote for the *Reprint*, but on the whole his work in this period was relatively uninfluential compared to that of the early period. I will show that his later work was dominated by the themes he had set out in the early period. I will then present a detailed discussion of his controversial reaction to Maxwell's theory, in which I claim that this reaction can only be fully understood on the background of the outlook Thomson had acquired during his early life, and

Ole Knudsen

which had developed into a rather rigid standard for electromagnetic theory before he began a close study of Maxwell's theory.

The main theme of this chapter as a whole, referred to in the title by the phrase 'mathematics and physical reality', will be that of analogical reasoning in Thomson's work.[1] I shall try to delineate Thomson's view of the role of analogy in relation to mathematical formalism, and the relation of both to 'true physical theory', that is to dynamical theories of the structure of ether and matter. One important topic will be what one might call the problem of demarcation between an analogy and a theory purporting to say something definite about the ultimate nature of the real world. It will appear that Thomson had rather strict views on this point, and that he was categorically opposed to developing an analogy further than warranted by its correspondence with the 'known laws' of the subject treated. This is, I think, one important root of his dissatisfaction with Maxwell's theory.

The historical significance of this aspect of Thomson's work is not limited to his having obtained certain propositions by means of an analogy, or to his views on the function of analogies having prevented him from accepting Maxwell's theory. Maxwell praised Thomson's discovery of the heat flow analogy to electrostatics as the invention of a whole new method for discovery in physical science, the method of 'physical analogy', as he called it (Maxwell 1890, 1:156; and 2:301-4). The historian may add that this method turned out to be eminently useful in many parts of physics (Klein 1972). Indeed, when a modern physicist, well aware that an atomic nucleus is a system very different from a charged drop of any liquid, nevertheless tries to account for some properties of nuclear fission by the liquid drop model, he is using a sophisticated research strategy which Thomson was the first to explore in a conscious manner.

Heat conduction and electrostatics: the discovery of an analogy

Any discussion of Thomson's early work must begin with Fourier's analytical theory of heat conduction, published in his famous book in 1822 (Fourier 1878).[2] The influence of this book on the young Thomson is well established and may perhaps best be indicated by pointing to the fact that the first dozen or so of Thomson's published scientific papers are all directly related to Fourier's work.

In his book, Fourier had created a mathematical description of the flow of heat from one part of a solid body to other parts, or to the body's surroundings. The basic quantities in his theory are the temperature, V, and the heat flux vector $\mathbf{F}$. Both quantities are functions of space and time, and the latter is defined so that the amount of heat passing in unit time through an infinitesimal surface element dS, with unit normal $\mathbf{n}$, is equal to $\mathbf{F} \cdot \mathbf{n} \, dS$.

Fourier set up the following two differential equations relating these two quantities

William Thomson's electromagnetic theory

and
$$\mathbf{F} = -K \nabla V \tag{1}$$

$$\nabla \cdot \mathbf{F} = -CD \frac{\partial V}{\partial t} \tag{2}$$

The first of these expresses 'the principle of communication of heat': that in a homogeneous body the flux is in the direction of decreasing temperature and proportional in magnitude to the temperature gradient. K is the conductivity of the body for heat. Fourier derived equation (1) by applying Newton's macroscopic law of cooling to infinitesimal elements of the body, and not from a microscopic theory of radiative heat exchanges between material molecules. For this reason he was severely criticised by a number of his colleagues, foremost among them Poisson. The debate between Fourier on one side, and Poisson, Laplace and Biot on the other, reflects a fundamental disagreement about the proper method in mathematical physics. While Poisson maintained that a differential equation like equation (1) must be derived in a rigorous way from a definite theory of molecular action, Fourier claimed that the validity of his equation was independent of hypotheses about the nature of heat and its interaction with the molecules of matter; and that it should be judged solely on the criterion as to whether it led to a correct representation of known macroscopic phenomena (Wise 1979, 51–8).

The second equation (2) was soon to be known as the continuity equation. It is a differential statement of the conservation of heat: that the total amount of heat flowing inwards through a closed surface equals the increase in the heat content of the volume enclosed by the surface, this increase being equal to the rise in temperature, multiplied by the heat capacity, here expressed as the product of the heat capacity per unit mass, C, and the mass density, D.

By eliminating $\mathbf{F}$ from the two equations, one obtains the following differential equation for the temperature distribution alone,

$$\frac{\partial V}{\partial t} = \frac{K}{CD} \nabla^2 V \tag{3}$$

This is Fourier's famous heat conduction equation. Combining it with appropriate boundary conditions, Fourier solved a number of particular problems, using his equally famous Fourier series technique to describe the solutions.

A special case of heat conduction will be of particular interest in the following. This is the case of steady flow, defined mathematically by the condition that the temperature at each point remains stationary, i.e.

$$\frac{\partial V}{\partial t} = 0 \tag{4}$$

In this case, the fundamental equation (3) reduces to

$$\nabla^2 V = 0 \tag{5}$$

151

that is, the temperature distribution satisfies Laplace's equation. Further, equation (2) leads in this case to

$$\nabla \cdot \mathbf{F} = 0 \qquad (6)$$

which means that heat is flowing as if it were an incompressible fluid. Physically, this case of steady flow is realised if, for instance, an infinite, homogeneous solid is submitted for a sufficiently long time to the action of given constant sources of heat. Equations (5) and (6) will then hold at all points except those in which the sources are placed.

Fourier also stated the conditions which must hold at an interface between two homogeneous bodies of different conductivities, K and K', say. These are simply that the temperature and the normal component of the heat flux must both be continuous across the interface. In mathematical terms this means that, $\mathbf{n}$ being a unit normal to the interface:

$$V = V'; \mathbf{n} \cdot \mathbf{F} = \mathbf{n} \cdot \mathbf{F}' \qquad (7)$$

or, using equation (1):

$$V = V'; K\mathbf{n} \cdot (\nabla V) = K'\mathbf{n} \cdot (\nabla V)' \qquad (8)$$

In the case of steady flow one must, accordingly, seek a solution to equation (5) which satisfies the interface conditions (8).

This is the basic structure of the theory which the 16-year-old Thomson made himself familiar with in May of 1840, when he went 'right through it' in a fortnight (Thompson 1910, 1:14). Only 15 months later, having already written two papers on Fourier's mathematics, he submitted his first important paper on the analogy between electrostatics and the steady flow of heat, 'On the uniform motion of heat in homogeneous solid bodies, and its connection with the mathematical theory of electricity' of 1842 (Thomson 1872a, 1–14).

This analogy can be stated very simply by noting that Laplace's equation (5), which Fourier derived for the temperature distribution, holds equally well for the electrostatic potential in regions which are free of electric charges. This is, however, not how Thomson did it. Instead, he took a particular solution to equation (5),

$$V = \frac{A}{r} \qquad (9)$$

which describes the case of an infinite solid heated by a constant source at the origin. From this solution he could easily deduce that if an isothermal, closed surface S' is covered by sources of heat of density equal to $4\pi\rho$, and no other sources are present, the stationary temperature of a point $\mathbf{x}$, not on S', will be given by

William Thomson's electromagnetic theory

$$V(\mathbf{x}) = \oint_{S'} \frac{\rho(\mathbf{x}')}{r} \, \mathrm{d}S' \tag{10}$$

where r is the distance between the point $\mathbf{x}$ and a point $\mathbf{x}'$ on S'.

On this basis Thomson described the analogy by noting that equations (9) and (10) are also correct expressions for the electrostatic potential from a point charge or, respectively, a conducting surface carrying a charge distribution proportional to ρ; and further, that the heat flux and the electrostatic attraction will be proportional to each other, because both of them will be proportional to the gradient of the function V. This allowed Thomson to translate propositions which are almost self-evident in heat conduction into not so evident propositions in electrostatics; and conversely, to convert theorems in the theory of the attraction of ellipsoids into statements about the heat flux in geometrically similar situations. In several cases Thomson succeeded, by ingenious use of the analogy, in proving propositions that are in fact applications of the theorems of Gauss and Green, theorems that were still unknown to him at this time. As Maxwell was to say much later, in his 'Address to the mathematical and physical sections of the British Association' in 1870, Thomson's paper illustrates that: 'the recognition of a formal analogy between two systems of ideas leads to a knowledge of both, more profound than could be obtained by studying each system separately' (Maxwell 1890, 1:219). It also shows not only that physics may benefit mathematics by providing mathematically interesting problems, but also that physical reasoning may lead to solutions or propositions which cannot yet be derived by mathematics proper. And it is always easier to invent a mathematical proof if the result is known beforehand.

To give just one example of this in Thomson's reasoning, consider an isothermal closed surface inside which there are no sources of heat. Then, in the stationary case, the net flux inwards from that surface must obviously be zero, hence the temperature of any isothermal surface in the interior must be equal to that of the outer one, and therefore the temperature must be constant and equal to the surface temperature all through the interior. In electrostatics this means that the potential must have a constant value inside a conducting surface, or, mathematically, that the only solution to Laplace's equation, with the boundary condition that V must have a constant value on a closed surface, is that V has the same constant value at every point in the interior — a result which is not so easy to derive mathematically if one does not have something like Green's theorem available. Another mathematical result which probably came out of the heat analogy was Thomson's invention of the method of images, which furnished a 'synthetic', geometrical way of constructing solutions for potential and electric force. The method consists in replacing conducting surfaces by certain systems of equivalent point charges, and is the inverse of the replacement of point sources of heat by isothermal surfaces which Thomson had used in his first paper (Thomson 1872a, 144–6; see also Maxwell 1890, 1:302–3 for an account of the invention of the method of images).

Three years later Thomson published a paper in which he again made use of the same analogy, but this time with a different purpose. The paper, entitled 'On the elementary laws of statical electricity' (1845), had been requested by the French mathematician and physicist Liouville, and was first printed in French in Liouville's journal. Liouville had been worried by some of Faraday's electrostatic discoveries, and even more by Faraday's description of them, a description which seemed to suggest that there was something fundamentally wrong with the usual mathematical theory of electrostatics. 'I was led to suspect', Faraday wrote, 'that common induction itself was in all cases an *action of contiguous particles*, and that electrical action at a distance never occurred except through the influence of the intervening matter' (Faraday 1839–55, 1:362). In support of this heterodox view, Faraday cited two experimental results. The first, and in his view the most important, was that, as he put it, induction may take place in curved lines, that is, that the lines of electric force are not necessarily straight ones. This led him to an explicit rejection of the very foundation of the usual theory. 'I do not see how the old theory of action at a distance and in straight lines can stand, or how the conclusion that ordinary induction is an action of contiguous particles can be resisted.' (Faraday 1839–55, 1:380.)

The second result was Faraday's discovery that the mutual action between two charged conductors depends on the intervening insulating medium. This result indicated once more that electrostatic action is not a direct action at a distance between electric charges, but something that is propagated through the intervening medium, which Faraday therefore termed a dielectric, and whose properties he characterised by a material constant called the specific inductive capacity of the dielectric.

It was therefore with good reason that Liouville, during Thomson's stay in Paris in 1845, asked him to write an exposé of Faraday's views and compare them to the usual theory. And it was an important achievement of Thomson's to prove in his paper that Faraday's results and statements, contrary to Faraday's own expectations, were in perfect agreement with mathematical electrostatics. In this proof, the analogy between electrostatics and heat conduction was of great importance. First, it provided the following general argument: Faraday's views of electrostatic action are very similar to the physical picture underlying Fourier's theory, in which heat likewise is conceived as being propagated from molecule to molecule. (This similarity was noted by Faraday himself, cf. Gooding 1980, 96.) Since this picture has led to a mathematical theory of heat conduction which is completely equivalent to action-at-a-distance electrostatics, there is every reason to expect that, despite the conceptual conflict between Faraday and his predecessors, their theories may well turn out to be mathematically equivalent. Secondly, the analogy furnished an easy way of discussing

William Thomson's electromagnetic theory

specific problems. Thus, Faraday's curved lines of force correspond exactly to lines of heat flux, and their curvedness is therefore immediately seen to follow also from the action-at-a-distance theory of electrostatics. Similarly, substances of different inductive capacities correspond exactly to substances of different heat conductivities, and it is therefore not difficult to reconcile Faraday's discovery with electrostatic theory. The latter point involves a consideration of Poisson's theory of magnetic polarisation together with the interface conditions (7) or (8); I shall pass over the details of this for the moment, but I return to this point below.

That this episode was an important event in the history of electromagnetic theory is well known. One reason for its importance lies, I think, in the fact that it was the first time an analogy of this kind had been discovered and used. It is true that the mathematical analogy between electrostatic forces, magnetic forces and gravitation had been recognised and applied, most extensively by Poisson; but in these cases the mathematical analogy was a simple reflection of the identity of the physical basis of these theories, the concept of particles interacting across space according to the inverse square law. What Thomson discovered was something of far greater consequence, namely a complete mathematical equivalence of theories based on opposing, mutually exclusive, physical concepts: action-at-a-distance as opposed to local action propagated in a field. It was Thomson's discovery which first made Faraday's field-theoretical approach respectable among mathematical physicists, and it was the further exploration of this approach by Thomson himself and his younger friend, Maxwell, which led to one of the finest achievements of the nineteenth century, the electromagnetic theory of light.

Put in slightly different words, what Thomson discovered was that one did not have to reject, or even modify, one single result derived from action-at-a-distance electrostatics in order to switch to a radically different way of thinking about electrostatic phenomena. It is important to note that this liberating discovery was essentially a mathematical one. Because the access to formal mathematics had been denied to Faraday, he was unable even to suspect that something like this would hold true. From this time onwards, mathematics would become much more than just a convenient language for stating physical laws; it would be increasingly involved in the very construction of physical theories.

However fruitful analogical reasoning has proved itself to be, its use is not without problems, and Thomson must have realised this, at least while working on his second paper. The fact is, as Wise has pointed out (Wise 1981, 49–50), that the analogy between media of different conductivities for heat and media of different dielectric constants is not quite perfect. It is true that the conditions (8) for the temperatures on each side of an interface also hold for the variation of the electrostatic potential across an interface between two media with dielectric constants K and K'. But the more basic equations (1) and (7) from which they are derived do not have exact analogues in electrostatics. Here, instead of equation (1), we have always, regardless of the value of K,

$$\mathbf{E} = -\nabla V \tag{11}$$

where $\mathbf{E}$ denotes the electric force. This is, in fact, the definition of the potential V in terms of the macroscopically measurable quantity, electric force. And from Poisson's theory of magnetic polarisation (Poisson 1826a, b), which formed the basis of Thomson's discussion of dielectrics, it follows that equation (7) must, in electrostatics, be replaced by

$$K(\mathbf{n} \cdot \mathbf{E}) = K'(\mathbf{n} \cdot \mathbf{E}') \tag{12}$$

In other words, while there is still a perfect analogy between temperature distribution and electrostatic potential, the analogy between heat flow and electric force breaks down, because heat will flow in a continuous manner across an interface, while the normal component of electric force jumps in value from one medium to the other.

It is hard to believe that Thomson did not see this flaw in the analogy between heat flow and electric force, because in his discussion of a dielectric body C, whose dielectric constant relative to air is k, he stated the interface condition between C and air in terms of electric force just as in equation (12). In a footnote he claimed, however, that 'From this it follows that, in the case of heat, C must be replaced by a body whose conducting power is k times as great as that of the matter occupying the remainder of the space ...' (Thomson 1872a, 33). As we have seen, this claim is rather dubious. It can probably only be explained by pointing to the fact that in his ensuing calculation of the effect of a dielectric on the capacity of a condenser, Thomson eliminated electric forces and stated his end result solely in terms of the potential, that is, in a form which does justify his claim.

However this may be, and whether or not Thomson really did have a precise notion of the limitation of his analogy, he was at least clearly aware that he had not developed a new explanation of electrostatic action. At the fundamental level, all the analogy had done — but this was no minor achievement — was to show that Faraday's view on the propagation of electrostatic action constituted a possible foundation for the mathematical theory, not that it was the only one possible. In his concluding paragraph he noted the possibility that electrostatic, magnetic and gravitational forces might be found to be propagated by the action of contiguous particles of intervening media, but he ended by a word of caution: 'We know nothing, however, of the molecular action by which such effects could be produced, and in the present state of physical science it is necessary to admit the known laws in each theory as the foundation of the ultimate laws of action at a distance.' (Thomson 1872a, 37.)

The phrase 'known laws' in this quotation refers to a problem Fourier had faced in his theory of heat, a problem that Thomson took up in a similar way.

William Thomson's electromagnetic theory

Fourier had wanted his theory to be valid independently of hypothetical notions of heat and its interaction with matter at the molecular level. He therefore formulated his mathematical laws solely in terms of macroscopic quantities: temperature, quantity of heat, thermal conductivity; and he took great pains to explain how these quantities could be defined by macroscopic experimental procedures. In a similar vein, Thomson began his 1845 paper by praising Coulomb for having given the elementary laws of electrostatics in a form 'which is independent of any hypothesis', so that his theory could 'only be attacked in the way of proving his experimental results to be inaccurate' (Thomson 1872a, 15). The second instalment of his series of papers 'On the mathematical theory of electricity', published in 1848, was, as its sub-title said, 'A statement of the principles on which the mathematical theory of electricity is founded'. It gave purely macroscopic, operational definitions of electric quantity, electric force, etc., and stated the fundamental laws without any reference to either the two-fluid model of Coulomb and Poisson or Faraday's idea of contiguous action. Speaking of the distribution of electricity on a surface, Thomson said that '... instead of the expression "the thickness of the stratum" [used by Poisson] Coulomb's far more philosophical term, *Electrical Density*, will be employed ...; a term which is to be understood strictly in accordance to the following definitions, without involving even the idea of a hypothesis regarding the nature of electricity' (Thomson 1872a, 48).

Thus, despite his willingness to speculate, and to use analogies as heuristic tools, Thomson felt a need to formulate the 'known laws' or, as he was later to say, 'the positive parts' of a theory in a non-hypothetical manner, in order to have a foundation from which the theory could be extended. Although Thomson first learned this attitude from Fourier, it also agreed well with the views of his other model, Faraday. As Gooding has pointed out, Faraday's and Thomson's agnosticism had by 1845 become mutually reinforcing (Gooding 1980, 108–11).

The elastic solid analogy; hydrodynamics; field energy

The heat analogy could not be said to give any information about the deeper nature of electrostatic action. One reason for this was that the primary electrostatic quantity, electric force, was represented by the flux of heat, a quantity which had nothing in common with the concept of force, neither in the material theory of heat nor in the theory of heat as molecular motion. In 1846, Thomson discovered a new analogy, which looked more promising as a heuristic means of penetrating deeper into the nature of things. This was the 'mechanical representation of electric, magnetic, and galvanic forces', as he entitled his paper describing the new analogy, 'galvanic' being his way of denoting what we would call 'electromagnetic' (Thomson 1882–1911, 1:76–80).

The analogy itself can be stated as follows. Thomson noted that the mathematical expressions for the electric force from a point charge, the magnetic force from a small dipole, and the electromagnetic force from an infinitesimal current element, represented three different solutions to the equations of equilibrium of an elastic solid strained by forces acting on its bounding surface. In the first case the analogy held between the electric force and the elastic displacement itself, while the magnetic and electromagnetic forces were analogous to the *curl* of the displacement, i.e. to the differential rotation of a volume element of the solid. Thomson was quite pleased with this discovery, as can be seen both from his notebook entries and from a letter to Faraday (Wise 1981, 52). First, this analogy was *mechanical*, i.e. it represented the forces by mechanical states in an elastic solid, so that it suggested more directly the propagation of force by mechanical processes in the ether. Secondly, it encompassed not only electrostatic action, but magnetic and electromagnetic action as well, and Thomson had been led to it by pondering Faraday's recent discovery of magneto-optic rotation, which pointed strongly towards a relation of magnetic theory with the elastic theory of the propagation of light. Nevertheless, caution was still needed, as Thomson explained to Faraday:

> What I have written is merely a sketch of the mathematical analogy. I did not venture even to hint at the possibility of making it the foundation of a physical theory of the propagation of electric and magnetic forces, which, if established at all, would express as a necessary result the connection between electrical and magnetic forces, and would show how the purely *statical* phenomenon of magnetism may originate either from electricity in motion, or from an inert mass such as a magnet. If such a theory could be discovered, it would also, when taken in connection with the undulatory theory of light, in all probability explain the effect of magnetism on polarized light. (Thompson 1910, 1:203–4.)

The analogy might be suggestive of great things to come, but Thomson felt he was still a long way from having even a beginning of the physical theory he was hoping for. Quotations of a similar kind are found quite frequently all through Thomson's later writings. In the published paper the only mention of the ideas behind the mathematics is a reference in the introduction to Faraday's theory of electrostatic action and his discovery of the magneto-optic rotation, and the laconic statement at the end of the paper: 'I should exceed my present limits were I to enter into a special examination of the states of a solid body representing various problems in electricity, magnetism, and galvanism, which must therefore be reserved for a future paper.' (Thomson 1882–1911, 1:80.)

The elastic solid is one of the recurrent themes in Thomson's thinking. Another is the hydrodynamical analogy which developed out of the early heat analogy. Hydrodynamics began to figure prominently in Thomson's writings in 1847 when, in collaboration with Stokes, he started publishing a series of 'Notes on hydrodynamics'. In the first of these, entitled 'On the equation of continuity' (Thomson 1847), Thomson derived, from conservation of mass, the equation of continuity in the form

William Thomson's electromagnetic theory

$$\nabla \cdot (\rho \mathbf{v}) + \frac{\partial \rho}{\partial t} = 0 \tag{13}$$

where ρ is the mass density, and $\mathbf{v}$ the velocity, of an infinitesimal element of any fluid. The analogy with the continuity equation (2) for heat is, of course, obvious; and this indicated that the heat analogy for electrostatics could be transformed into a hydrodynamical analogy.

The fifth note is essentially a hydrodynamical interpretation of a mathematical theorem published one year earlier (Thomson 1872a, 139–43). The theorem deals with that solution to Laplace's equation

$$\nabla^2 V = 0 \tag{14}$$

which satisfies the boundary condition that, in every point of a closed surface S, with unit normal $\mathbf{n}$,

$$\mathbf{n} \cdot (\nabla V) = F \tag{15a}$$

where F is a given, arbitrary function. The theorem holds just as well, and is even simpler to prove, if the 'Neumann condition' (15a) is replaced by the 'Dirichlet condition'

$$V = F \tag{15b}$$

in every point on S. Now consider a quantity Q, defined by the following integral over all space exterior to S,

$$Q \equiv \int (\nabla V)^2 d^3\mathbf{x} \tag{16}$$

The theorem then states that, among all functions V which satisfy

$$\oint_S V F \, dS = A \tag{17}$$

there is one which, when inserted in (16), makes Q a minimum; and this particular function V is a solution to (14) — unique up to an additive constant — which satisfies the boundary condition

$$\mathbf{n} \cdot (\nabla V) = c \, F$$

on S. By a suitable choice of the arbitrary constant A in (17), one can obtain $c = 1$, so that the boundary condition (15a) is satisfied. (In the case defined by (15b) the restriction (17) is unnecessary.) This theorem is a fine piece of mathematics which became known in the mathematical literature as 'Dirichlet's principle' and led to interesting discussions by, among others, Casorati, Weierstrass, and Hilbert (Klein 1926, 98–9; Neuenschwander 1978, 23–33). More important for our discussion is the fact that, as Thomson stated, the theorem, and the analysis behind it, 'possesses very important applications in the theories of heat, electricity, magnetism, and hydrodynamics, which may form the subject of future communications'. (Thomson 1872a, 141.)

Ole Knudsen

In his note referred to above, 'On the vis-viva of a liquid in motion' (Thomson 1882–1911, 1:107–112), Thomson applied this analysis to the motion of an incompressible fluid enclosed within a flexible envelope. If the motion is irrotational, the velocity **v** can be derived from a potential, and Thomson could then apply his theorem to showing that the integral over all space

$$Q = \rho \int \mathbf{v}^2 \, d^3\mathbf{x} \tag{18}$$

would be a minimum for the actual motion of the fluid. But since Q is seen to be the 'vis-viva', or twice the kinetic energy, of the motion described by the velocity field **v**, this meant that the fluid would automatically 'choose', among all possible motions consistent with the given boundary conditions, that motion which minimised its kinetic energy.

The application to electrostatics is even more interesting. Consider a number of closed, conducting surfaces, S_i. Let q_i and V_i denote the charge, respectively the constant value of the potential, on S_i. Then, as Helmholtz showed in 1847 (Helmholtz 1882–95, 1:41–6), the energy of the system is given by

$$Q = \tfrac{1}{2} \sum_i q_i V_i = \tfrac{1}{2} \sum_i \oint_{S_i} \sigma V \, dS \tag{19}$$

where σ denotes surface density of electric charge. Using the well-known relation between σ and the electric force **E**,

$$4\pi\sigma = \mathbf{n} \cdot \mathbf{E} = -\mathbf{n} \cdot (\nabla V) \tag{20}$$

equation (19), which expresses Q in terms of the states of the conducting surfaces alone, may be transformed into an integral over all space:

$$Q = \frac{1}{8\pi} \int (\nabla V)^2 d^3\mathbf{x} = \frac{1}{8\pi} \int \mathbf{E}^2 d^3\mathbf{x} \tag{21}$$

The application of the theorem is now straightforward. The potential V satisfies Laplace's equation (14) — we have already presupposed this in deriving equation (21) from equation (19) — and the boundary condition (15b) is of course trivially fulfilled by the requirement that the potential must have the constant value V_i on S_i. The theorem then says that the actual potential is distributed in the space between the conductors in such a way that the energy of the system, Q, given by equation (21), is a minimum. Equally or perhaps more important is the physical interpretation of the mathematical transformation of the expression (19) for the energy into the expression (21). Equation (19) describes the energy of the system as a sum of contributions from each of the charged surfaces, that is, as being confined to the places where free electricity is found. It is therefore a natural expression of the energy in terms of action-at-a-distance between electric charges. On the other hand, equation (21) gives the energy as an integral over all space outside the charged surfaces, every volume element contributing an amount proportional to the square of the electric force

William Thomson's electromagnetic theory

in that part of space. Thus, it can only be interpreted as stating that the energy is localised, not on the charged surfaces, but in every part of the field of electric force.

This interpretation was, then, another important step in the development of the field-theoretical viewpoint. Precisely at what time Thomson realised this cannot be established from his publications, but the suggestive analogy between the expression (21) for electrostatic energy and the expression (18) for hydrodynamical kinetic energy — which undisputedly resides in the velocity field — would certainly not have been lost on him. As we shall see, he introduced energy considerations in his theory of magnetism already in 1850.

'A mathematical theory of magnetism'

In 1849, while his sequence of papers 'On the mathematical theory of electricity in equilibrium' was still being printed in the *Cambridge and Dublin Mathematical Journal*, Thomson began to publish a new series of papers in the *Philosophical Transactions*. Bearing the common title, 'A mathematical theory of magnetism', these were intended to form a comprehensive treatise on the 'two distinct kinds of magnetic action — the mutual forces exercised between bodies possessing magnetism, and the magnetization induced in other bodies through the influence of magnets' (Thomson 1872a, 341). This treatise was clearly meant to do for magnetism what Fourier had done for heat and he himself had done for electrostatics. In an abstract, written for the Royal Society's *Proceedings*, he discarded the hypothesis of two magnetic fluids, and stated his view of how a proper mathematical theory should be constructed:

> No physical evidence can be adduced in support of such a hypothesis; but on the contrary, recent discoveries, especially in electromagnetism, render it extremely improbable. Hence it is of importance that all reasoning with reference to magnetism should be conducted without assuming the existence of those hypothetical fluids.
>
> The writer of the present paper endeavours to show that a complete mathematical theory of magnetism may be established upon the sole foundation of facts generally known, and Coulomb's special experimental researches. The positive parts of this theory agree with those of Poisson's mathematical theory ... (Thomson 1872a, 340).

This might sound as an arch-positivist speaking. Even the word 'positive' itself is used in its Comtean meaning. But Thomson did not wish to restrict scientific research to 'the positive parts' of a theory. In the paper itself he began with insisting on the primacy of macroscopic force as the instrument by which magnetism is recognised and its properties explored. He went on to mention magneto-optic rotation as another type of physical effect and said,

> ... however interesting such other phenomena may be in themselves, however essential a knowledge of them may be for enabling us to arrive at any satisfactory ideas regarding the physical nature of magnetism, and its connexion with the general properties of matter, we must still consider the investigation of the laws, according to

Ole Knudsen

which the development and the action of magnetic force are regulated, to be the primary object of a Mathematical Theory ... (Thomson 1872*a*, 341).

The programme outlined here, and executed in great detail on the following many pages, is clearly to begin by developing as far as possible a mathematical, macroscopic theory of magnetism, based on the concept of magnetic force defined with reference to Coulomb-type measurements of the attractions and repulsions between magnets. In this part of the programme all reference to such hypothetical entities as the magnetic fluids, or Ampère's molecular currents, should be avoided. Thus Thomson took great pains to free such concepts as 'pole', 'polarity', and 'magnetism', from any hypothetical microscopic connotations, and in a characteristic passage he wrote: 'However different are the physical circumstances of magnetic and electric polarity, it appears that the positive laws of the phenomena are the same, and therefore the mathematical theories are identical. Either subject might be taken as an example of a very important branch of physical mathematics, which might be called "A Mathematical Theory of Polar Forces"'. (Thomson 1872*a*, 347.) Only when the mathematical theory had been completed would the time be ripe for attempts to penetrate deeper into 'the physical nature of magnetism'. For such attempts Thomson expected the Faraday effect to be essential and, as we shall see, this would indeed turn out to be the case.

I shall not go into a detailed discussion of these magnetic papers, although they contain a number of new technical results, not least in the additions which Thomson wrote in 1871–72 when the papers were reprinted. Two points, however, deserve to be mentioned. One is that in his analysis of conditions inside a magnet Thomson discovered an ambiguity in the very concept of magnetic force. It was possible to define this concept in two equally reasonable ways, corresponding to the quantities denoted **B** and **H** in modern notation. These two definitions led to different properties of magnetic force. One of them, **H**, corresponded to representing a magnetised body, as Poisson had done, by an equivalent distribution of 'imaginary magnetic matter' over the body's surface; the force could then be derived from a potential, but its normal component would be discontinuous across the surface. The other definition, **B**, corresponded to the representation of magnetisation by electrical currents; the normal component would then be continuous, but the force could not be derived from a potential. Maxwell would later come to regard **B** and **H** as two distinct physical quantities, **B** being of the nature of a flux and **H** of the nature of a force. To Thomson, however, both were valid definitions of the single concept, magnetic force; and it was simply a matter of convenience which definition one would choose in a given investigation. After all, the ambiguity only concerned the unobservable force *inside* a magnetised body; the choice had no effect on external, observable force. Later, when he came to believe in the reality of Ampèrean molecular currents, he would tend to regard **B** as the appropriate

William Thomson's electromagnetic theory

representation of magnetic force; and this had the further advantage that **B**, unlike **H**, because of its continuity, was easy to represent by a hydrodynamical analogy (cf. Wise 1981, 61–7).

Another point worth mentioning is that Thomson inserted a short discussion of the energy of a system of two permanent magnets. He denoted this energy by Q, the same letter he had used for the kinetic energy of a fluid, and for the minimum quantity in his mathematical theorem (cf. equation (16)). Having written an expression for the magnetic Q in terms of the magnetisations of the two magnets, he went on to emphasise its importance,

> The mechanical value [his phrase for potential energy in this period] of a distribution of magnetism, although it has not, I believe, been noticed in any writings hitherto published on magnetism, is a subject of investigation of great interest, and, as I hope on a later occasion to have an opportunity of showing, of much consequence, on account of its maximum and minimum problems ... (Thomson 1872a, 378).

Clearly, Thomson had realised the importance of the energy concept for unifying magnetic theory with electrostatics and hydrodynamics. Another important use of energy considerations from this period is found in a memorandum on the energy of electromagnets, written in 1851, but only published in 1872 (Thomson 1872a, 441–2). In his *Über die Erhaltung der Kraft* Helmholtz had given a faulty analysis of the energy of two current-carrying circuits, which had led him to the erroneous conclusion that the existence, and the mathematical law, of electromagnetic induction was a necessary consequence of energy conservation, given the mathematical law of electromagnetic force between the two circuits (Helmholtz, 1882–95, 1:61–5). Thomson in his memorandum was the first to give the correct analysis of this very tricky problem and to prove that if the circuits are displaced so that the electromagnetic forces during a time interval dt perform an amount of work wdt, the galvanic elements must supply an additional energy equal to twice this amount, so that while mechanical work is being performed by the system, the energy of the system is being *increased* by the same amount. This shows that a system of two circuits cannot, as Helmholtz had done, be treated in the same way as, say, a system of two point charges, where the potential energy *decreases* by precisely the amount of work performed by the electrostatic force during a displacement of the charges. This is almost the only point where one could argue on the basis of what was known around 1850 that field theory was not just as good as, but decidedly superior to, action at a distance, because the only natural answer to the question as to what becomes of this additional energy that the galvanic elements must supply is to say that it is spent in increasing the energy residing in the magnetic field surrounding the two circuits. Whether Thomson realised this fully in 1851 is not known, but it cannot have been much later that he transformed his expression for the magnetic energy Q into the form appropriate for field theory, namely

Ole Knudsen

$$Q = \frac{1}{8\pi} \int H^2 d^3x \qquad (22)$$

an expression he published in *Nichol's Cyclopædia* in 1860 (Thomson 1872a, 443).

The reality of microscopic rotation

In February 1856, Maxwell finsished reading his first electromagnetic paper, 'On Faraday's lines of force', to the Cambridge Philosophical Society (Maxwell 1890, 1:155–229). More than anything else, this paper represented a systematic exposition, albeit with some further developments, of the main points of Thomson's work during the previous decade. Not only in its technical aspects and its use of a hydrodynamical analogy was this paper an exercise in the Thomsonian approach to electricity and magnetism. Also, in his more philosophical discussion of the value of 'physical analogies' and their relation to mathematical formalism and 'true physical theory', Maxwell demonstrated his complete understanding of the subtleties of Thomson's reasoning, and his ability to formulate his mentor's views more clearly than the latter had ever done. Maxwell particularly warned his readers against taking an analogy too seriously and against believing that a true physical theory of electromagnetic action was at all within sight. At the same time, of course, the paper was, and was meant to be, an attack on action-at-a-distance electrodynamics as developed by Wilhelm Weber, and a piece of propaganda for the field theory of electromagnetism which Maxwell and Thomson were hoping for. This 'true physical theory' would furnish a microscopic explanation of electric and magnetic force as mechanical states in the luminiferous ether, and would therefore have to be based on a hypothesis of the microscopic constitution of the ether. Unfortunately, neither Maxwell nor Thomson dared believe that they had any clue to this problem.

However, only a few months later the situation had improved tremendously. In June 1856, Thomson for the first time allowed himself to publish a categorical statement about the ultimate nature of magnetism (Thomson 1856); it consisted, he claimed, in rotatory motions of microscopic constituents of matter or ether. As Thomson had predicted in 1849 (cf. pages 161–2, *supra*) it was indeed the Faraday effect that provided him with this insight. By comparing the magnetic rotation of the plane of polarisation with the 'natural' rotation occurring in certain substances, he concluded that the only possible explanation of the former would be to assume that the vibrations constituting light enter into combinations with circular motions, around axes parallel to the magnetic field, of parts of the medium having linear dimensions comparable to the wavelength of the light wave. This was more than just an analogy: 'I think it is not only impossible to conceive any other than this dynamical explanation of the fact ... but I believe it can be demonstrated that no other explanation of that fact is possible.' (Thomson 1904, 570–1.)

William Thomson's electromagnetic theory

Important consequences could be drawn from this discovery. First, it afforded 'a demonstration of the reality of Ampère's explanation of the ultimate nature of magnetism'. Previously, Thomson had shared Faraday's scepticism towards Ampère's molecular currents; he had, in fact, had little more faith in their existence than in that of Poisson's magnetic fluids (Thomson 1872*a*, 419). Secondly, it suggested a dynamical explanation, according to which magnetic moment would be the effect of, and would be measured by, the resultant angular momentum of the microscopic rotations. This would, in turn, lead to a complete dynamical foundation for electromagnetic theory:

> The explanation of all phenomena of electro-magnetic attraction or repulsion, and of electro-magnetic induction, is to be looked for simply in the inertia and pressure of the matter of which the motions constitute heat. Whether this matter is or is not electricity, whether it is a continuous fluid interpermeating the spaces between molecular nuclei, or is itself molecularly grouped; or whether all matter is continuous, and molecular heterogeneousness consists in finite vortical or other relative motions of contiguous parts of a body; it is impossible to decide, and perhaps in vain to speculate, in the present state of science. (Thomson 1904, 571.)

Whether in vain or not, Thomson did not refrain from speculating further along these lines. On 6 January 1858, he put down in his notebook the results of his investigation of one of these alternatives, that of the continuous fluid interpermeating the spaces between molecular nuclei.[4] First, he noted his reservations and doubts:

> It does not seem probable that a complete theory of physical science can be founded on such a hypothesis. For what qualities are we to give this fluid? Is it to be perfectly incompressible or is it to have compressibility, & if so perfect elasticity of course? Are the motes or molecules to be perfectly elastic, & if not *atoms*, are they to be worn down and altered in shape, or to be occasionally broken? How is gravitation to be accounted for, & how the law of proportionality between gravitation and inertia for all different substances?

He went on to speculate on the equally impossible question of how to account for the properties of matter, particularly that of stability, in terms of 'a particular form & order of motions or eddies in a fluid'. Again, his conclusion was negative: 'I see no possibility however of explaining the constancy of the qualities of particular substances on this hypothesis, and I see no opening for a successful investigation on dynamical principles of any of the motion that would result from the supposed circumstances'.

This meant that the system of fluid and 'motes' could not be the ultimate truth about the constitution of the universe. Still, it had its value as 'a temporary mechanical illustration of some of the agencies hitherto looked upon as among the most inscrutable phenomena of inorganic physics'. First of all, it could account for the generation of heat by mechanical action or by electrical currents as well as for heat conduction. That is, it would serve as a basis for the dynamical theory of heat, if heat was assumed to consist in rotating motions of the motes. (Already in the 1856 paper Thomson had related his new insight to Rankine's vortex theory of heat.) Secondly, it could probably explain at least one mechanical property of matter, macroscopic elasticity, as repulsion by fluid pressure between two rotating motes. This is, then, the origin of one of Thomson's most cherished ideas, that of 'elasticity as a mode of motion' (W. Thomson 1891, 1:149–53), which from now on would provide a unifying link between the hydrodynamical and the elastic solid analogies for electromagnetism and light. Further, it would, of course, fully explain the Faraday effect which had 'really brought on the whole attack'. And finally, because there would be less repulsion in the fluid along, than perpendicular to, the direction of axes of rotation, magnetic attraction would be 'perfectly illustrated'. His enthusiasm over this result made him forget himself to the extent that he wrote, 'A complete dynamical theory of magnetism and electromagnetism seems not at all difficult or far off.' On a second reading he had sobered enough to delete the word 'theory' and replace it by 'illustration'.

This substitution of one word for another, when seen in conjunction with the first part of the entry, shows in a nutshell one of the most striking aspects of Thomson's work, namely the impossibly high standard he set for theoretical physics. With modern hindsight one might say that he reserved the designation 'dynamical theory' or 'complete physical theory' for a theory that would encompass as much as quantum mechanics and the general theory of relativity combined, or even as much as the unified field theory that Einstein was trying to create. In one characteristic passage he wrote of the possibility of conceiving 'that all the phenomena of matter might be explained by the consequences of contractility in a universal fluid constituting the material world, and created with such a distribution as to density (and possibly also with motion ...) that the present and past phases of dead matter may have followed from it in accordance with constant mechanical laws'. As long as this goal had not been achieved, or at least was within sight, one was merely possessing a 'dynamical illustration' or 'mechanical illustration'. On the other hand, a dynamical illustration was more than just an analogy of the kind he had applied in electrostatics. In an analogy a given type of macroscopic force might be represented by a macroscopic flux of heat or matter, or a potential by a temperature distribution: a dynamical illustration, on the other hand, would consist in a possible microscopic model of the constitution of ether or matter from which magnetic force, say, could be derived by purely mechanical reasoning. Thus, at least in 1858, and I believe for the rest of his life, Thomson operated with a hierarchy of three types of theoretical construction. At the lowest level there was analogy, later sometimes to be called 'mathematical analogy'; at a higher level there was dynamical illustration, later sometimes denoted 'physical analogy' (Thomson 1882–1911, 3:498–501) or 'mechanical representation'; and finally at the supreme level there was the never attained

William Thomson's electromagnetic theory

ideal of a complete dynamical theory of the inorganic universe. I think that Maxwell's phrases 'physical analogy' and 'physical theory' correspond pretty closely to the first two of Thomson's categories (Knudsen 1976, 248–50).

There is one aspect of Thomson's subsequent work on electromagnetic theory which is probably related to this ideal of a complete dynamical theory, and distinguishes him from Maxwell. Maxwell's electromagnetic theory was primarily a theory of the electromagnetic field in vacuum and there appear to be very few connections between this part of his work and his researches on the kinetic theory. In contradistinction, Thomson was unwilling to concentrate for any length of time on electromagnetism, and forget about the constitution of matter and the interaction of matter with light and electromagnetic forces. Indeed, we shall see that this aspect formed the basis of one of Thomson's explicit criticisms of Maxwell's electromagnetic theory.

For all his reservations, Thomson undoubtedly felt that he had gained a real insight into the nature of magnetism and its interaction with light. He pursued this advance in a characteristic way by working out the equations of motion of a macroscopic solid moving through a liquid. These equations were only published in 1870 as part of a larger paper which is no doubt a valuable, minor contribution to hydrodynamics, but contains no indication of the origin and fundamental significance of its subject matter (Thomson 1872b). It was Maxwell who was to develop Thomson's ideas into the first version of the electromagnetic theory of light, in his 'On physical lines of force', which appeared in 1861–62 (Maxwell, 1890, 1:451–513). Maxwell's success with the vortex theory of magnetism underlines the problem of understanding why Thomson did not himself attempt to develop a theory of this sort. One obvious reason is, of course, simply lack of time. By 1858, Thomson was already deeply involved in the Atlantic cable enterprise which was to absorb so much of his effort until the final success in the summer of 1866. Naturally his work on the problem of telegraphic signalling diverted his attention from fundamental to applied aspects of electromagnetism; but even if this had not been the case, it is extremely unlikely that Thomson would have moved in the direction Maxwell did. Although the first two parts of Maxwell's paper can be seen as a Thomsonian 'dynamical illustration' of the known macroscopic laws of magnetic and electromagnetic force and electromagnetic induction, the two subsequent parts are written in a style that most of all resembles Poisson's 'physical mechanics' which Thomson, following Fourier, had rejected from the beginning of his career onwards. In these later parts there is no longer any clear distinction between microscopic hypotheses and known macroscopic laws, and this is particularly true of the famous concept of electric displacement. Although Maxwell tried to argue for the introduction of this concept by macroscopic considerations, the relation between displacement and electric force was mainly derived from an analogy between dielectrics and elastic media. And Maxwell's addition of displacement current to the macroscopic equation relating current and magnetic force could not be justified by reference to any macroscopic electromagnetic phenomenon. Maxwell's only argument for this step consisted in his extension of the vortex illustration of magnetism by the additional mechanism of 'idle wheel' particles, a mechanism he himself described as 'somewhat awkward' and as not to be taken as an 'electrical hypothesis'. To put it bluntly, what Maxwell did was to postulate a new, hitherto unsuspected and experimentally unrecognised, source of macroscopic magnetic force, solely on the basis of a microscopic hypothesis which even he himself could not take quite seriously. (For a closer discussion of Maxwell's procedure, see Siegel's chapter in this volume.) Even if this postulate allowed Maxwell to identify the luminiferous and the electromagnetic medium and to derive a relation between the velocity of light and electromagnetic quantities, the whole procedure must have seemed to Thomson an example of all that a physical theory should not be. He never accepted the notion of displacement current, and as late as 1890 he made a determined effort to reach Maxwell's result without using his unacceptable procedure. I return to this point below.

The hydrokinetic analogy

In 1872, Thomson published his *Reprint of Papers on Electrostatics and Magnetism*. It contained not only his published papers, but also a number of unpublished manuscripts from the early period, and the whole was intermingled, in a chronologically confusing although systematically understandable manner, with footnotes and additional sections, all freshly written for the occasion. This new material is of interest to the historian mainly because it sheds a good deal of light on the early papers, and particularly because it abounds in applications of the hydrodynamical analogy — which Thomson now called the hydrokinetic analogy because it represented force by flux, not by force — and in explanations of how it had helped Thomson to derive this or that result in the early papers.

However, Thomson also gave a substantial new contribution to the theory of induced magnetisation, which he had begun to develop in 1851. It contained, among other things, a clear definition of magnetic susceptibility and permeability, and an explanation of the relation between them. In giving the physical meaning of the concept of permeability, Thomson used the analogy between four different branches of physics to explain that permeability was his word for Faraday's 'conducting power for lines of force', and that its role in magnetic theory was analogous to that of thermal conductivity in the theory of heat, that of 'specific inductive capacity' in electrostatics, and of 'the specific quality of a porous solid, according to which, when placed in a moving frictionless liquid, it modifies the flow' (Thomson 1872*a*, 484). This is a typical example of a general feature of Thomson's later writings, namely that the hydrokinetic analogy became increasingly a pedagogic aid to be called upon whenever a divergence-free vector field appeared in the exposition. In fact, when discussing Thomson's

William Thomson's electromagnetic theory

addiction to analogies, one should not forget that for more than 50 years his main job — and one he took very seriously — was to teach physics to students having a less than adequate mathematical background.[5] Similarly, it is a reasonable guess that he would have found this technique helpful in making Faraday grasp the gist of his mathematical ideas.

This is not, however, to imply that Thomson had given up all attempts to develop the hydrokinetic analogy into a dynamical illustration. An example of this is the very last section of the volume, written in 1872. Here Thomson used the analogy to give the magnetic energy Q its mathematical expression appropriate for the case of induced magnetisation:

$$Q = \frac{1}{8\pi} \int \frac{1}{\mu} \mathbf{B}^2 d^3\mathbf{x} \qquad (23)$$

where μ is the permeability. He tried to extend the dynamical aspect of the analogy by relating the forces between a permanent magnet and an inductively magnetised body to the forces between porous solids immersed in a flowing liquid. He found that '... the force required to balance [the solid] A in this case of the hydrokinetic system will be equal and *opposite* to the force required to balance a rigid body corresponding to A in the magnetic analogue.' (Thomson 1872a, 587, italics added.) Thus, as Maxwell said in his review of the book, this is merely an illustration, not an *explanation* of magnetic force, 'for in fact the forces are of the opposite kind to those of magnets' (Maxwell 1890, 2:306). That is, the attempt to develop the hydrokinetic analogy into a (hydro-) dynamical illustration, in Thomson's sense, had been unsuccessful.

The elastic solid revisited

In his later life, Thomson would often refer to his early paper 'On a mechanical representation ...' (1847) as the origin of his most persistent efforts to understand the dynamical constitution of the electromagnetic ether. Thus, in 1889 in his Presidential Address entitled 'Ether, electricity, and ponderable matter' he told the Institution of Electrical Engineers, after quoting the introduction and the final paragraph of this paper: 'I may add that I have been considering the subject for forty-two years ... I have been trying, many days and many nights, to find an explanation, but have not found it.' (Thomson 1882–1911, 3:502.) And in 1896, he referred to the same paper in a letter to Fitzgerald: 'I have not had a moment's peace or happiness in respect to electromagnetic theory since Nov. 28, 1846 (see vol. i. p. 80 of M.P.P.). All this time I have been liable to fits of ether dipsomania, kept away at intervals only by rigorous abstention from thought on the subject.' (Thompson 1910, 2:1065). In the same letter he went on to describe some of the obstacles he had been struggling to overcome:

> The greatest of my difficulties to get something towards a physical theory out of the 'mechanical representation' (M.P.P. Art. XXVII) up to now has been the steel

Ole Knudsen

magnet. Next greatest perhaps has been electrostatic stress: next greatest perhaps magnetic induction of currents. The greatest of all *was* the mobility of magnets and electrified bodies, showing the ponderomotive forces experienced by them in virtue of the rigidity of the ether in which they are embedded. But this difficulty (?) is annulled by the ether of Article XCIX, which acts as an incompressible liquid except in so far as its virtual rigidity is called into play by frictionality between it and ponderable matter. (Thompson 1910, 2:1066.)

This is not the place for a detailed discussion of these problems, of which the above quotation is by no means a complete list, or of Thomson's many attempts at solving them. It should be mentioned, however, that the 'ether of Article XCIX' refers to Thomson's gyrostatic ether whose equations of motion he described in a paper published in 1890 (Thomson 1882–1911, 3:436–65). This was an 'ideal substance' having a peculiar type of elastic reaction to absolute, as opposed to relative, rotation in space. In another paper (Thomson 1882–1911, 3:466–72) Thomson showed that such a substance could be realised by a horrendously complicated structure of liquid gyrostats, and therefore represented a possible hypothesis for the constitution of the real ether. In a previous paper Thomson had shown that an ether of this type could account for optical reflection and refraction (Thomson 1888), and he now showed that it could also give a 'complete mechanical representaton of an electromagnet consisting of any distribution whatever of closed electric currents'. The gyrostatic ether was perhaps Thomson's nearest approach to a theory fulfilling at least some of his hopes: 'It need scarcely be said that the 'ether' which we have assumed is a merely ideal substance. It seems to me highly probable however, that the assumed dependence of its forcive on absolute rotation, is at all events analogous to the truth of real ether.' (Thomson 1882–1911, 3:464). Another advantage of the gyrostatic ether was that it realised Thomson's old idea, which we have met already in the 1858 notebook entry (Thomson 1891, 1:149–53), of explaining elasticity by motion. In 1881 he had lectured at the Royal Institution on 'Elasticity viewed as possibly a mode of motion' (paraphrasing the title of Tyndall's popular book, *Heat, a Mode of Motion*). In this lecture he had demonstrated how a gyrostat, an endless chain, a limp disc of india rubber, and even water, acquired stiffness and elasticity when set in rapid rotation; and he had expressed his belief that 'the elasticity of every ultimate atom' might be similarly explained. He had now effected this for the ether, and thereby succeeded in unifying the hydrodynamical analogy with the elastic solid representation.

Still, Thomson did not allow himself or his readers to forget that this was only a partial success. He ended his paper by the usual recital of problems yet to be solved:

> But to give anything like a satisfactory material realisation of Maxwell's electromagnetic theory of light, it is necessary to show *electro-static force* in relation to the

William Thomson's electromagnetic theory

forcive (X, Y, Z) of my formulas; to explain the generation of heat according to Ohm's law in virtue of the action of this forcive when it causes an electric current to flow through a conductor; and to show how it is that the velocity of light *in ether* is equal to, or perhaps we should rather say, *is*, the number of electro-static units in the electro-magnetic unit of electric quantity. All this essentially involves the consideration of ponderable matter permeated by, or imbedded in ether, and a *tertium quid* which we may call electricity, a fluid go-between, serving to transmit force between ponderable matter and ether and to cause by its flow the molecular motions of ponderable matter which we call heat. I see no way of suggesting properties of matter, of electricity, or of ether, by which all this, or any more than a very slight approach to it, can be done, and I think we must feel at present that the triple alliance, ether, electricity, and ponderable matter is rather a result of our want of knowledge, and of capacity to imagine beyond the limited present horizon of physical science, than a reality of nature. (Thomson 1882–1911, 3:465.)

We note again that the interaction of ether and molecular matter was, as always, the crucial problem for Thomson. It is also interesting to note how close Thomson had come to Maxwell's ether of 1861–62. The structure of liquid gyrostats is not so different from Maxwell's system of vortices, and electricity as a 'fluid go-between' bears some resemblance, at least in its functions, to Maxwell's system of idle wheel particles.

Thomson and Maxwell

An important purpose in studying Thomson's work on electromagnetic theory must be to reach an understanding of his attitude towards Maxwell's theory. One difficulty in this respect is to acquire a precise impression of what that attitude actually was. In the 1880s and 1890s Thomson made a number of statements which at first sight seem impossible to fit into a consistent picture, ranging as they do from a labelling of the electromagnetic theory of light as 'rather a backward step' to a praise of the same theory as marking 'a stage of enormous importance in electromagnetic doctrine'. In the following I shall present Thomson's judgements on some of the main elements of Maxwell's theory and, I hope, show that they fall into an intelligible pattern and that Thomson's seemingly conflicting statements refer to different aspects of the theory.

One clue to a more detailed study of Thomson's reactions is the following passage from a letter to Fitzgerald, written in 1885: 'I have never yet felt any satisfaction in Maxwell's §§783, 784, 790, 791, 792, 645, 646, 794, 797, 798, 824 ... 829. I have never yet met anyone who understood a definite dynamical foundation for §783'. (Thompson 1910, 2:1038.) These sections in Maxwell's *Treatise* deal with the propagation of electromagnetic disturbance (note the omission of §§785–789 containing the electromagnetic interpretation of the velocity of light), the electromagnetic stress tensor, and the theory of the Faraday effect. I shall consider these points in due order below, but first I must discuss one of the most basic elements of Maxwell's theory, the electric displacement concept, and its role in the electromagnetic theory of light. Earlier I gave some reasons for believing that Maxwell's introduction of this concept, and his postulate about the magnetic effects of displacement currents, would have been difficult for Thomson to accept. In 1896, Thomson explained to Fitzgerald his persistent troubles with this concept, and contrasted this with his belief in the vortex theory of magnetism:

> Electric force (X, Y, Z) cannot be a mere displacement, because mere displacement does not in an elastic solid or in any conceivable 'ether' give rise to energy equal to $R^2/8\pi$ per unit volume of field. I could not in Nov. 1846, nor have I, ever since that time, been able to regard 'displacement' as anything better than a mere 'mechanical representation of *electric* force'. But I have always from that time till now felt, and I now still feel, that somehow or other we shall find rotation of a medium to be the reality of *magnetic* force. (Thompson 1910, 2:1069.)

In other words, displacement was only an illustration of electric force. But Thomson, as we have seen, would not permit analogies or illustrations to be used as arguments for the postulation of new macroscopic laws. Representing electric force by displacement might be valuable for describing the 'known laws' of dielectrics; but it could not justify Maxwell's doctrine of the true current as the sum of conduction and displacement current (cf. Wise 1981:50).

However, even as a 'mechanical representation', Maxwell's theory was not very successful. We noted earlier that Maxwell's stress tensor was one of the objectionable points; and in this quotation we see that the reason was that it was inconsistent with Thomson's old expression, $E^2/8\pi$, for the density of electrostatic field energy. In fact, Maxwell himself was probably referring to difficulties of this sort when he wrote: 'I have not been able to make the next step, namely, to account by mechanical considerations for these stresses in the dielectric. I therefore leave the theory at this point ...' (Maxwell 1873a, 1:132).

In 1888 Thomson published a discussion of Maxwell's theory of the propagation of electromagnetic disturbance. Maxwell's fundamental equation for this problem is, in slightly modernized notation

$$\mu\left(4\pi C + K\frac{\partial}{\partial t}\right)\frac{\partial}{\partial t}\mathbf{j}_T = \nabla^2 \mathbf{j}_T \tag{24}$$

where $\mathbf{j}_T$ is Maxwell's total current, C is the conductivity and K is the dielectric constant of the medium. The second term on the left-hand side represents Maxwell's postulate on the magnetic effect of displacement current. It is this term that ensures the existence of wave solutions for C equal to zero, i.e. in free ether and in insulating media. Having quoted this equation, Thomson said that it

> ... cannot be right, I think, according to any conceivable hypothesis regarding electric conductivity, whether of metals, or stones, or gums, or resins, or wax, or

William Thomson's electromagnetic theory

shellac, or india-rubber, or gutta-percha, or glasses, or solid or liquid electrolytes; being, as seems to me, vitiated for complete circuits by the curious and ingenious, but as seems to me not wholly tenable hypothesis which he introduces, in §610, for incomplete circuits. (Thomson 1882-1911, 4:543.)

The meaning of this seems to be that Thomson could find no justification for Maxwell's hypothesis in the known electrical properties of material substances. Instead, he proposed to take the known equation for complete circuits and use it in unchanged form also for incomplete circuits. This amounts simply to omitting the displacement current term from Maxwell's equation, and thus to destroying the foundation for Maxwell's theory of light. Thomson went on to say that his own 'simple' hypothesis led to

... simple and natural solutions, with nothing vague or difficult to understand, or to believe when understood, by their application to practical problems, or to conceivable ideal problems, such as the transmission of ordinary or telephone signals along submarine telegraph conductors and land lines, electric oscillations in a finite insulated conductor of any form, transference of electricity through an infinite solid, &c.&c. This, however, does not prove my hypothesis. Experiment is required for informing us as to the real electro-magnetic effects of incomplete circuits; and, as Helmholtz has remarked, it is not easy to imagine any kind of experiment which would decide between different hypotheses which may occur to anyone trying to evolve out of his inner consciousness a theory of the mutual force and induction between incomplete circuits. (Thomson 1882-1911, 4:544.)

Thomson's criticism of Maxwell's hypothesis was, of course, entirely justified. For the types of problem that he mentioned, which all have to do with the propagation of electrical effects in *conductors*, the effects of displacement currents are indeed negligible compared with those of ordinary conduction currents. It was, however, also rather beside the point. The justification for Maxwell's hypothesis lay in the fact that it opened the way for understanding light as an electromagnetic phenomenon, not in its application to 'practical' problems in telegraphic engineering, which could be solved just as well without it. In fact, the most important result of Maxwell's theory was the identification of the ratio v between the electromagnetic and electrostatic units of charge with c, the velocity of light in free ether.

In Maxwell's derivation of the equality of v and c, the equation (24) of electromagnetic disturbances, with the displacement current term, was of essential importance, because the derivation consisted in showing that equation (24), applied to empty space, i.e. with $C = 0$, $K = 1$, and $\mu = 1/v^2$, admitted of plane wave solutions propagated with velocity v. These waves of electromagnetic disturbance could then be identified as plane waves of light. As we have seen, Thomson explicitly rejected equation (24), and it is therefore at first sight surprising to find him expressing his enthusiasm for Maxwell's identification of v and c, as in the following quotation from his Presidential Address to the Institution of Electrical Engineers, in 1889,

Ole Knudsen

Heaviside points out that electro-magnetic induction causes a less great difference in the attenuation of signals of different periods than there is without it; and that electro-magnetic induction (as we knew forty years ago) tends to reduce the retardation of phase to the same for all different notes — that is, to the retardation equal to what would depend on a velocity not very different from the velocity of light — if the signals have but sufficient frequency. That velocity was then and is still known as the velocity which is the conductance in electrostatic measure, and the resistance in electro-magnetic measure of one and the same conductor. But its relationship to the velocity of light was brought out in a manner by Maxwell to make it really a part of theory which it never was before. Maxwell pointed out its application to the possible or probable explanation of electric effects by the influence of a medium, and showed that that medium — the medium whose motions constitute light — must be ether. Maxwell's 'electro-magnetic theory of light' marks a stage of enormous importance in electro-magnetic doctrine, and I cannot doubt but that in electro-magnetic practice we shall derive great benefit from a pursuing of the theoretical ideas suggested by such considerations. (Thomson 1882–1911, 3:489–90.)

To understand this passage one must remember that the electromagnetic significance of the velocity of light had been hinted at long before Maxwell's theory came into existence. Thus in 1856, Kirchhoff had pointed out explicitly that under certain conditions electric signals might be propagated through a wire with a velocity equal to v, and that the numerical value of this velocity was very close to that of light (Kirchhoff 1857, 209–10). Thomson had known the same result already from about 1854 when he had himself worked out a theory of the propagation of signals (Thomson 1904, 45 and 688–94). In Thomson's view Maxwell had made it 'part of a theory' by pointing out that this result about the propagation of telegraphic signals should probably be taken as an indication that electrical effects in general are propagated through the ether with the same velocity as the waves of light. In other words, Thomson accepted the general moral of Maxwell's result, but rejected the specific theory behind it.

Thomson also discussed this point in his famous *Baltimore Lectures*. In the fourth lecture, dealing with waves in the ether, he used the occasion to talk about one point where he felt that Maxwell's theory must be in error, namely in its claim that electric waves in vacuum are strictly transverse. Thomson stated his conviction that, if two conducting globes are connected by a wire subjected to a rapidly alternating e.m.f., the ensuing electrical waves will be 'condensational [i.e. longitudinal] waves in the luminiferous ether'. He went on to mention the 'so-called Electro-magnetic theory of light' and the theory of electric signalling 'which I worked out myself about the year 1854, and in which I found a velocity comparable with the velocity of light', and he continued: 'That is a very different case from this very rapidly varying electrification I have ideally put before you: and I have waited in vain to see how we can get any justification of the way of putting the idea of electric and magnetic waves in the so-called electro-magnetic theory of light'. (Thomson 1904, 45.) He ended this discussion by referring to his 'Mechanical representation' paper of 1847:

William Thomson's electromagnetic theory

It is shown in that paper that the static displacement of an elastic solid follows exactly the laws of the electro-static force, and that rotatory displacement of the medium follows exactly the laws of magnetic force. It seems to me that an incorporation of the theory of the propagation of electric and magnetic disturbances with the wave theory of light is most probably to be arrived at by trying to see clearly the view that I am now considering.

Thomson clearly felt that, while Maxwell's idea of the meaning of the equality of v and c was basically right, his identification of light waves with transverse electromagnetic waves represented an oversimplification derived from an erroneous foundation. He had already made it clear to his audience at Baltimore in the first, introductory lecture, that he did not believe in transverse electric waves: 'It seems to me that when we have an electro-magnetic theory of light, we shall see electric displacement as in the direction of propagation, and simple vibrations as described by Fresnel with lines of vibration perpendicular to the line of propagation, for the motion actually constituting light'. (Thomson 1904, 9.)

It should be evident by now that Thomson's reaction to Maxwell's theory cannot be described adequately as an outright rejection, although it undoubtedly was far from being an acceptance. Thomson certainly believed in the possibility of an electromagnetic theory of light. This theory would, like Maxwell's purported to do, embrace both electromagnetism and light within a common dynamical framework. It would resemble Maxwell's in accounting for magnetic force by microscopic rotation and in explaining why the velocity of light and the velocity of electric signals are both related to the ratio between the two different units of charge. But it would differ from Maxwell's theory in several respects. First, it would be based on a satisfactory and intelligible dynamical foundation, that is, a consistent set of assumptions about the dynamical constitution of the ether. In this respect Maxwell's theory was, to Thomson, 'rather a backward step from an absolutely definite mechanical notion that is put before us by Fresnel and his followers' (Thomson 1904, 9); that is, it was inferior to the old elastic theory of light. Secondly, it would avoid unwarranted hypotheses, such as that of displacement currents having magnetic effects never seen in any experiments. And thirdly, in a number of specific cases, like that of electric vibrations, it would lead to more reasonable predictions.

The above discussion does not exhaust the subject. It has not touched on the most substantial of Thomson's objections to Maxwell's theory, namely its failure to provide more than the merest beginning of an attack on the problem of the interaction of light with matter. We have already seen that Thomson had the problem of the constitution of matter in mind while speculating on the vortex theory of magnetism; and its continued importance for him is shown by the fact that his *Baltimore Lectures* were nothing less than a complete survey of the state of the physical theory of the interaction of light and matter. On this problem Thomson could find very little in Maxwell's *Treatise*, and the most detailed discussion of this sort, the chapter on the Faraday effect, he did not like, as we have seen (cf. Knudsen 1976, 255–61). It is, then, quite natural that Thomson made one of his sharpest factual criticisms of Maxwell during one of his lectures at Baltimore: 'Maxwell's electro-magnetic theory of light was essentially molar; and therefore not in touch with the dynamics of dispersion essentially involved in metallic reflection and translucency; though, outside his electro-magnetic theory, he was himself one of the foremost leading molecularists of the nineteenth century ....' (Thomson 1904, 376.)

With this passage Thomson gave a succinct statement of a problem that was generally recognised at the time. But while many of his colleagues sought to solve it by modifying or extending Maxwell's theory (Buchwald 1981), Thomson was alone in rejecting much of Maxwell's fundamental doctrine and advocating a return to the old elastic theory of light as the foundation for optical theory.

Concluding remarks

In the preceding section I have described Thomson's reactions to Maxwell's theory as conditioned by the views he had formed in the early part of his career on the proper relation between analogy, mechanical representation and macroscopic mathematical theory. This description presupposes a claim that Thomson stuck to these views also in the later period. I do not wish to state such a claim completely without qualifications, however. First, there is no doubt that Thomson's engineering experiences led to his insistence that Maxwell's theory should be judged by its application to 'practical problems', and to his understanding of the physical meaning of the constant v as primarily related to the velocity of signals transmitted through a telegraph cable. Secondly, there seems to be a significant difference between Thomson's early and later views on the value of a mathematical formalism.

We have seen that Thomson, in the 1840s and 1850s, stressed the importance of obtaining a secure theoretical foundation for macroscopic electrostatics and magnetism in the form of mathematical theories formulated without reference to underlying microscopic hypotheses. In the 1880s, Heinrich Hertz, despairing of reaching a full understanding of Maxwell's conception of the nature of electricity, made an attempt to build a mathematical theory of electromagnetism and light on Maxwell's equations in their simplest form, and to show how this theory could be verified by experiment (Hertz 1893, 195–240). Hertz was consciously trying to make the validity of the electromagnetic theory of light independent of 'Maxwell's peculiar conceptions or methods' (Hertz 1893, 21), just as Thomson had freed electrostatics and magnetism from any dependence on the hypotheses of electric and magnetic fluids. In particular, Hertz wanted to remove from the theory such 'rudimentary ideas of a physical nature' as 'that of electric displacement in free ether, as distinguished from the force which produces it' (Hertz 1893, 196). Hertz's endeavour was an example of a rather

William Thomson's electromagnetic theory

general tendency which one might have expected Thomson to welcome. On the contrary, already in the *Baltimore Lectures* he objected to 'the so-called Electromagnetic theory of light in the way it has been taken up by several writers of late' (Thomson 1904, 9), and in 1896 he explained the meaning of this statement to Fitzgerald: 'It is mere nihilism, having no part or lot in Natural Philosophy, to be contented with two formulas for energy, electromagnetic and electrostatic, and to be happy with a vector and delighted with a page of symmetrical formulas'. (Thompson 1910, 2:1065.) Thomson was undoubtedly referring to Hertz here. Three years earlier he had written a preface to the English edition of Hertz's electrical papers in which, while extolling Hertz's experimental work, he had managed not to say one word about Hertz's theoretical papers (Hertz 1893, ix–xv). Clearly, Thomson had either changed his opinion of the importance of mathematical theory, or Hertz's was not the right kind of theory. I think that both possibilities hold. Hertz's theory does differ from Thomson's early theories in one important aspect. Thomson's mathematical electrostatics was based on Coulomb's law which related macroscopic force to macroscopic distance and was capable of direct experimental verification. By contrast, Hertz's axioms were the field equations for free ether which could not be verified in the same direct way, and Hertz explained this point quite explicitly:

> Explanations will be added to the formulæ; but these explanations are not to be regarded as proofs of the formulæ. The statements will rather be given as facts derived from experience; and experience must be regarded as their proof. It is true, meanwhile, that each separate formula cannot be specially tested by experience, but only the system as a whole. But practically the same holds good for the system of equations of ordinary dynamics. (Hertz 1893, 197.)

Hertz's sophisticated view of a theory as a system which is testable as a whole although based on non-testable axioms, may be contrasted with the following passage from the preface to Thomson and Tait's famous textbook. 'The Laws of Motion, the Law of Gravitation and of Electric and Magnetic Attractions, Hooke's Law, and other fundamental principles *derived directly from experiment*, lead by mathematical processes to interesting and useful results, for the full testing of which our most delicate experimental methods are as yet totally insufficient'. (Thomson and Tait 1867, vi. Italics added.) Here it is rather the other way around: the axioms are directly testable whereas their consequences are not fully so. It is, then, not hard to imagine that Hertz's philosophy of science would have appeared 'nihilistic' to Thomson.

However, I think it is also true that Thomson in his later life did not value mathematical theory quite as highly as he had done in the early period. In one of his *Baltimore Lectures* he said:

> I never satisfy myself until I can make a mechanical model of a thing. If I can make a mechanical model I can understand it. As long as I cannot make a mechanical model all the way through I cannot understand; and that is why I cannot get the

Ole Knudsen

electromagnetic theory. I firmly believe in an electromagnetic theory of light, and that when we understand electricity and magnetism and light we shall see them all together as parts of a whole. But I want to understand light as well as I can, without introducing things that we understand even less of. That is why I take plain dynamics. I can get a model in plain dynamics; I cannot in electromagnetics. (Thompson 1910, 835–6.)

Apparently, even if electromagnetism and light could be unified in a consistent mathematical theory, such a theory would be inferior in value to the elastic solid theory of light, as long as it did not admit of a mechanical representation. Even though he had not succeeded, Maxwell had at least attempted to construct a dynamical theory; by renouncing mechanical representations some of the younger theoreticians were moving away from, rather than towards the type of theory Thomson wanted. He certainly did not see Hertz's work as analogous to his own early efforts; instead it epitomised a 'nihilistic' tendency to deny even the possibility of dynamical explanation.

If Thomson seemingly changed his attitude towards mathematical theory, it was no doubt because he felt that the situation had altered. He had developed his mathematical theories of electrostatics and magnetism at a time when, so he believed, absolutely nothing was known with certainty about the deeper nature of things and when the existing notions were untenable. However, since 1856 he had felt convinced that he possessed at least one ultimate truth, namely that magnetic effects were produced by microscopic rotation. Furthermore, the doctrine that light consisted in elastic waves in the ether had been a secure foundation for optics ever since Fresnel's time. To attempt, then, to develop a mathematical theory which did not encompass these profound results seemed to him like sacrificing true physical knowledge in order to obtain a mathematical formalism without much physical meaning.

By thus clinging to what he felt to be real advances in the knowledge of the nature of magnetism and light, Thomson was forced to reject the new style in electromagnetic science which was emerging in the 1880s and which found a clear expression in Hertz's work. Physics had progressed in such a way that Thomson's revolutionary style of the 1840s became in the 1880s a reactionary attitude out of touch with the real development in electromagnetic theory.

Notes

1. Aspects of this topic are treated in Buchwald 1977, Gooding 1980 and 1982, Heimann 1970a, Moyer 1977 and 1978, Siegel 1981, Smith 1978, and Wise 1979, 1981 and 1982. Wise's papers have been particularly influential in shaping my ideas for this chapter. I wish to thank P.M. Harman for proposing the topic, and D. Gooding, P. Lervig, J. Lützen, and K.M. Pedersen for valuable criticism. I am grateful to the Danish Natural Science Research Council for a grant enabling me to attend the conference at Grasmere.
2. My discussion in the following sections owes much to the treatment given in Wise 1979 and Wise 1981.
3. The above is largely a paraphrase of Maxwell's discussion in Maxwell 1873a, 2:246–7.

4. Published in Knudsen 1972, pp. 47–50, unfortunately under a wrong date (page 47). The correct date is 'Jan. 6, 1858'.
5. See, for instance, Thomson 1872a, 52, footnote, for an early instance of Thomson's concern about solving electrostatic problems 'without the explicit use of the differential or integral calculus'.

7

Mechanical image and reality in Maxwell's electromagnetic theory

DANIEL M. SIEGEL

Paradigmatic of Cambridge mathematical physics in the nineteenth century was the work of James Clerk Maxwell, who made pivotal contributions to the two broad areas of mathematical physics that matured in the nineteenth century: thermodynamics and statistical mechanics; and electromagnetism and optics. In the field of electromagnetism and optics, it was Maxwell who demonstrated the unity of the field and systematised its foundations. In doing this, he made extensive use of mechanical representations of electromagnetic phenomena — that is, mechanical pictures or models taken to correspond in some way to electromagnetic phenomena. This use of mechanical images was characteristic of Cambridge mathematical physics in the nineteenth century, and to this circumstance attaches a fundamental question: to what extent were these mechanical images taken as literal representations of reality? To a modern eye, the mechanical representations of electromagnetic phenomena proposed by Maxwell and his followers may seem outlandish; the historian, however, must attempt to approach these conceptual artifacts with an open mind, considering them in the context of the nineteenth-century commitments and goals that they reflected and served.

If one preserves an open mind and a sensitivity to nuance, one finds that the situation is in fact extremely complex: in various situations, and in the hands of various investigators, mechanical representations were used in various ways, and with varying degrees of commitment to their reality. The complexity is not much reduced by focusing on Maxwell, for his opinions and commitments were rich and nuanced in this area as in others. His pronouncements on this matter were also at times cryptic and ambiguous. In this situation there can be no easy answers, but neither is the situation hopeless. In what follows, I approach the problem by giving particular attention to the development, over time, of Maxwell's attitudes towards and use of mechanical representation, as reflecting his responses to various influences and the shifting needs of his developing research programme. The picture that emerges from this analysis is one of a nineteenth-century mechanical philosopher, methodologically perceptive and adventurous within that framework — using mechanical representations in a variety of ways — while maintaining throughout his basic allegiance to the

Maxwell's electromagnetic theory

notion that reality is ultimately mechanical and that the highest aim of mechanical representation is to be faithful, as far as possible, to that reality.

Thomson, Maxwell, and the uses of analogy

Maxwell's teacher in the use of mechanical representation in an analogical vein was William Thomson, a fellow Scotsman and fellow Cambridge product seven years Maxwell's senior; relatedly, Thomson's work and tutelage provided a vehicle for Maxwell's initiation into the arcana of the field approach to electricity and magnetism pioneered by Michael Faraday (Siegel 1981, 239–46). The *locus classicus* of Thomson's application of the method of analogy to Faraday's lines of force was an analogy Thomson had developed in the 1840s, comparing Faraday's lines of force to lines of heat flux. Then, in a paper of 1847 entitled 'On a mechanical representation of electric and magnetic forces' (Thomson 1882–1911, 1:76–80), Thomson considered states of linear and rotational mechanical strain in an elastic solid, showing that their distributions in space would be analogous respectively to distributions of electric force and magnetic force. Although Thomson regarded these as mere analogies, he thought they might be suggestive concerning the real basis of electrical and magnetic phenomena: what he had published was a 'mathematical analogy'; what he looked forward to was something different — a 'physical theory' (Thompson 1910, 1:203–4). Maxwell was particularly impressed by Thomson's analogical method: 'Have you patented that notion with all its applications?' asked Maxwell in a letter of 1855, 'for I intend to borrow it for a season' (Larmor 1936, 705). Borrow it he did, along with the notion that its primary value was as a temporary expedient, as a prelude to something better.

Maxwell saw himself in a situation where he could make good use of just such a temporary expedient. 'The present state of electrical knowledge', he felt, was 'peculiarly unfavourable to speculation'. Much was known about electricity, but that knowledge was scattered and fragmentary, and Maxwell could not yet see his way through to a theoretical structure that would unify all the ramified phenomena of electricity and magnetism. One might be tempted, in such a situation, to 'adopt a [working] hypothesis', which might at least lead to a 'partial explanation' of electromagnetic phenomena. In Maxwell's opinion, however, this would be dangerous, because it might foster premature commitment: 'If [in this situation], we adopt a physical hypothesis, we see the phenomena only through a medium, and are liable to that blindness to facts and rashness in assumption which a partial explanation encourages'. Along with British methodologists from Francis Bacon to John Herschel, Maxwell was much concerned to 'avoid the dangers arising from a premature theory', to avoid being 'carried beyond the truth by a favourite hypothesis'; aligned with the Scottish Common Sense philosophical tradition, Maxwell was particularly sensitive to the comparative and relational aspects of knowledge. In this situation and against this background, Thomson's analogical approach seemed

Daniel M. Siegel

particularly well suited as a tool of investigation, for it would allow for the use of a mechanical representation, which would aid in the task of 'simplification and reduction of the results of previous investigation to a form in which the mind can grasp them, . . . without being committed' in any way to the truth of that mechanical representation (Maxwell 1890, 1:155–9).[1]

Merely as an analogy, then, Maxwell proposed a mechanical representation in which an incompressible fluid was pictured as flowing through a porous medium. In a paper entitled 'On Faraday's lines of force' (1855–56), this image was applied to the elucidation of electric fields, magnetic fields, and distributions of electric current; the flow lines of the incompressible fluid were taken to correspond to magnetic lines of force, electric lines of force, or lines of electric current, depending on the particular problem being analysed. The flow analogy could provide no image of the coexistence and interaction of electric fields, magnetic fields, and electric currents; rather, the flow analogy provided segmented, compartmentalised understanding of each of the three electromagnetic phenomena, considered each in isolation from the other. This compartmentalisation would have been intolerable from the point of view of providing a theory: 'No electrical theory can now be put forth, unless it shews the connexion not only between electricity at rest and current electricity, but between the attractions and inductive effects of electricity in both states'. The flow representation, however, did not have to be measured against this high standard, for it was intended as a mere analogy, with no claim to either comprehensiveness or truth value. Maxwell explicitly and insistently cautioned the reader that the incompressible fluid referred to was an 'imaginary fluid', and 'not even a hypothetical fluid'. The mathematical isomorphism between the equations of percolative streamline flow and the equations describing electric or magnetic lines of force was nothing more than that, and no 'physical theory', no specification of the actual 'physical nature of electricity' or magnetism, was implied (Maxwell 1890, 1:155–229, especially 155–60).

There were further reasons for Maxwell's diffidence. Wilhelm Weber, a German electrical experimenter and theorist, had developed, in the 1840s, a 'professedly physical theory of electro-dynamics', which, in Maxwell's own admission, was 'so elegant [and] so mathematical' that it could not be ignored. Weber, working in the action-at-a-distance tradition, had apparently not found the 'state of electrical science' quite so 'unfavourable to speculation' as had Maxwell; Weber had in fact been able to develop a theory, based on the interaction of charged particles through distance forces depending on their relative positions, velocities and accelerations, that gave a coherent and connected account of the basic phenomena of electricity, electromagnetism, magnetism and electromagnetic induction — basically all of the phenomena of electricity and magnetism. Maxwell was not able at this point to propose an alternative theory of comparable range, and he apparently felt that to make strong claims for a partial theory of his own, as against Weber's elegant,

Maxwell's electromagnetic theory

mathematical, and very comprehensive theory, would be to invite criticism and perhaps ridicule. Maxwell therefore presented his mechanical picture merely as an analogy — a heuristic device and 'temporary' expedient — justifying his effort on grounds of theoretical pluralism: 'It is a good thing to have two ways of looking at a subject, and to admit that there *are* two ways of looking at it'. (Maxwell 1890, 1:207–8; Woodruff 1976.)

There was another ground on which Maxwell felt he might be criticised: he was presuming to contribute to the theory of electricity and magnetism, while having made no contribution whatever to the experimental side of that field. Only in the twentieth century has the theoretical physicist *per se* had an acknowledged role; given nineteenth-century norms, Maxwell felt he had to be both apologetic and moderate in his claims: 'By the [analogical] method I adopt', Maxwell wrote, 'I hope to render it evident that I am not attempting to establish any physical theory of a science in which I have hardly made a single experiment'. Maxwell's deference, one surmises, was primarily towards Faraday, who was eponymously honoured in the title of Maxwell's paper; beyond this, Maxwell's main competitor in the realm of mathematical theory — Wilhelm Weber — also had impressive experimental credentials. Having no such credentials himself, Maxwell did not presume to present a 'true solution' to the problems of electrical science; instead, he offered a heuristic analogy, defining for himself an auxiliary role *vis-à-vis* the experimental philosophers, helping but not usurping:

> If the results of mere speculation which I have collected are found to be of any use to experimental philosophers, in arranging and interpreting their results, they will have served their purpose, and a mature theory, in which physical facts will be physically explained, will be formed by those who by interrogating Nature herself can obtain the only true solution of the questions which the mathematical theory suggests. (Maxwell 1890, 1:157, 159.)

This statement expresses not only the modesty of Maxwell's aims in 'Faraday's lines' but also the hopes he had for the future. He did look forward to something better than a mere analogy: he looked forward to 'a mature theory, in which physical facts will be physically explained'. Thus, while his aim in 'Faraday's lines' had 'not [been] to establish [a] theory', his hope for the future was that 'a ... theory ... will be formed'; while the fluid-flow analogy had 'not [been] introduced to explain actual phenomena', the hoped-for theory would be one 'in which physical facts will be physically explained'. In 1855 Maxwell was a young man without reputation, diffident and deferent with respect to both Faraday and Weber; he was just beginning to make headway in the task of understanding electromagnetic phenomena and wanted to avoid premature commitment; and he found himself able to devise only a segmented and compartmentalised mechanical representation of electromagnetic phenomena. In these circumstances, he made no strong claims for the flow represent-

Daniel M. Siegel

ation, putting it forward merely as an illustrative and heuristic analogy; he looked forward, however, to a better time.

Towards a realistic, comprehensive and explanatory theory

The signal that the time had come to go beyond mere analogy and begin to talk in earnest about the nature of things came from William Thomson. Thomson had pioneered the use of physical analogies in electromagnetic theory, and Maxwell had followed him; Thomson had, from the outset, felt that these analogies were merely preliminary steps along the way towards a hoped-for 'physical theory', and Maxwell had followed him in this; and finally, in 1856, Thomson decided that the time had come to talk about the nature of things in electromagnetic theory, and Maxwell was to follow him in this. Maxwell was no blind follower — we have seen that he had his own good reasons for his use of the method of analogy — but he was none the less a devoted follower of Thomson's, and we shall not be surprised to find Maxwell following his mentor in this last step as well.

Thomson's route to the conclusion that he was now able to propose a description of the 'reality' underlying the magnetic field, a specification of the 'ultimate nature of magnetism', is described in detail in Knudsen's chapter in this volume. In brief, the general background for Thomson's bold step was supplied by the principle of conservation of energy and the associated kinetic theory of heat, which together dominated the scientific agenda of the 1850s. The general opinion, especially in Britain, was that the new view of the nature of heat represented the extirpation of ancient error and the achievement of true insight into the real nature of things. Applied to other phenomena — and the postulated universality of the energy law certainly suggested broad applicability — the hope was that this basic insight could be parlayed into an understanding of the real physical natures of those other phenomena as well. The particular form of the kinetic theory of heat that Thomson adopted stemmed from the work of W.J.M. Rankine, who assumed that the motions constituting heat are rotatory motions of atmospheres surrounding individual material molecules — 'molecular vortices,' in Rankine's terminology. Applying this picture to an understanding of the magnetic field, by way of an argument based on the behaviour of light in magnetic fields (i.e. the Faraday rotation), Thomson was able to conclude that the actual mechanical condition characterising a region traversed by magnetic lines of force would be one where the axes of the molecular vortices were all aligned in one direction, this being the direction of the line of force. Starting with the firm conviction that Rankine's molecular vortices were real, and regarding the logic of the argument based on the Faraday rotation as airtight, Thomson felt justified in asserting that 'a certain alignment of axes of revolution in this [vortical] motion IS *magnetism* [Thomson's emphasis]. Faraday's magneto-optic experiment makes this not a hypothesis, but a demonstrated conclusion'.[2]

Maxwell's electromagnetic theory

'Professor Thomson has pointed out that the cause of the magnetic action on light must be a real rotation going on in the magnetic field', Maxwell wrote approvingly; the time had come to go beyond the analogical approach and begin constructing, on the basis of Thomson's picture of molecular vortices oriented along magnetic field lines, something like the 'mature theory' to which Maxwell had looked forward. Maxwell may have begun thinking seriously along these lines in 1857 (Campbell and Garnett 1882, 199n), and the results of his deliberations were published in a series of instalments — constituting Parts I to IV of a paper entitled 'On physical lines of force' — over a period of eleven months in 1861-62. The promise of Thomson's suggestion for resolving the difficulties that had hitherto hindered the development of a serious theory was twofold: the theory of molecular vortices promised comprehensiveness, and it promised explanatory power. One of the major limitations of the flow analogy in 'Faraday's lines' had been its inability to give any account of the connections and interactions between electric fields, magnetic fields, and electric currents. The molecular-vortex picture, on the other hand, gave promise of just that comprehensive and connected coverage which the flow picture lacked: 'If, by the [molecular-vortex] hypothesis', Maxwell wrote, 'we can connect the phenomena of magnetic attraction with electromagnetic phenomena and with those of induced currents, we shall have found a theory which, if not true, can only be proved to be erroneous by experiments which will greatly enlarge our knowledge of this part of physics'. A theory having comprehensive coverage thus was not guaranteed to be true, but it was, by virtue of this comprehensive coverage, to be regarded as a candidate for truth; it was to be regarded in any case as having truth value, whether true or false, in contradistinction to the flow analogy of 'Faraday's lines', which was to be regarded as neither true nor false, but only illustrative (Maxwell 1890, 1:451-513, on 452).

The molecular-vortex representation was to be distinguished from the fluid-flow picture also in that the molecular-vortex representation was to be regarded as having explanatory power, whereas the fluid-flow representation did not. The issue of explanatory power was a central one for Maxwell, and is worth discussing at some length. Consider, for example, the case of two unlike magnetic poles, exerting mutual attractive forces on each other (see Figure I). As Faraday had conceptualised this situation in a paper entitled 'On the physical character of the lines of magnetic force' (1852) — this is clearly the referent of the title of Maxwell's own paper — the lines of magnetic force behave as if they have a tendency to contract along their lengths and also to repel each other; acting in this way, they tend to pull the unlike magnetic poles together. The attribution of this behaviour to the magnetic lines of force provides an explanation or account of the attraction between the unlike poles, in terms of the system of lines of magnetic force existing in the space surrounding the magnets. Faraday distinguished between a merely *geometrical* treatment of the lines of force, dealing descriptively with their distribution in

Daniel M. Siegel

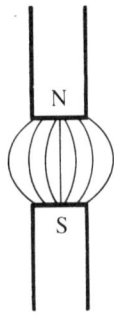

Figure I

space, and a *physical* treatment of the lines of force, dealing with their dynamical tendencies that give rise to the actual forces exerted. In the former context, 'the term *magnetic line of force*' applied; in the latter context, one spoke of a '*physical line of [magnetic] force*'. Clearly, the titles of Maxwell's two papers reflect this usage, indicating at the outset the merely geometrical and descriptive character of the flow representation as opposed to the physical and explanatory character of molecular vortices (Faraday 1839–55, 3:407–37, especially 419, 435–7).

In 'Faraday's lines', then, Maxwell had presented a mechanical representation of magnetic lines of force that adequately modelled the geometrical distribution of magnetic lines of force in space — this was given by the flow lines — but gave no purchase for understanding the forces of attraction or repulsion between magnetic poles. No tendencies of the magnetic lines to repel each other and contract along their lengths were derivable from the flow picture, and magnetic forces hence were not explained or accounted for by that picture. Maxwell had been explicit and insistent on this point: 'By referring everything to the purely geometrical idea of the motion of an imaginary fluid, I hope to ... avoid the dangers arising from a premature theory professing to explain the cause of the phenomena.' Reviewing the matter in the introduction to 'Physical nes', Maxwell stated again that in 'Faraday's lines' he had been 'using mechanical illustrations to assist the imagination, but not to account for the phenomena'. Referring in a similar vein to Thomson's paradigmatic mechanical analogies of 1847, Maxwell observed that 'the author of this method of representation does not attempt to explain the origin of the observed forces ... but makes use of the mathematical analogies ... to assist the imagination'. Maxwell's avowed purpose in 'Physical lines', by contrast, was 'to examine magnetic phenomena from a mechanical point of view, and to determine what tensions in, or motions of, a medium are capable of *producing* the mechanical phenomena observed'. He was seeking for a way in which 'the observed resultant forces may be *accounted for*'. This was where the molecular-vortex representation showed its superiority: assuming that the magnetic line of force

Maxwell's electromagnetic theory

represented the axis of a molecular vortex, it was easy to demonstrate that centrifugal forces would tend to make each vortex tube expand in thickness, thereby tending to increase the spacing between magnetic lines; at the same time, owing to the incompressibility of the fluid in the vortex tubes, those tubes would tend to shrink in length, making the magnetic lines have a corresponding tendency to contract along their lengths. This *physical* behaviour of the magnetic lines was what was needed to *explain*, to *account for*, magnetic forces (Maxwell 1890, 1:159, 452–5, 467, emphasis mine).

The contrast between the illustrative mechanical analogy of 'Faraday's lines' and the explanatory mechanical theory of 'Physical lines' can be developed in a more formal manner. Consider a physical system described by a set of variables $\{F_i, G_i\}$, where the F_i are variables that represent observable mechanical forces, and the G_i are the other variables describing the system. A mechanical representation of that physical system would refer to a mechanical system represented by variables $\{f_j, g_j\}$, whose interrelationships were isomorphic to those of some subset $\{F_j, G_j\}$ of the $\{F_i, G_i\}$. (The larger the subset, the more complete the mechanical representation.) Whether that mechanical representaton were to be construed as illustrative or explanatory would depend upon the nature of the f_j: if the set $\{f_j\}$ were empty, or if the f_j were not themselves forces, then the mechanical representation would be merely illustrative; if, on the other hand, there were some f_j and they were forces, then the mechanical representation could be said to have explanatory power. In the molecular-vortex representation of 'Physical lines', for example, there were forces f_m, produced by the centrifugal forces of the rotating vortices, that were isomorphic to the forces F_m exerted in magnetostatic situations. In this case, the system of molecular vortices could be *identified with* the magnetic field, the forces f_m then being identified with the forces F_m — which is possible because they are variables of the same kind; the behaviour of the molecular vortices, which explains the f_m, then also explains the F_m — that is, the theory of molecular vortices *explains* magnetic forces. In the fluid-flow representation of 'Faraday's lines', on the other hand, the set $\{f_j\}$ was empty: the variables relevant to magnetostatics, for example, were all of the g_j class — representing pole strength, field strength, etc. — and no forces f_m corresponding to magnetostatic forces F_m were exhibited; there could therefore be no explanation of magnetic forces by this mechanical representation — it could only by illustrative. In general, because the set $\{f_j\}$ was completely empty in the fluid-flow representation, this mechanical representation could have no explanatory power.[3]

Maxwell's assertion that the mechanical representation based on molecular vortices was explanatory rather than merely illustrative was thus no mere rhetorical device, but rather had a precise technical meaning. Because the mechanical representation based on molecular vortices was explanatory, and also because it provided comprehensive and connected coverage of the whole range of electromagnetic phenomena, Maxwell felt justified in referring to it as 'the theory of molecular vortices'. By calling it a theory, Maxwell indicated that it was definitely something more than a 'mechanical illustration ... to assist the imagination': it was at least a candidate for reality — perhaps 'true' and perhaps 'erroneous', but in any case not merely illustrative (Maxwell 1890, 1:451–2).

On the reality of molecular vortices

Although the theory of molecular vortices was a candidate for reality, the strength of its candidacy was subject to vicissitudes. The four instalments of 'Physical lines', taken together with other evidence from the period, furnish a rich record of the variations and nuances in Maxwell's views during the period in 1861–62 when he was working on the theory. Parts I and II of the paper were published in March to May of 1861, and the nuances in Maxwell's views contained therein would appear to represent primarily a range of essentially coexisting sentiments, rather than a development in time. As discussed above, the tone of Part I was enthusiastic concerning the promise of the theory of molecular vortices, and guardedly optimistic concerning the theory's candidacy for reality. In Part II, however, there were cross-currents. On the one hand, the basic 'hypothesis of vortices' was characterised as a 'probable' hypothesis. Also classified as probable was Maxwell's judgement concerning the sizes of the vortices: 'The size of the vortices is ... probably very small as compared with that of a complete molecule of ordinary matter'. Maxwell went on to observe that although the precise sizes of the vortices could not be specified by electromagnetic measurements alone, they could be determined if one were able to measure directly the mechanical angular momenta of the vortices. This in turn 'might be detected by experiments on the free rotation of a magnet'. Maxwell had indeed already 'made experiments to investigate this question', but he had 'not yet fully tried the apparatus', and did not report results. It is clear, however, that in Maxwell's estimation these vortices could well have been real enough to have detectable angular momentum. He therefore continued to work on the experiment, but, as he complained to Faraday some months later, he had 'not yet overcome the effects of terrestrial magnetism in marking the phenomenon'. Indeed, a successful measurement of the effect continued to elude Maxwell, as he reported retrospectively in the *Treatise on Electricity and Magnetism* (1873). His continuing efforts to detect the vortices, however, testified to their continuing candidacy for realistic status.[4]

Maxwell was not optimistic, on the other hand, concerning another part of the theory. In extending the theory to include electric currents, Maxwell had postulated that between the vortex cells there were interposed monolayers of small spherical particles, which rolled without slipping on the surfaces of the vortices (see Figure II), thus coupling the vortex rotations in the manner of '"idle wheel[s]"'; these were, moreover, to be regarded as *moveable* idle wheels, and, 'according to our hypothesis, an electric current is represented by the

Maxwell's electromagnetic theory

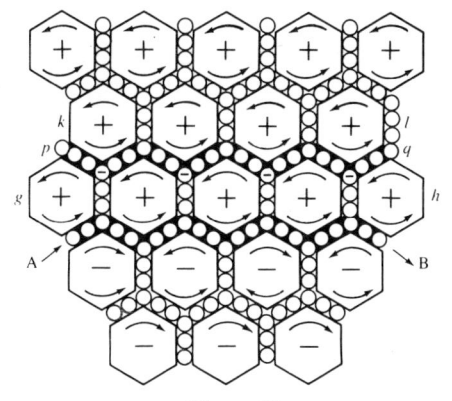

Figure II

transference of the[se] moveable particles'. The mathematics of these 'moveable particles' turned out very neatly, but Maxwell nevertheless had to admit that this hypothesis could be regarded only as a 'provisional' one, of 'temporary character'. Indeed, the 'conception' was 'awkward', and Maxwell did 'not bring it forward as a mode of connexion existing in nature, or even as that which [he] would willingly assent to as an electrical hypothesis'. (Part of Maxwell's distaste for the moveable particles probably stemmed from the fact that they constituted a sort of electrical fluid in his theory, and Maxwell, as a good follower of Faraday, found the notion of electrical particles or fluids repugnant.) Maxwell, then, was careful to distinguish between a part of the theory that he regarded as a 'probable' hypothesis and a good candidate for reality, and another part, which he regarded as a 'provisional' hypothesis (a kind of place-holder in the theory) and a very poor candidate for reality.[5]

Part II of 'Physical lines' ended on a down-note. As I have argued elsewhere, Maxwell encountered difficulty in extending his theory to embrace electrostatics, and he did not anticipate, when he submitted Part II of the paper for publication, the triumphant further extension of the theory that he was to publish as Parts III and IV after a nine-month hiatus. (It would appear that the new breakthrough was accomplished in the summer of 1861, months after the publication of Part II.) He thus closed Part II with what was clearly intended to be a final conclusion to the paper, and he was clearly not too happy with what he had wrought. Through the omission of electrostatics, the theory fell short of the comprehensive coverage Maxwell was striving for, and he remained still at a disadvantage *vis-à-vis* Weber's more comprehensive theory. In a somewhat cynical closing statement, Maxwell observed that 'those who [had] been already inclined' towards the field point of view might find his paper worth while. Others might be predisposed to 'look in another direction for the explanation of the facts' — clearly the reference was to Weber's action-at-a-distance theory — and Maxwell knew he was not going to win them over with the incomplete

189

Daniel M. Siegel

theory he had presented. Unable, in this situation, to take a strong stand on behalf of his own theory, Maxwell had to content himself with taking a shot at Weber's theory: 'Those who look in a different direction for the explanation of the facts, may be able to compare this theory with that ... which supposes electricity to act at a distance with a force depending on its velocity, and therefore not subject to the law of conservation of energy'. Continuing in this dyspeptic vein, Maxwell summed up his own contribution as follows: 'We have now shewn in what way electro-magnetic phenomena may be imitated by an imaginary system of molecular vortices'. This statement is clearly out of tune with Maxwell's stance in the rest of 'Physical lines'; it echoes instead Maxwell's tone of diffidence with respect to Weber in 'Faraday's lines'. Maxwell had hoped to construct a theory that would rival Weber's in comprehensiveness; he had failed and he was disappointed. Apparently, however, his disappointment spurred him on to further efforts, which were crowned with spectacular success.[6]

Part III of 'Physical lines' was published in January 1862, and there, by extending the theory to electrostatics and by providing an explanation of electrical forces in terms of stresses in the medium, Maxwell finally achieved the full comprehensiveness and explanatory character that he had sought. He could now 'explain the condition of a body with respect to the surrounding medium when it is said to be "charged" with electricity, and account for the forces acting between electrified bodies, [thereby] establish[ing] a connexion between all the principal phenomena of electrical science' (Maxwell 1890, 1:490). The extension to electrostatics was accomplished by assigning elastic properties to the molecular vortices, making the system of vortices in space — the 'magneto-electric medium' — now capable of sustaining elastic waves (Maxwell 1890, 1:489). Using values of electrical parameters measured by Wilhelm Weber and Rudolph Kohlrausch — Weber was now being enlisted as an ally — Maxwell was able to calculate the velocity of elastic waves in the magneto-electric medium, arriving at a value that agreed precisely with (in fact was bracketed by) existing measurements of the speed of light in air or vacuum. Maxwell's conclusion was that 'we can scarcely avoid the inference that *light consists in the transverse undulations of the same medium which is the cause of electric and magnetic phenomena*' (Maxwell 1890, 1:497–500). Maxwell's italics were well justified: he, and soon his colleagues, judged this to be a result of immense consequence, and historical perspective has reinforced that assessment (Everitt 1975, 111–30).

The result was also of the greatest importance as concerned the status of the theory of molecular vortices. First, given the widespread belief in the real existence of the luminiferous medium, the identification of the luminiferous and magneto-electric media became an argument for the real existence of the magneto-electric medium, with its vortex structure. At the level of particulars, certain features of the vortices — their elasticity, for example — were now no

Maxwell's electromagnetic theory

longer *ad hoc* to electromagnetic theory, but rather could be seen as the natural extension of a theoretical structure already in place, namely, the wave theory of light. Relatedly, the generality and broad range of the resulting unified theory of electromagnetism and optics argued powerfully for the physical significance of the theory of molecular vortices. Beyond this, as Maxwell stressed in a letter announcing the new results to Faraday, the new, unified theory predicted various relationships between electromagnetic and optical phenomena, the experimental test of which would help to establish the truth value of the theory: 'The conception I have hit on has led, when worked out mathematically, to some very interesting results, capable of testing my theory, and exhibiting numerical relations between optical, electric, and electromagnetic phenomena, which I hope soon to verify more completely.' In a parallel letter to Thomson, Maxwell acknowledged Thomson's seminal role in the development of the theory of molecular vortices; sketched the application of the theory of molecular vortices to electricity and magnetism, making clear the comprehensiveness and explanatory character of the theory; and discussed the experimental basis and implications of the unification with optics. In all three announcements of the new results — the public one in 'Physical lines' and the private ones to Faraday and Thomson — Maxwell's confidence in the theory of molecular vortices, now with a vastly enhanced range of applicability, was in marked contrast to his earlier vacillations. Further arguments for the reality of molecular vortices were to be presented in the final instalment of 'Physical lines', which appeared one month later.[7]

The mathematics and physics of linear and rotatory vectors

Maxwell's most focused argument in favour of the reality of molecular vortices was given in Part IV of 'Physical lines'. It was a mathematical argument, but it was not presented as pure and abstract mathematics; it was, rather, according to Maxwell's enduring commitment, 'embodied mathematics' — mathematics represented in mechanical examples (Olson 1975, 302–3). The central mathematical relationship that Maxwell focused on was the one exemplified in the relationship between electric current and magnetic field, i.e. Ampère's circuital law in differential form:

$$p = \frac{1}{4\pi}\left(\frac{d\gamma}{dy} - \frac{d\beta}{dz}\right)$$
$$q = \frac{1}{4\pi}\left(\frac{d\alpha}{dz} - \frac{d\gamma}{dx}\right) \quad (1)$$
$$r = \frac{1}{4\pi}\left(\frac{d\beta}{dx} - \frac{d\alpha}{dy}\right)$$

where α, β, γ are the cartesian components of the vector representing magnetic field intensity; p, q, r are the components of the vector representing electric

Daniel M. Siegel

current density; and the differential operators d/dx, d/dy, d/dz represent partial differentiation with respect to cartesian coordinates x, y, z. In order to illustrate the deeper meaning of this equation, Maxwell listed a series of mechanical examples to which this equation could be applied:

(1) If α, β, γ represents linear displacement or change of location, then p, q, r represents rotatory displacement or change of location.
(2) If α, β, γ represents linear velocity, then p, q, r represents rotational velocity.
(3) If α, β, γ represents a force or push, then p, q, r represents a torque or twist.

Common to these examples is the circumstance that α, β, γ has '*linear* ... character', representing a motion or a thrust in a certain direction, while p, q, r has '*rotatory* character', representing a rotation or a twist about a certain axis. Further mechanical examples exhibited the inverse relationship:

(4) If α, β, γ represents rotatory displacement or motion in a continuous medium, then p, q, r represents linear displacement or relative motion in that medium.
(5) If α, β, γ represents the rotational velocities (angular velocities) of the vortices in Maxwell's theory, then p, q, r represents the averaged linear flow density of the idle-wheel particles postulated in that theory.

In these examples, α, β, γ represents a rotary motion about some axis, while p, q, r represents an associated linear motion. Maxwell's conclusion from the two sets of instances was that equations (1) in general represent a kind of relationship that obtains 'between certain pairs of phenomena, of which one has a *linear* and the other a *rotatory* character'; if α, β, γ is linear, then p, q, r is rotatory, and if α, β, γ is rotatory, then p, q, r is linear (Maxwell 1890, 1:502–3, emphasis Maxwell's).

The content and character of Maxwell's argument can be highlighted by comparison and contrast with a modern treatment of the same issue. Modernly, and in fact building on Maxwell's continuing work in this area (Maxwell 1890, 2:257–66, especially 263), one maintains his distinction between two kinds of vectors: Maxwell's linear quantities are now denoted true vectors or polar vectors, and Maxwell's rotatory quantities are now designated pseudovectors or axial vectors. The distinction between these two kinds of vector quantity is, however, established in a modern treatment on the basis of their transformation characteristics — in particular with respect to spatial reflection — rather than by reference to mechanical examples (Morse and Feshbach 1953, 1–44). Maxwell, conversely, relied exclusively on mechanical examples to establish the distinction and to illustrate the relationship between the two kinds of vectors, as in equations (1). Maxwell was, after all, a mechanical philosopher: not a mechanical philosopher of the eighteenth-century type, with their qualitatively distinct subtle matters, but a characteristic British mechanical philosopher of the post-1850 period, who espoused an ontology of matter and motion, in which

Maxwell's electromagnetic theory

'all matter must in itself be the same, and can be modified only by differences of arrangement and motion and by being actuated by different systems of force'. Given this ontology, which Maxwell discussed, for example, in his Inaugural Lectures at Marischal College, Aberdeen, and King's College, London (1856 and 1860, respectively), it followed that 'Electrical and Magnetic Science', and indeed all 'Physical Sciences', 'treat[ed] of ... phenomena ... depending on conditions of matter', conditions of 'arrangement and motion and ... being actuated by different systems of force' (Jones 1973, especially 73). The reality underlying the electromagnetic field thus had to be mechanical, and the equations describing electromagnetic relationships, ultimately, were the expression of mechanical conditions. Mechanically, equations (1), by virtue of their mathematical structure, had to represent the relationship between a linear mechanical motion or force and a rotational motion or torque. Either magnetism was rotational and electric current linear, or magnetism was linear and electric current rotational. The mathematical relationship of equations (1), taken together with the mechanical ontology, guaranteed this; all that remained was to make a decision between the two possibilities.

The way to tell whether a given phenomenon had linear or rotatory character, Maxwell suggested, was to look at its effects: 'All the direct effects of any cause which is itself of a longitudinal character, must themselves be longitudinal, and ... the direct effects of a rotatory cause must themselves be rotatory.' Maxwell proceeded to inventory the effects of electric currents, to judge whether they were to be classified as linear or rotatory in character. In the first place, Maxwell observed, 'Electric currents are known to produce effects of transference in the direction of the current.' In the electrolysis of water, for example, hydrogen is moved in one direction along the current line, and oxygen in the other; this indicates a linear character for the electric current, and there was no known rotational effect of electric current (Faraday had in fact searched for such an effect but found none), so that Maxwell felt confident in characterising the electric current as linear. Given the previous argument concerning equations (1), this result was sufficient to demonstrate not only that electric current was linear in character, but also that magnetism was rotatory; any additional information bearing directly on the rotatory character of magnetism would then introduce a reassuring redundancy into the argument (Maxwell 1890, 1:503–4).

As directly concerned the nature of magnetism itself, Maxwell first argued that magnetism produced no known linear effects. (The magnetic lines of force and their actions on magnetic poles appear to be linear, but, as Maxwell argued, some phenomenon like electrolysis in the electrical case, where the opposite ends of the line of action are physically distinguished, is needed to establish that the line is not merely an axis of rotation; no such phenomenon, however, had been found for the magnetic line of force. This is a subtle point, which still bedevils students.) Magnetism did, however, produce a rotatory effect, namely, the Faraday rotation — 'the rotation of the plane of polarized light when transmitted along the lines of magnetic force'. This, of course, was the phenomenon that had formed the basis for the whole line of thinking that Maxwell was pursuing. Maxwell acknowledged Thomson's role in 'point[ing] out that the cause of the magnetic action on light must be a real rotation going on in the magnetic field'. Also invoked was Thomson's argument concerning the handedness of the Faraday rotation, leading to the conclusion that 'the direction of rotation is directly connected with that of the magnetic lines, in a way which seems to indicate that magnetism is really a phenomenon of rotation' (Maxwell 1890, 1:504–6).

What Maxwell had added to Thomson's original argument for the reality of magnetic rotations was a fourfold redundancy: given Maxwell's argument concerning the mathematical character of equations (1) and their associated mechanical significance, four kinds of evidence converged on the conclusion that magnetism was rotational:

(1) linear effects of electric current, as in electrolysis;
(2) lack of rotatory effects of electric current, as generally established and further tested in Faraday's experiments;
(3) lack of linear effects of magnetism, as generally established and as supported by Maxwell's argument concerning the lack of any physical distinction between the two ends of a magnetic line of force; and
(4) rotatory effects of magnetism, as experimentally established by Faraday and interpreted theoretically by Thomson.

Maxwell added further redundancy to the argument in favour of the rotational character of magnetism by enlisting even the views of the action-at-a-distance theorists — namely André-Marie Ampère and Wilhelm Weber — to the effect that electric currents involved linear transport of electric charge, whereas magnetism was a manifestation of current loops (Maxwell 1890, 1:505–6).

The strength and weakness of this argument in Part IV of 'Physical lines' for the rotatory character of magnetism was its generality, its independence of the specifics of the theory of molecular vortices. The details of the theory of molecular vortices were not supported by this argument, and where those details were questionable — as in the case of the idle-wheel particles — they remained questionable. On the other hand, as this argument was not tied to those details, it could retain its general force for Maxwell even when he backed away from the details. In fact, as will become apparent in the sequel, Maxwell never did give up the belief that there was 'a real rotation going on in the magnetic field'. Given the mechanical ontology, the mathematics of linear and rotatory vector quantities, and the experimental evidence concerning the linear and rotatory character of electric currents and magnetic fields, respectively, the conclusion that there was 'a real rotation going on in the magnetic field' was simply unavoidable for Maxwell (Maxwell 1890, 1:505).

Maxwell's electromagnetic theory

The decline of the theory of molecular vortices

After the publication of 'Physical lines', Maxwell began a measured retreat from the mechanical concreteness and detail that characterised his presentation of the theory of molecular vortices there. His next major paper on electromagnetic theory, 'A dynamical theory of the electromagnetic field' (1864–65), exemplified this trend in his thinking. Clearly, dissatisfaction with the weakest link in the theory — the idle-wheel particles — was a crucial reason for this retreat. There were, however, other reasons, having to do with the broader development of Maxwell's research programme: his research on gas theory, in the 1860s, took a direction that had negative implications concerning molecular vortices; and his concern to develop the electromagnetic theory of light in a manner that would be acceptable to a broader audience led also to a de-emphasis on molecular vortices. The result of these was to redirect Maxwell's efforts in electromagnetic theory towards a more phenomenological emphasis, without, however, engendering either a return to the analogical approach or a complete loss of faith in the reality of molecular vortices (Maxwell 1890, 1:526–97).

The original stronghold of the hypothesis of molecular vortices had been the theory of heat and gases, and one of the most attractive features of Thomson's suggestion, in 1856, that the hypothesis of molecular vortices be applied to electromagnetic phenomena, was the broad unification this promised. In the later 1850s, however, most notably as a result of the work of Rudolph Clausius, the idea that the motion of gas molecules giving rise to gas pressure was translatory — characteristically in straight lines — rather than rotatory began to look more and more appealing. Maxwell's initial response to this situation, $c.$ 1860, was to maintain his primary allegiance to molecular vortices, employing them as the basis for a physical theory in 'Physical lines', while utilising the linear-motion picture as the basis of a physical analogy in gas theory, with no commitment to it as a realistic representation of nature. In this way, conflict between the respective requirements of electromagnetic theory and gas theory was minimised. This accommodation, however, proved to be unstable (Maxwell 1890, 1:377–409; Brush 1965; Siegel 1983, 416).

In the course of the 1860s, Maxwell's investment in and commitment to the linear picture — that is, to what has been ever since the standard kinetic theory of gases — increased substantially. Exhibiting the same kind of progression from an analogical stage to a theoretical stage that we have already seen in his electromagnetic theory, Maxwell in 1866 published what he was prepared to call a 'Theory of gases'; once again, just as in the electromagnetic case, the step to the theoretical stage was justified by the comprehensiveness and explanatory character of the given mechanical representation. The methodological progression in Maxwell's gas theory thus mirrored quite faithfully the methodological development of his electromagnetic theory; as far as content was concerned, however, the direction in which Maxwell's gas theory was developing in the first half of the 1860s tended to undermine the foundations of his electromagnetic theory. Not that there was any direct conflict between the gas theory and the electromagnetic theory — molecular vortices in the ether were perfectly consistent with translational motion of gas molecules — but the grand synthesis on the basis of molecular vortices as originally envisaged by Thomson now appeared to be ruled out, and some of the appeal of applying molecular vortices to electromagnetic phenomena was thereby lost (Maxwell 1890, 2:26–78, especially 27; Everitt 1975, 95).

In part, it was the very success of the theory of molecular vortices that led to its downfall. This theory had provided the context for a unified treatment of electromagnetism and optics on the basis of the mechanics of one universal medium or ether. The further development and consolidation of this unification of electromagnetism and optics then became the new focus of Maxwell's continuing research. On the experimental side, he became quite productively involved in work on electrical measurements and standards, oriented towards the more precise determination of the ratio of electrical units, on the basis of which the connection with light had been established. On the theoretical side, he worked towards the establishment of a more direct connection between electromagnetism and optics. The original connection between the two, as established in 'Physical lines', has aptly been designated an 'electromechanical' theory of light, rather than an electromagnetic theory of light (Bromberg 1967a, 219): in 'Physical lines', Maxwell had argued from electromagnetic phenomena, by way of the theory of molecular vortices, to the mechanical properties of the magnetoelectric medium; then, from the mechanical properties of that medium, Maxwell had deduced the propagation of transverse elastic waves in it, with all the characteristic properties of light waves. From the outset, however, Maxwell had had the feeling that a more direct and theoretically parsimonious establishment of this result should be possible, and a more direct argument, 'cleared ... from all unwarrantable assumption', would certainly be more palatable to the Continental action-at-a-distance electricians, who were sceptical of the whole Faraday–Thomson–Maxwell approach. Maxwell was in fact successful in devising such an argument, which proceeded directly from the electromagnetic equations — as appropriately modified — to the calculation of electromagnetic waves propagating at the velocity of light. This, finally, was truly an 'electromagnetic theory of light', and it formed the centrepiece of Maxwell's paper 'On a dynamical theory of the electromagnetic field'.[8]

In this paper, then, in consonance with the exigencies of his research programmes in both gas theory and electromagnetic theory, Maxwell retreated from the specifics of the theory of molecular vortices, but not from the general framework. He still insisted on the existence of a mechanical medium in space, which was both the carrier of light waves and the seat of electric and magnetic fields; he was still willing to propose, as a 'very probable hypothesis', that magnetic and electric fields were manifestations respectively of 'motion' and 'strain' in that medium; and he still judged that the Faraday rotation gave

Maxwell's electromagnetic theory

'reason to suppose that th[e] motion [underlying the magnetic field was] one of rotation, having the direction of the magnetic force as its axis'. Beyond this Maxwell was not willing to go, and he did not even use all of this in developing the mathematical theory. Basically, all that Maxwell used was the equations of electromagnetic phenomena as established by experiment, together with the assumption that these reflected conditions in a connected mechanical medium pervading space and capable of storing, exchanging and transmitting kinetic and potential energy. This warranted Maxwell in treating the field variables (and other electromagnetic variables) as generalised mechanical variables, in the sense of the Lagrangian formalism. The result was a 'dynamical theory' of the electromagnetic field, which was still a mechanical theory, but abstract and general rather than concrete and pictorial.[9]

Molecular vortices in the *Treatise on Electricity and Magnetism*
Maxwell's primary methodological commitment in the later 1860s and 1870s was to the dynamical approach, which was fully developed and centrally positioned in the *Treatise on Electricity and Magnetism* (1873). The *Treatise*, however, was intended to be a comprehensive work, treating all aspects of electromagnetic phenomena, and some of these were not amenable to a treatment by macroscopic dynamical theory, without any assumption as to mechanical details. In particular, an entire chapter was devoted to the Faraday rotation (important because of its bearing on the electromagnetic theory of light), and here molecular vortices played a central role. Maxwell began his analysis — characteristically for the *Treatise* — by 'consider[ing] the dynamical conditions[s]' attendant upon the Faraday rotation, that is, by applying the Langrangian formalism; he arrived thereby at the following by now familiar result, stated in the abstract terminology characteristic of that formalism: 'the consideration of the action of magnetism on polarized light leads ... to the conclusion that in a medium under the action of magnetic force something belonging to the same mathematical class as an angular velocity, whose axis is in the direction of the magnetic force, forms a part of the phenomenon'. He then proceeded to interpret this result in the light of his mechanical ontology: given a universe of matter and motion, the 'something belonging to the same mathematical class as an angular velocity' must in fact *be* an angular velocity of some rotating portion or portions of the medium filling space. Experiment indicated that no sizeable angular momenta were associated with these rotations, so the rotating portions of the medium had to be small, and the conclusion was that 'we must therefore conceive the rotation to be that of very small portions of the medium, each rotating on its own axis'. 'This', once again, was 'the hypothesis of molecular vortices'; Maxwell used the hypothesis basically in this unadorned form, not fleshing it out as he had in 'Physical lines'.[10]

In a final summary of his ultimate views concerning the reality of molecular vortices, Maxwell again invoked the distinction that he had made in 'Physical lines' between the status of the vortices themselves — they had been a 'probable' hypothesis — and the status of the system of idle-wheel particles that coupled the motions of the vortices — this had been a 'provisional,' 'temporary,' and 'awkward' hypothesis. Concerning the vortices themselves, the 'probable' hypothesis, Maxwell now had this to say: 'I think we have good evidence for the opinion that some phenomenon of rotation is going on in the magnetic field, that this rotation is performed by a great number of very small portions of matter, each rotating on its own axis, this axis being parallel to the direction of the magnetic force, ...'. Concerning this part of the theory of molecular vortices, then, Maxwell was fully as sanguine in the *Treatise* as he had been in 'Physical lines'; his language in the *Treatise* — 'I think we have good evidence for the opinion that' — was if anything a bit stronger than in 'Physical lines' (Maxwell 1890, 1:468, 486; Maxwell 1873a, 2:415–17).

Concerning the *existence* of a mechanism coupling the motions of the individual vortices, there was also good evidence: 'I think we have good evidence ... that the rotations of the ... different vortices are made to depend on one another by means of some kind of mechanism connecting them'. The *particular* connecting mechanism envisaged in 'Physical lines' — that is, the system of idle-wheel particles — was, on the contrary, not to be taken seriously:

> The attempt which I then made to imagine a working model of this mechanism must be taken for no more than it really is, a demonstration that mechanism may be imagined capable of producing a connexion mechanically equivalent to the actual connexion of the parts of the electromagnetic field. The problem of determining the mechanism required to establish a given species of connexion between the motions of the parts of a system always admits of an infinite number of solutions.

This agnostic statement notwithstanding, Maxwell observed that certain things were more definitely known even about the connecting mechanism: 'Electromotive force arises from the stress on the connecting mechanism[, and] electric displacement arises from the elastic yielding of the connecting mechanism.' Thus, certain *general* mechanical properties of the connecting mechanism were known, along with their relationships to electromagnetic phenomena (Maxwell 1873a, 2:416–17).

An infinite number of different mechanisms could be imagined that would fulfil these specifications, and Maxwell clearly did not entertain hopes that one could ever determine which of these was the 'actual connexion' existing in nature. The effort that Maxwell had made in 'Physical lines' to envisage a concrete example of such a mechanism had not, however, been a worthless exercise: it had provided a 'demonstration that mechanism may be imagined capable of' fulfilling the given specifications. Such a concrete mechanism, not realistically intended, but intended instead to show that a mechanism of the sort required was *possible*, was called by Maxwell a 'working model'. Now such a working model is similar to a physical analogy in that it is a concrete and

Maxwell's electromagnetic theory

pictorial mechanical representation, with imaginary rather than realistic status. There is, however, an important difference: the working model must be able to *produce* the mechanical effect in question — that is, it must be a model that really *works*, in the sense of accomplishing the effect. The working model furnishes a possible explanation of the effect, just because it is able to produce the effect. One may judge it highly improbable that the given working model faithfully represents the details of the actual situation, either because the working model is manifestly awkward or artificial, or simply because one knows that there are an infinite number of possible working models, so that the *a priori* probability that a given one is the true one is vanishingly small. Nevertheless, the working model is 'capable of producing' the observed effects, and in that sense represents a *possible* explanation. A physical analogy, however, will in general not represent even a possible explanation, because the mechanical system envisaged is not capable of producing the phenomenon in question. (In the formal language introduced earlier, a working model is described by variables $\{f_j\}$ that are forces, whereas a physical analogy is not.) Thus, even as concerned the connecting mechanism — the weakest part of the theory of molecular vortices — Maxwell was not retreating back to the physical analogy stage.[11]

To sum up Maxwell's final position concerning the theory of molecular vortices, he regarded part of the theory as a hypothesis for which 'we have good evidence', and part of the theory as a 'working model'. In order to get a comprehensive theory of electromagnetic phenomena, one had to put the two parts together; evaluating the resultant theory by its weakest link, one would have to characterise the whole as merely a 'working model'. Maxwell, however, chose to maintain the separation of the two parts in his characterisation of the status of the theory, thus highlighting the strong and continuing commitment he had to the reality of the central core of the theory, the vortices themselves. Beyond this, Maxwell judged certain general mechanical features of the theory to be firmly established throughout, including both the vortices and the connecting mechanism:

> The following results of the theory, however, are of higher value:
> (1) Magnetic force is the effect of the centrifugal force of the vortices.
> (2) Electromagnetic induction of currents is the effect of the forces called into play when the velocity of the vortices is changing.
> (3) Electromotive force arises from the stress on the connecting mechanism.
> (4) Electric displacement arises from the elastic yielding of the connecting mechanism (Maxwell 1873a, 2:416–17).

Conclusion

Maxwell remained faithful to the ontology of matter and motion, and to the hypothesis of molecular vortices, to the end. A new articulation of the mechanical worldview, based on a new form of the hypothesis of molecular vortices, was developed by Thomson in 1867 and the years following: material atoms were to be regarded as nothing but patterns of vortex motion — closed and knotted vortex filaments — in a universal material substratum, and the luminiferous and magnetoelectric ether was to be regarded as the locus of linearly extended vortex filaments in the universal substratum. A programme based on Thomson's vortex atom thus promised full realisation of the mechanical programme of the British natural philosophers of the second half of the nineteenth century, in which 'all matter must in itself be the same, and can be modified only by differences of arrangement and motion and by being actuated by different systems of force' — thus Maxwell in his Aberdeen Inaugural. Maxwell therefore endorsed the programme based on Thomson's vortex atom with the greatest enthusiasm in his article 'Atom' for the ninth (1875) edition of the *Encyclopaedia Britannica*: 'The difficulties of this method are enormous', he observed, for horrendous calculations in fluid mechanics were involved, 'but the glory of surmounting them would be unique'. A nineteenth-century mechanical philosopher speaks here, and the achievements of Maxwell's mathematical physics came in the context of his sincere belief in, and development of, the mechanical ontology. There were rich and interesting variations and nuances in his use of mechanical representations and equations, and some of these were to have significance for the development of the demechanised mathematical physics of the twentieth century, but Maxwell himself remained a nineteenth-century mathematical physicist.[1,2]

Acknowledgements

Conversations with my colleagues at the conference that gave rise to this volume, especially Peter Harman, and with my students in a recent seminar on Maxwell at the University of Wisconsin, including Michael Boersma and Derrick Mancini, have been of great value to me.

Notes

1. The Scottish Common Sense nexus is discussed especially in Olson 1975; see also Everitt 1975, 87–8, and Harman 1982*b*, 130–4. An alternative view concerning the philosophical roots of Maxwell's use of analogies was presented in Paul Theerman, 'Maxwell and method: Cultural resonances in the philosophy of physics', History of Science Society Annual Meeting, Norwalk, October 1983; and in a recent seminar of mine on Maxwell at the University of Wisconsin, Michael Boersma called attention to the emphasis on analogical thinking by James David Forbes, Maxwell's teacher in natural philosophy at the University of Edinburgh. For an overview of Bacon to Herschel, see, for example, Blake, Ducasse and Madden 1960. Further, on Maxwell and analogy see his 'Are there real analogies in nature?' (1856) in Campbell and Garnett 1882, 235–44. For a recent survey of the literature on Maxwell and analogies, see Wise 1982. Cf. Siegel 1975, 364–7; 1981, 239–44.
2. Thomson 1856, 151; Rankine 1881, 16–48, especially 16–18; Thomson 1877*a*, 208–26, on 224–5; Siegel 1981, 244–6.
3. My thinking about representations, analogies, models and theories has been much facilitated by my reading in Achinstein 1968, 203–25.

Maxwell's electromagnetic theory

4. Maxwell 1890, 1:468, 485, 485–6n. Maxwell to Faraday, 19 October 1861, in reprint of Campbell and Garnett 1882, xx–xxii, on xxii (appended by the editor from the second, 1884, edition). Maxwell 1873a, 2:202–4. Everitt 1975, 106–8. The experiments are to be discussed in greater detail, utilising additional evidence from manuscript materials, in Paul Theerman, 'James Clerk Maxwell and the mechanical philosophy: optics and electrodynamics in the 1860s'.
5. Maxwell 1890, 1:468–71, 486, and Plate VIII facing page 488. Finding the phrase 'molecule of electricity' useful in a discussion of electrolysis in the *Treatise*, Maxwell noted that 'this phrase, gross as it is, and out of harmony with the rest of this treatise' nevertheless had heuristic value (Maxwell 1873a, 1:312).
6. Maxwell 1890, 1:488; Larmor 1936, 728–9; Everitt 1975, 98–9; Siegel 1981, 250–1.
7. Maxwell 1890, 1:489–90; reprint of Campbell and Garnett 1882, xx–xxii; Larmor 1936, 728–9. Cf. Maxwell 1873a, 2:383; Bromberg 1967a, 227–9.
8. Everitt 1975, 99–101; Bromberg 1967a, 227–30; reprint of Campbell and Garnett, 1882, xxii, 329, 340; Maxwell 1890, 1:577–80. Cf. Maxwell's 'Note on the electromagnetic theory of light' (1868), Maxwell 1890, 2:137–43.
9. Maxwell 1890, 1:529–30, 564. The meaning and significance of 'dynamical theory' in the nineteenth century is a large issue, treated only in passing herein, where the central concern is with Maxwell's use of concrete and pictorial mechanical representation rather than with his avoidance of it through the use of abstract and generalised dynamics. Concerning dynamical theory, see Siegel 1981 and the works cited there; cf. Wise 1982 and the works cited there. Maxwell did make use of a macroscopic mechanical analogy for inductive circuits, which was in some sense — although not the usual sense — a mechanical representation, beginning in 'Dynamical theory' and fully developed later (Everitt 1975, 102–5).
10. Maxwell 1873a, 2:399–417, especially 406, 408; see, in addition, a much condensed version of the redundancy argument of Part IV of 'Physical lines', in Maxwell's 'On the mathematical classification of physical quantities' (1870), Maxwell 1890, 2:257–66, on 263. Cf. Knudsen 1976, especially 273–81, and Bromberg 1967b, 145.
11. Maxwell 1873a, 2:416–17. Cf. the discussion of 'imaginary models' and the contrast with 'analogue models' in Achinstein 1968, especially 218–25.
12. Maxwell 1890, 2:466–77, on 472 and 792; Siegel 1981, 254–64; cf. Bromberg 1967b, 145.

= 8 =

Edinburgh philosophy and Cambridge physics: the natural philosophy of James Clerk Maxwell

P.M. HARMAN

Introduction
Ever since his lifelong friend and biographer Lewis Campbell drew attention to Clerk Maxwell's close regard for his philosophy professor at Edinburgh, William Hamilton, and to Maxwell's approval of Hamilton's Logic class as '[by] far the most solid' he was attending in his first year at Edinburgh University in 1847 (Campbell and Garnett 1882, 108, 116), commentators have drawn attention to peculiarly Scottish features in Maxwell's mature scientific achievement. It can hardly be doubted that Hamilton did indeed stimulate Maxwell's lifelong interest in the general problems of philosophy, in metaphysics and ethics; and there are evident traces in Maxwell's early scientific writings of the traditional Scottish philosophical approach which came to him by way of Hamilton, notably Maxwell's development of the principle that knowledge involves the comparison of experiences, in his emphasis on the role of mathematical and physical analogy in the enunciation of scientific theories (Davie 1964, 192–4).

The extent of the influence of Scottish philosophy in shaping Maxwell's scientific thought should not, however, be exaggerated. It has been claimed that Maxwell's scientific development was determined by philosophical regulative principles, that 'the changing theoretical bases of Maxwell's successive papers on electricity and magnetism depended upon a conscious attempt to vary his theoretical focus ... [which] demonstrates the psychologizing tendency of the whole Scottish Common Sense tradition' (Olson 1975, 295–7). Maxwell's distinction between a physical model of the electromagnetic field and a mechanical theory grounded on the Lagrangian formalism of dynamics has been interpreted as being derived from Hamilton's distinction between empirical knowledge obtained from perception and *a priori* knowledge originating in the modes of our consciousness: that for Maxwell 'dynamical thinking is certain, necessary and rooted in the very nature of our thought processes', providing a 'mechanical theory, whose structure is known *a priori*' (Simpson 1970, 255, 258).

The natural philosophy of James Clerk Maxwell

In contesting claims that the fundamental principles of Maxwell's physics were derived from the categories of Hamilton's metaphysics, I will argue that Maxwell appeals to philosophical argument to provide a *justification* for his physics. My aim in this essay is to clarify the foundational status of geometrical, analogical and dynamical principles in Maxwell's mathematical physics, and to analyse some of the issues that arise from any claim that Maxwell's science was shaped by Hamilton's metaphysics. Maxwell's abiding concern to establish the conceptual rationale of his physics by an appeal to philosophical argument, which distinguishes his work from that of his older contemporaries William Thomson and George Gabriel Stokes, does suggest the influence of Hamilton's Edinburgh lectures in nurturing and forming his distinctive intellectual style. But his appeal to geometrical representation, his emphasis on the role of physical analogies, and his commitment to the fundamental status of dynamical explanation in physics have their source in the physical and mathematical problems of elasticity, hydrodynamics, heat, electricity and the ether in the work of Stokes and Thomson, who, like Maxwell, were Cambridge-educated mathematicians, and whose work of the 1840s shaped Maxwell's style of mathematical physics in an enduring fashion.

There are therefore two connected themes in this chapter. The first is concerned to chart the status of philosophical argument (specifically Hamilton's metaphysics) in Maxwell's physics; the second theme is concerned with the relation between Maxwell's mathematical and physical modes of representation of nature, an issue which was fundamental to the mathematical physics of Stokes and Thomson. These two themes are connected, because much of the case for Maxwell's supposed indebtedness to Scottish philosophy rests on the claim that he gave primacy to geometry and in this echoed the attitudes of Scottish philosophers (Davie 1964, 192). The clarification of the relation between Hamilton's metaphysics and Maxwell's science is therefore of considerable importance for an understanding of Maxwell's thought, for it relates to the problem of characterising the threads of mathematical, physical and philosophical argument which make up the distinctive weave of Maxwell's natural philosophy.

In contesting exaggerated claims of the regulative status of Hamilton's ideas in shaping Maxwell's physics, I will, however, emphasise that Maxwell's construal of philosophical issues, and his mode of expression, are frequently indebted to styles of argument which were probably imbibed from Hamilton's Edinburgh lectures. Hamilton cannot, however, be regarded as the sole source for Maxwell's philosophy. Maxwell was philosophically sophisticated and notably well read, and any account of the status of philosophical principles in his scientific thought must recognise the breadth of influences which helped to shape his reasoning. I will note the influence of the writings of the Cambridge scholar William Whewell (who was Master of Trinity during Maxwell's time at the College) in shaping some of Maxwell's philosophical arguments, notably on the relation between theory and reality and on the status of mechanical laws, thus questioning interpretations which stress the unique and formative influence of Hamilton's Edinburgh lectures, and of the Scottish intellectual milieu, on Maxwell. While Maxwell's physical and mathematical work cannot be interpreted as being derived from any overarching metaphysical programme obtained from either Hamilton or Whewell, their arguments probably lie behind his concern to provide a justification of his physical and mathematical worldview by appeal to philosophical argument.

Maxwell's discussions have their source in physical and mathematical problems which shaped his arguments in fundamental ways, and cannot be interpreted as being derived from prior metaphysical commitments. The attempt to trace Maxwell's emphasis on geometrical–physical modes of representation to the stress by Scottish philosophers and mathematicians on the primacy of geometry rests on a misconception. Following Colin Maclaurin, Scottish philosophers and mathematicians argued for the foundational status of a geometrical theory of limits for the fluxional calculus, arguments that Maxwell does not repeat. These foundational issues must, however, be construed as quite distinct from Maxwell's espousal of a geometrical–physical approach to mathematical physics, where the influence of Stokes and Thomson is apparent.

The thesis that Maxwell's appeal to a physical geometry of lines of force and to a dynamical explanation of the electromagnetic field rests on the pillars of presumed metaphysical foundations will therefore be challenged. This interpretation misrepresents the status of philosophical principles in Maxwell's physics; ignores the breadth of his philosophical sources; misrepresents the basis of his appeal to geometrical argument; and fails to characterise the depth and subtlety of his application of dynamical principles to physical phenomena.

Edinburgh: geometry and physics

Maxwell's first scientific paper, written when he was a schoolboy of fourteen, was devoted to a geometrical subject, an extension of the theory of conic sections. Entitled 'Observations on circumscribed figures having a plurality of foci, and radii of various proportions', and published in abstract by J.D. Forbes, shortly to be Maxwell's professor in the natural philosophy class at Edinburgh University (Maxwell's father having shown Forbes the manuscript), Maxwell's work was based on the method of drawing an ellipse by means of a cord of any given length being fixed by the ends to the foci. Maxwell argued that just as the sum of the focal distances of an ellipse was a constant quantity, the essential condition for all circumscribed figures was that the sums of the focal distances were constant, and that circumscribed figures could have any number of foci and focal distances of various proportions. Maxwell went on to demonstrate the construction of curves by wrapping thread around pins, producing curves with up to five foci and varying the distances of the foci. As

The natural philosophy of James Clerk Maxwell

J.D. Forbes observed in his published abstract of the paper, Descartes had described a mechanical method of describing ovals, but — Forbes added — Maxwell's new method was an improvement (Maxwell 1890, 1:1–3). Forbes drew attention to the optical analogies to the geometrical properties of ovals as discussed by Newton in *Principia*, a treatment of the surfaces which cause light diverging from one focus to converge to another focus (Newton 1972, 1:344–5), and in a later manuscript Maxwell himself sketched curves (circles, ovals, hyperbolas and ellipses) converging rays (Campbell and Garnett 1882, 97,104).

These very early papers illustrate Maxwell's interest in geometrical problems and their physical implications, which was to lead to a stream of publications on aspects of geometry throughout his life. These include his geometrical treatment of the calculation of stresses in frameworks (1864), his extension of the method of reciprocal diagrams to continuous media (1870), and his paper 'On the cyclide' (1867), a study of the geometry of wave surfaces as the foci of lenses, based on a treatment of the surfaces generated as the envelope of a sphere which touches three fixed spheres continuously. Maxwell illustrated the four forms of cyclides he analysed by stereoscopic figures (Maxwell 1890, 2:144–59).

Maxwell's mathematical method in these papers does of course contrast with the approach in the youthful study of ovals, which was based on geometrical construction. The paper on the cyclide exhibits a sophisticated application of analysis to geometry, the use of equations to represent and analyse the relations between the geometric variables of curves and surfaces. Maxwell had already acquired a knowledge of analytical geometry while at Edinburgh University, certainly by 1848,[1] and these methods were fundamental to his paper 'On rolling curves' (1849), an analytical treatment of the geometry of the cycloidal curves generated by one curve rolling on another. The study of cycloidal curves, as Maxwell was aware, goes back to the Greeks, and he himself referred to work by de la Hire and Euler among others, though not to Newton's discussion of cycloids in *Principia* (Newton 1972, 1:245–51; see Whiteside 1967–81, 6:387), as well as to the discussion of the cycloid in Robert Willis' recent *Principles of Mechanism* (1841), a work which he studied during his Edinburgh years (Campbell and Garnett 1882, 137). The paper is notable for its systematic analytical and geometrical treatment of the cycloid.

In Maxwell's third paper, again dating from his Edinburgh years, 'On the equilibrium of elastic solids' (1850), his first paper on a physical topic, as in his later major paper 'On Faraday's lines of force' (1856), his method was based on the elaboration of a mathematical formalism which represented the phenomena, but which was not grounded on a physical hypothesis about the constitution of material substances. Maxwell wished to emphasise the gap between mathematical representation and a physical theory of the constitution of elastic solids. Nevertheless, the paper developed important physical ideas, an investigation of the phenomenon of induced double refraction in strained glass, leading to a development of Brewster's method for determining the strain in an elastic solid by means of the colour-fringes produced when the solid is viewed by polarised light. Maxwell demonstrated that the pressures acting in the interior of transparent elastic solids could be determined by photoelastic techniques. He exhibited coloured patterns which provided a graphical solution of stress distributions, arguing that the axes of pressure corresponded to the axes of double refraction (Maxwell 1890, 1:30–73).

The mathematical method which Maxwell employed is of especial significance. The method is presented as a development of George Gabriel Stokes's paper 'On the theories of the internal friction of fluids in motion, and of the equilibrium and motion of elastic solids', presented to the Cambridge Philosophical Society in 1845. Stokes had based his theory of the elasticity of solids on the assumption of the independence of the elasticity of shape and volume of solids, claiming that his was an axiomatic approach based on a 'geometrical' model which assumed the independence of the shape and volume of a system of points moving in an element of space (Stokes 1880–1905, 1:80). Stokes thus emphasised the independence from physical hypothesis of his theory of elasticity (Wise 1982, 186). Similarly, Maxwell rejects any theory of an elastic solid based upon physical assumptions about 'the laws of molecular action' of elastic solids (Maxwell 1890, 1:71). Maxwell rejects the central force theories of the Laplacian school, which sought to explain elasticity by a 'physical mechanics' (Poisson 1829, 361) grounded on molecules acting at a distance, in favour of a theory derived from two 'axioms' which he claims are the 'results of experiments'. These experiential axioms state relations between the pressure and compression of elastic media, pressure being defined in terms of the ratio of units of weight to the unit of surface, and compression as the proportional change of any dimension of the solid caused by pressure. The equations, Maxwell states, 'agree with all the laws of elasticity which have been deduced from experiments' (Maxwell 1890, 1:31).

The mathematical method here possibly reflects Maxwell's reading of Fourier as well as Stokes; Maxwell had acquired a copy of Fourier's *Théorie analytique de la chaleur* (1822) in 1849 (Campbell and Garnett 1882, 134). Fourier had set the study of heat in the tradition of rational mechanics, avoiding physical hypotheses and deriving his theory from empirical laws of the temperature distribution in bodies. These 'primary facts' (Fourier 1878, 6) could be tested by experiment, and were subjected to mathematical analysis, leading to the formulation of differential equations that characterised the transmission of heat, equations that were independent of all physical hypotheses. This was a method analogous to Maxwell's mathematical theory of elastic solids, which was based on experiential axioms relating pressure and compression, neither of these concepts being defined in terms of a physical hypothesis about the molecular structure of elastic media.

The natural philosophy of James Clerk Maxwell

Cambridge: 'mixed mathematics' and physical analogy

Maxwell's papers on rolling curves and elastic solids, written during his years at Edinburgh University, were congruent in approach with the work of the Cambridge mathematicians Willis and Stokes, which belies any claim that Maxwell's early mathematical work was shaped by Scottish cultural values. Maxwell's early work is congruent with the geometry and 'mixed mathematics' of contemporary Cambridge mathematics (Becher 1980*a*). At Edinburgh, Maxwell had attended classes in mathematics and natural philosophy with Philip Kelland and Forbes (both prominent spokesmen of Cambridge mathematical education), who fostered his early papers. When Maxwell entered Peterhouse (before migrating to Trinity) in 1850, he arrived at Cambridge as an accomplished mathematician. His considerable private reading, of Fourier, Monge, Cauchy, Poisson and Willis, among others, had clearly provided him, to quote his friend Peter Guthrie Tait, with 'a mass of knowledge which was really immense for so young a man, but in a state of disorder appalling to his methodical private tutor ... William Hopkins' (Campbell and Garnett 1882, 133).

By 1850 the reforms in Cambridge mathematical education of the Analytical Society, formed in 1812 by Babbage, Herschel and Peacock, who had attempted to divorce the foundations of analysis from geometric and physical representations, had been considerably modified. The Analytical Society had aimed to replace Newtonian fluxions, justified (as by Maclaurin) by the method of first and last ratios of nascent and vanishing finite quantities approaching their limits. The reformers aimed to avoid both infinitesimals and the concept of the limit by appealing to Lagrange's theory of analytical functions in which the derivatives of a function are defined as the coefficients of the terms in its expansion in a Taylor series. Although the Cambridge analysts failed to keep up with continental developments on the rigorous definition of limits, notably the limit-avoidance arguments of Cauchy, in the 1830s and 1840s analytic geometry, algebra, and calculus texts which incorporated continental techniques dominated the work for the Mathematical Tripos. Maxwell's own Cambridge undergraduate notes on the differential calculus reflect these developments. A discussion of Taylor's theorem leads to the definition of a derivative as the coefficients of the terms in the Taylor series, and to a further definition of a derivative by a limiting ratio, yet without a definition of limits, reflecting a partial absorption of Cauchy's methods.[2]

In the late 1840s under Whewell's influence the Mathematical Tripos was reformed, its analytical focus being replaced by a more traditional concern with geometry and 'mixed mathematics', the problems of mathematical physics, here construed as the mechanics deriving from *Principia* but including geometrical and physical optics though excluding topics of current physical research such as electricity and heat (Becher 1980*a*). Maxwell's undergraduate notes reflect this

P.M. Harman

concern with 'mixed mathematics'. Notebooks on mechanics, hydrodynamics, astronomy and optics — these latter topics being given special emphasis by his tutor William Hopkins (see Wilson's chapter in this volume) — give evidence of Maxwell's study of the mathematical aspects of these physical subjects.[3] In 1854, the year Maxwell took the Mathematical Tripos (graduating Second Wrangler), 'mixed' physical questions outnumbered 'pure' analytic questions by three to two, with some emphasis on geometric problems (Becher 1980a, 42). In the Smith's Prize examination, set by Stokes (which included a question that required a proof of Stokes's theorem), where Maxwell was placed equal Smith's Prizeman, the questions had a similar orientation (Stokes 1880–1905, 5:320–2).

The Mathematical Tripos provided Maxwell with a systematic mathematical education but hardly discouraged his interest in physical and geometrical problems, issues which were of central concern to Cambridge mathematicians. It is no surprise to find Maxwell informing William Thomson in February 1854 that he wished 'to attack Electricity', a subject excluded from the Tripos but which was open to incorporation within 'mixed mathematics', nor that he was also engaged in a geometrical study of the bending of surfaces (Larmor 1936, 697).

Maxwell's paper 'On the transformation of surfaces by bending' (1856) represented the bending of surfaces by 'lines of bending', and there are clear links between this paper and his geometrical representation of 'lines of force' in his first paper on the theory of the electromagnetic field 'On Faraday's lines of force' (1856). He gave a geometrical discussion of the bending of surfaces, without reference to the elasticity and other mechanical properties of material lamina. Maxwell defines bending as 'a continuous change of the form of a surface, without extension or contraction of any part of it', and he introduced the concept of 'lines of bending' to provide a representation of the change of form of a surface. Describing a surface as 'the limit of the inscribed polygon', Maxwell supposes that the lines of bending are formed by the edges of the inscribed polygons, and determine the forms of the polygons, and hence provide a representation of the kind of bending performed on the surface (Larmor 1936, 699; Maxwell 1890, 1:81, 87).

On writing to Thomson in November 1854 about his developing interest in the science of electricity, Maxwell observed that 'I have heard you speak of "magnetic lines of force" & Faraday seems to make great use of them'. Maxwell declared that he aimed to explain magnetism as 'a property of a "magnetic field" or space' by 'developing the geometrical ideas [of lines of force] according to this view' (Larmor 1936, 701–2). His attempt to formulate 'a geometrical model of the physical phenomena [of electricity]' in his paper 'On Faraday's lines of force' was no doubt in part suggested by the analogy between lines of bending and lines of force. Lines of force provide a geometrical model of the electromagnetic field. Thus drawing lines of force to represent the direction of the force acting, 'till we had filled all space with curves indicating by their direction that of the force at any assigned point', would provide a geometrical model of lines of force based on propositions about lines and surfaces, and did not imply any physical hypothesis about the nature of electrical action (Maxwell 1890, 1:158). In a draft of the paper, Maxwell observes that 'Faraday treats the distribution of forces in space as the primary phenomenon, and does not insist on any theory as to the nature of the centres of force round which these forces are generally but not always grouped'.[4] The contrast between a purely geometrical representation of forces and a physical hypothesis about the nature of electricity echoes his contrast, in his early paper on elastic solids, between the axiomatic method (grounded on experiential concepts) and the appeal to physical hypotheses based on a theory of molecular action, in theorising about elastic media.

Maxwell's mathematical method in 'On Faraday's lines of force' again suggests his reading of Fourier. Indeed, Maxwell emphasises Thomson's demonstration of the 'analogy between the formulae of heat and [electrical] attraction' (Maxwell 1890, 1:157). The analogy showed that a mathematical formalism analogous to the mathematical theory of heat distribution of Fourier's analytical theory of heat could be employed to provide a representation of electrostatic attraction. Writing to Thomson in September 1855, Maxwell acknowledged the importance of Thomson's 'allegorical representation of the case of electrified bodies by means of conductors of heat' in shaping the development of his theory of lines of force (Larmor 1936, 711).

The crucial mathematical concept in Fourier's theory of heat flow is his elaboration of a differential equation for the diffusion of heat. Fourier considered that the conduction of heat was the result of the communication of heat between the molecules of a substance, but he did not base his mathematical theory on a physical model of radiating molecules. The flux of heat was related to the temperature difference between molecules; assuming that heat was conserved during its flow, Fourier obtained the 'continuity equation' (derived by Euler for the flow of a fluid), a differential equation relating the flow of heat and the temperature gradient between contiguous molecules. In the conduction of heat in solid bodies, heat was communicated between contiguous points in the solid, and the flow of heat was represented by a differential equation which expressed empirical observations about the temperature gradient, rather than being calculated by integration, as the summation of heat radiated at a point by numerous radiating molecules (Fourier 1878, 454; Wise 1979; Harman 1982*a*, 27–30).

This model of heat flow had important implications for the field theory of electromagnetism that Maxwell was seeking to illuminate. The essence of Faraday's concept of the 'field' was to suppose that the forces between bodies were *mediated* by some property of the ambient space or field. The field could be characterised as a field of force and represented in terms of the spatial distribution of forces (as in Faraday's theory of lines of force); or the mediating

P.M. Harman

property of the field could be characterised in terms of an intervening ether or medium (as in Faraday's theory of action between contiguous particles or Maxwell's famous ether model developed in his paper 'On physical lines of force' of 1861–62), the ether particles being supposed as acting only on neighbouring contiguous particles. In his *Treatise on Electricity and Magnetism* (1873), Maxwell argues that the mathematical language of partial differential equations provides an expression of the theory of the physical field. The equations of the electromagnetic field involve quantities that are continuous functions of their variables, and these equations express the continuous propagation of force or energy between neighbouring infinitesimal elements of space in the field. Similarly, in Fourier's theory of heat, the conduction of heat in a solid is characterised in terms of the flux of heat between contiguous infinitesimal elements of the solid, and is represented by differential equations of heat flow. Thus, Maxwell observes that 'the integral, therefore, is the appropriate mathematical expression for a theory of action between particles at a distance, whereas the differential equation is the appropriate expression for a theory of action exerted between contiguous parts of a medium'. In this latter method, characteristic of a field theory, 'the partial differential equations are supposed given, and we have to find the potential and the distribution of electricity' (Maxwell 1873a, 1:98–9).

William Thomson's paper 'On the uniform motion of heat in homogeneous solid bodies and its connexion with the mathematical theory of electricity', written when Thomson was an undergraduate at Peterhouse and published in the *Cambridge Mathematical Journal* in 1842, provided a theory of electrostatics which employed a mathematical formalism analogous to the mathematical theory of heat distribution of Fourier's analytical theory of heat, a method later amplified in his exploration of the analogy with fluid flow (see Knudsen's chapter in this volume). The distribution of electricity is represented by a flux of electric force, analogous to the representation of heat by a flux of heat. The continuity equation for fluid flow could therefore be applied to the theories of heat and electricity: the continuity or conservation of the flux of heat implied the conservation of the flow of electric force. The analogy is one of mathematical form, rather than a physical analogy between thermal conduction and electrostatic attraction. The analogy is geometrical, between the flux of heat and the flow of electric force, flowing continuously across isothermal or equipotential surfaces. The differential equations for the flux of heat and the flow of electric force (continuity equations) thus express a system of flow lines, but the analogy with fluid flow was merely geometrical, not physical (Thomson 1872a, 1–14; see Wise 1981). As Maxwell observes in a draft of his paper 'On Faraday's lines of force', 'while the mathematical laws of the conduction of heat derived from the idea of heat as a substance are admitted to be true, the theory of heat has been so modified that we can no longer apply to it the idea of substance'.[5]

Maxwell's theory of lines of force is based on a mathematical theory of an incompressible fluid moving in tubes formed by lines of force. This is not intended as a physical representation of the field; the fluid was 'not even a hypothetical fluid' but 'merely a collection of imaginary properties'. The value of a 'geometrical model of the physical phenomena' over a purely analytical formalism was that geometrical propositions would present 'the mathematical ideas to the mind in an embodied form, as systems of lines or surfaces, and not as mere symbols'. The appeal to a model of the flow of an incompressible fluid provides the basis for a 'physical analogy', as Maxwell terms it, the 'partial similarity between the laws of one science and those of another which makes each of them illustrate the other'. Bearing in mind Thomson's analogy between heat and electricity, the 'purely geometrical idea of the motion of an imaginary fluid' provided the basis for a 'physical analogy', a visual, geometrical representation, of lines of force. The appeal to a physical analogy was 'more applicable to physical problems than that [method] in which algebraic symbols alone are used'. Maxwell emphasised that he was not proposing a physical hypothesis about the constitution of the field; his theory was not intended as 'a mature theory, in which physical facts will be physically explained', and he declared that the 'use of the word "Fluid" will not lead us into error, if we remember that it denotes a purely imaginary substance' (Maxwell 1890, 1:156–60, 187).

As Maxwell's drafts show, the argument of the paper 'On Faraday's lines of force' was first formulated as a purely hydrodynamical study, a treatment of the equations of motion of an incompressible fluid.[6] Writing to Thomson in September 1855, Maxwell remarked that his theory was 'in itself a collection of purely geometrical truths embodied in geometrical conceptions of lines, surfaces &c', a physical geometry of 'lines of force &c which may be *afterwards* applied to Electricity, Heat or Magnetism' (Larmor 1936, 711). Maxwell thus developed a mathematical theory of fluids, based on a concept of 'lines of fluid motion' (Maxwell 1890, 1:175) which, construed as a physical analogy, he then applied to the science of electricity: the change in pressure at a point in a fluid corresponds to the electrical attraction exerted by electrical particles on one another, and the intensity of electric force is represented by the velocity of the fluid in a tube of force.

The hydrodynamical analogy is especially significant. Little of the mathematical argument in Maxwell's drafts, concerning the conditions of stability of an incompressible fluid, remained in the published paper, but the hydrodynamical argument is clearly shaped by his undergraduate study of the pressure, motion and equilibrium of fluids.[7] Maxwell established that the mathematical theory of electricity could be incorporated within the explanatory framework of 'mixed mathematics'.

'On Faraday's lines of force' thus represents the culmination of Maxwell's early scientific work. The approach followed, the development of a non-hypothetical mathematical theory rather than a physical hypothesis, is consistent with the method employed in his youthful paper 'On the equilibrium of elastic solids', a method which was especially prominent in the papers of the 1840s by Stokes and Thomson, work which shaped the form of Maxwell's mathematical physics. Maxwell's theory of lines of force, grounded on the mathematical theory of fluids (on which Stokes lectured at Cambridge), is rooted in Cambridge 'mixed mathematics'. The geometrical representation of lines of force, enabling differential equations to be represented by lines, surfaces and tubes, is congruent in approach with the renewed stress on geometry in the Mathematical Tripos in the 1850s.

The problem of Scottish philosophy
Maxwell's method of physical analogy was directly indebted to Thomson's work on the analogy between electrostatics and heat, though Forbes's stress on analogical physics may well have shaped Maxwell's outlook. Claims that Maxwell was influenced by Scottish philosophy rest, however, on the assumption that his appeal to geometrical representation and physical analogy displays the influence of the lectures delivered at Edinburgh by William Hamilton. In his lectures Hamilton emphasised that science could only aspire to characterise relations between phenomena, and that analogical argument was the key to the understanding of such relations. There are indeed some correlations between Hamilton's ideas and Maxwell's arguments, but these connections require further analysis, in relation to problems Maxwell probably derived from Whewell and Maclaurin, for their significance to be gauged.

In a class essay 'On the properties of matter' (written between 1848 and 1849), which Hamilton himself preserved, Maxwell developed Hamilton's and Dugald Stewart's discussions of the primary qualities, extension and solidity, as 'mathematical affections of matter' (Hamilton 1859, 2:112). Maxwell argued that the 'geometric properties of matter', the properties of position, extension and figure, were more fundamental than those properties which did not 'belong both to matter and to imaginary geometrical figures'. Maxwell went on to claim that the geometric properties were 'forms of thought and not of matter' (Campbell and Garnett 1882, 109–10). In his *Lectures* (as later published) Hamilton argued that extension is 'one of our necessary notions, — in fact, a fundamental condition of thought itself'. Hamilton went on to assert that 'the analysis of Kant ... has placed this truth beyond the possibility of doubt' (Hamilton 1859, 2:113). Although Maxwell did later declare his intention to Lewis Campbell, on arriving in Cambridge in 1850, to read Kant's *Critique of Pure Reason* 'with a determination to make it agree with Sir W. Hamilton' (Campbell and Garnett 1882, 135), any direct influence of Kant on the shaping of Maxwell's views on the status of geometry can be discounted.

Maxwell consistently emphasises the status of mathematics, concerned with the realm of thought, as contrasted with physics, concerned with fact and reality. The relationship between thought and things had been discussed by

The natural philosophy of James Clerk Maxwell

William Whewell in his *Philosophy of the Inductive Sciences* (1840, 2nd edn 1847). Maxwell was familiar with Whewell's philosophy, certainly by 1855 (Campbell and Garnett 1882, 215), and in an essay 'Are there real Analogies in Nature', which he read to the Cambridge Apostles' Club in February 1856, he sought to resolve the problems raised by Whewell in discussing the 'fundamental antithesis of philosophy' as Whewell termed it (Whewell 1847, 1:16), by appealing to the relationship between mathematical and physical representation.

The argument of this early essay is of especial interest and relevance because it has clear connection to the philosophical discussion in Maxwell's paper 'On Faraday's lines of force'. The mathematical embodiment of physical concepts does not, in Maxwell's view, imply that physical concepts are thereby granted an *a priori* status. Mathematical physics unified the poles of geometry, concerned with geometric figures which (as he had expressed it in his essay for Hamilton) were 'forms of thought', and the elaboration of physical models. The geometric structure of crystals leads Maxwell to conclude, in his essay on 'Analogies', that 'the only laws of matter are those which the mind must fabricate, and the only laws of mind are fabricated for it by matter', an argument that may reflect Whewell's view that thoughts and things were inseparable components of knowledge. Maxwell emphasises the relation between thoughts and things in terms of the link between geometrical models and physical phenomena. In Maxwell's view this link was to be expressed in terms of geometrical analogies of physical phenomena. Maxwell places emphasis on the importance of analogical reasoning: 'although pairs of things may differ widely from each other, the *relation* in the one pair may be the same as that in the other. Now, as in a scientific point of view the *relation* is the most important thing to know, a knowledge of one thing leads us a long way towards a knowledge of the other' (Campbell and Garnett 1882, 243–4). This construal of the role of analogical argument, amplifying his published discussion of the role of physical analogies in 'On Faraday's lines of force', no doubt does echo Hamilton's view that knowledge involves the comparison between experiences.

Maxwell's statement on analogies as relational seems also to be a development of a discussion by Colin Maclaurin in his *Treatise of Fluxions* (1742). Maclaurin had claimed that:

> the mathematical sciences treat of the relations of quantities to each other . . . [hence] . . . we enquire into the relations of things rather than their inward essences. Because we may have a clear conception of that which is the foundation of a relation, without having a perfect or adequate idea of the thing it is attributed to, our ideas of relation are often clearer and more distinct than those of the things to which they belong, and to this fact we may ascribe, in some measure, the peculiar evidence of the mathematics. (Maclaurin 1742, 1:51–2.)

Maclaurin's argument helped to shape the doctrine of Scottish philosophers that knowledge was relational, and probably influenced Maxwell's discussion of analogies (directly or indirectly); but Maclaurin's most fundamental claim in his book, of the foundational status of a geometrical theory of limits for the fluxional calculus, which was of crucial importance for the emphasis by Scottish philosophers and mathematicians on the primacy of geometry (Davie 1964, 138–47), is nowhere echoed by Maxwell. Maxwell's debt to specifically Scottish philosophical and mathematical traditions of geometry would therefore seem to be limited in scope.

Maxwell's insistence on the disjunction between mathematical and physical theory is in the spirit of Newton's emphasis in *Principia* that in the first two books of his treatise he was concerned merely with 'quantities and mathematical proportions' (Newton 1972, 1:298); as Maclaurin expresses it, 'it is the business of geometry to enquire into the measures, rather than unfold the hidden essences of things' (Maclaurin 1742, 1:54). This distinction between mathematical representation and physical hypothesis was characteristic of the work of Maxwell's scientific mentors Stokes and Thomson, and its role in Maxwell's mathematical physics does not imply the unique influence of Scottish philosphy. The distinctive feature of Maxwell's work is his concern to provide a justification of his mathematical practice by an appeal to metaphysical argument; and the role of Hamilton's lectures on Maxwell's intellectual development is apparent from the kind of philosophical argument that Maxwell deployed, the appeal to the relational character of analogical argument.

Geometry and metaphysics: the 'Dublin Hamilton'

In commenting on the systems of thought of the Edinburgh philosopher William Hamilton and the Dublin mathematician William Rowan Hamilton, in a letter to Mark Pattison in April 1868, Maxwell observed that 'the Edinburgh & the Dublin Hamilton differ in their metaphysical power in the direct ratio of their physical knowledge (not with the inverse as most people suppose)'. Maxwell himself therefore does not seem to have granted high status to the philosophy of his Edinburgh professor. In Maxwell's view, as he informed Pattison, 'the practical relation of metaphysics to physics is most intimate', because 'metaphysicians differ from age to age according to the physical doctrines of the age and their personal knowledge of them'.[8] While ascribing some merit to the metaphysical speculations of William Rowan Hamilton, Maxwell was disinclined to regard philosophical argument as shaping the enunciation of physical theory. The role of metaphysical foundations was to provide justificatory sanction for a physical and mathematical worldview.

Maxwell's view of the limited role and status of philosophical argument in relation to physical theories is apparent in his discussion of William Rowan Hamilton's mathematics; for despite his assessment of the metaphysical profundity of Hamilton's work, Maxwell showed no sympathy with Hamilton's philosophical doctrines. Maxwell's interest in Hamilton's mathematical work is most apparent in his discussion of Hamilton's calculus of quaternions. Maxwell was impressed with quaternions as a method of providing the conceptual framework for a geometrical representation of physical phenomena. The calculus of quaternions expressed the direction of physical quantities in space, providing the geometrical embodiment of algebraic symbols, a mode of representation which Maxwell had sought in 'On Faraday's lines of force'. Maxwell was not, however, attracted by Hamilton's philosophical justification of quaternions, and was critical of the application of the symbolism of quaternions to physical questions.

Hamilton's calculus of quaternions developed from his earlier work on algebra in the 1830s. Hamilton was concerned with the concept of number and the definition of complex numbers, attempting to give an algebraic definition of complex numbers and seeking to base algebra on the ordinal character of numbers. In his 'Essay on algebra as the science of pure time' (1835), Hamilton aimed to establish algebra as a 'SCIENCE, *in some sense* analogous to Geometry'. Appealing to Kant's concept of space and time as forms of intuition, Hamilton claims that 'the intuition of TIME is such a rudiment' to provide a basis for considering algebra a science, because the 'notion or intuition of ORDER IN TIME is not less but more deep-seated in the human mind, than the notion or intuition of ORDER IN SPACE', and hence algebra was 'not less an object of *a priori* contemplation than Geometry' (W.R. Hamilton 1931–67, 3:4–7). While this view of the *a priori* status of geometry is in accord with Maxwell's statement (in 1848) that geometric figures were 'forms of thought', and while Maxwell once jotted down the phrase 'Algebra as the science of Time' among a draft list of topics in mathematics and dynamics,[9] he did not pursue these philosophical issues in his discussions of mathematical reasoning, or indeed in his development of Hamilton's quaternion methods.

From his study of complex numbers, Hamilton had aimed to extend the complex number system to three dimensions, leading in 1843 to his invention of quaternions, hypercomplex numbers with one real and three (imaginary) complex parts. Hamilton interpreted the three imaginary numbers as 'vectors' directed along three mutually perpendicular lines in space. The fourth and real part of the quaternion corresponded to a line in space of only one dimension: this was therefore the 'scalar' part of the quaternion. Quaternions were therefore considered as the sum of their own vector and scalar parts, and the quaternion calculus '*selects no one direction in space as eminent* above another, but treats them all as equally related to that *extra-spatial* or simply SCALAR direction' (W.R. Hamilton 1931–67, 3:355–9). Quaternions thus provided a form of analytical geometry which required no prior selection of coordinates. Hamilton devoted much attention to the interpretation of the meaning of quaternions conceived as the sum of quantities of different dimensions. In his papers 'On symbolical geometry' (1846–49) and in his *Lectures on Quaternions* (1853) Hamilton appealed to an argument that he termed 'partly geometrical, but partly also metaphysical', elaborate geometrical illustrations of the algebraic properties of quaternions (W.R. Hamilton 1931–67, 3:145; see Hankins 1980, 311–15).

In a review of Kelland and Tait's *Introduction to Quaternions* (1873) Maxwell drily observed that Hamilton's tendency 'to become fascinated with the metaphysical aspects of the method' led the student of quaternions to become 'impressed with the profundity, rather than the simplicity of his doctrines' (Maxwell 1973*b*, 137). Maxwell was interested in the application of the concept of vectors to mathematical physics, their use in the geometrical representation of physical quantities. By 1870 he had become seriously interested in quaternions, making some use of these methods in his *Treatise on Electricity and Magnetism* (1873), encouraged by Tait. In his paper 'On the mathematical classification of physical quantities' (1871), Maxwell introduced the terms 'convergence' and 'curl': 'convergence' represented the effect of a vector function 'in carrying its subject inwards towards a point', and 'curl' represented 'the direction and magnitude of the rotation of the subject matter carried by the vector' (Maxwell 1890, 2:265; see Crowe 1967, 127–39). Maxwell was interested in the conceptual role of quaternions for the geometrical representation of physical quantities. He emphasised the value of quaternions as a 'mathematical method ... of thinking', because 'it calls upon us at every step to form a mental image of the geometrical features represented by the symbols' (Maxwell 1873*b*, 137). Vectorial representation was valuable as a method of representing directed quantities in space: 'the invention of the calculus of Quaternions', that is the fundamental 'ideas of this calculus, as distinguished from its operations and symbols', was 'a step towards the knowledge of quantities related to space which can only be compared with the invention of triple coordinates by Descartes' (Maxwell 1890, 2:259). In a letter to Tait in December 1871 he went further, contrasting quaternions and Cartesian coordinates: 'the one is a flaming sword which turns every way; the other is a ram, pushing westward and northward and (downward?)' (Knott 1911, 150).

In the *Treatise* Maxwell points out that the physical quantities of electrodynamics, the electric and magnetic forces, the electrotonic state (which he renamed the vector potential), and the displacement current were vector quantities, and he urges the use of vectorial representation as a means of fixing 'the mind at once on a point of space ... and on the magnitude and direction of a force'. This vectorial 'mode of contemplating geometrical and physical quantities is more primitive and natural' a method than the use of Cartesian coordinates. This geometrical method provided a direct representation of physical quantities congruent with 'physical reasoning' (Maxwell 1873*a*, 1:8–9), involving 'the continued construction of mental representations' (Maxwell 1873*b*, 137). The use of vectors thus satisfied Maxwell's desire to provide a geometrical representation of physical phenomena, a geometrical embodiment of algebraic symbols. While Maxwell appeals to criteria of spatial and physical

The natural philosophy of James Clerk Maxwell

simplicity, his use of philosophical argument is intended only to provide justificatory sanction for vectors.

The fundamental science of dynamics

The mathematical and physical method that dominates Maxwell's mature work was the application of mechanical principles to physical phenomena, a programme of mechanical explanation that was of seminal importance in nineteenth-century physics (Klein 1972; Harman 1982*a*). This method received its clearest expression in his work on the theory of the electromagnetic field. The method of his paper 'On physical lines of force' (1861–62) was foreshadowed in his indication, in 'On Faraday's lines of force', that a physical representation of lines of force required the formulation of a 'mechanical conception' of the ambient ether as the substratum of the field (Maxwell 1890, 1:188). In 'On physical lines of force' Maxwell advances from his early discussion of the physical geometry of lines of force to a treatment of the electromagnetic field 'from a mechanical point of view'. Maxwell's physical model of vortex tubes and 'idle-wheel' particles (Maxwell 1890, 1:452, 468), suggested by William Thomson's explanation of the action of magnetism on polarised light in terms of the rotation of molecular vortices (Heimann 1970*a*, 188–9; see the chapter by Knudsen in this volume), and by C.W. Siemens' use of differential gears in his governor for steam engines (Mayr 1971), provides a mechanical analogy of the electromagnetic field in which electromagnetic quantities had mechanical correlates (the angular velocity of the vortices corresponding to the intensity of the field, the electrotonic state being represented as the rotational momentum of the vortices).

Maxwell emphasises that his mechanical model of idle-wheel particles was only a 'provisional and temporary' hypothesis; he observes that the model might appear 'awkward', but states that he did 'not bring it forward as a mode of connexion [in the ether] existing in nature'. The supposition of a 'mechanically conceivable' model of the ether demonstrated the possibility of a mechanical explanation of the field, a theory which explained the mediation of action in the field by the postulation of a mechanism of ether particles (Maxwell 1890, 1:486). Faraday's discovery of the rotatory effect of magnetism on polarised light indicated, as he observes in the *Treatise*, that 'some phenomenon of rotation is going on in the magnetic field' and therefore implied the rotation of molecular vortices and the existence of 'some kind of mechanism' (Maxwell 1873*a*, 2:416; see Heimann 1970*a*, 190). Thus Maxwell's model of idle-wheel particles in 'On physical lines of force' was merely suggestive and illustrative (Siegel 1981 and his chapter in this volume; Harman 1982*a*, 89–93).

In a letter to Tait in December 1867, Maxwell remarks that his ether model was 'built up to show that the phenomena can be explained by mechanism. The nature of the mechanism is to the true mechanism what an orrery is to the Solar System' (Knott 1911, 215). The mechanical model of ether particles was therefore purely illustrative, and also demonstrated the possibility of a mechanical explanation of electromagnetism. In the *Treatise* he emphasises that his earlier attempt 'to imagine a working model of this mechanism [of the ether] must be taken for no more than it really is, a demonstration that mechanism may be imagined capable of producing a connexion mechanically equivalent to the actual connexion of the parts of the electromagnetic field' (Maxwell 1873*a*, 2:416–17). The mechanical ether model had the status of an analogy, like the analogy of the flow of an incompressible fluid which provided a geometrical model of lines of force. The mechanical analogy provided an embodiment of a mathematical representation, a link between the physical concept of the electromagnetic field (here expressed mechanically) and a mathematical formalism.

In his 'A dynamical theory of the electromagnetic field' (1865), Maxwell develops a theory of the field from analytical equations of mechanical systems, without employing a specific mechanical model to represent the structure of the ether. This method of analytical dynamics assumed that 'motion is communicated from one part of the [ethereal] medium to another by forces arising from the connexions of those parts', and he now disclaims any mechanical model of 'a particular kind of motion and a particular kind of strain' in the ether (Maxwell 1890, 1:532–3, 563). As he explains in the *Treatise*, there was in principle no limit to the number of mechanical models that could be proposed: 'the problem of determining the mechanism required to establish a given species of connexion between the motions of the parts of a system always admits of an infinite number of solutions' (Maxwell 1873*a*, 2:417). Maxwell's appeal in the *Treatise* to the Lagrangian form of analytical dynamics is constrained by his continued commitment to the aim of providing a physical expression of the symbols of analytical dynamics. His dynamical theory of the electromagnetic field assumes that electromagnetic phenomena were produced by particles of 'matter in motion' constituting 'a complicated mechanism capable of a vast variety of motion' (Maxwell 1890, 1:533), and in the *Treatise* he stresses his aim of formulating a dynamical theory of the field which would emphasise the primacy of concepts of momentum, velocity and energy. He therefore declares that he seeks to translate the results of the Lagrangian method 'from the language of the calculus into the language of dynamics, so that our words may call up the mental image, not of some algebraical process, but of some property of moving bodies' (Maxwell 1873*a*, 2:185).

To achieve this aim in the *Treatise* Maxwell developed the method of dynamical explanation which had been articulated by Thomson and Tait in their *Treatise on Natural Philosophy* (1867). Maxwell contrasts Lagrange's method, which he considers to be a formalism of generalised equations of motion conceived as 'pure algebraical quantities' in a manner 'free from the intrusion of dynamical ideas', with the method of Thomson and Tait which sought 'to cultivate our dynamical ideas' (Maxwell 1873*a*, 2:184–5). Whereas

The natural philosophy of James Clerk Maxwell

the Lagrangian method provided a purely mathematical formalism that avoided reference to the concepts of momentum, velocity and energy after they had been replaced by symbols in the generalised equations of motion, Thomson and Tait placed the emphasis on dynamical concepts. Thomson and Tait's treatment of dynamics was based on a mathematical theorem discovered by Thomson, which relates the variation in a system by impulsive forces (which act in an infinitesimal time-increment) to the kinetic energy of the system, enabling a generalised equation of motion to be derived (Moyer 1977). In Maxwell's view, while this method 'kept out of view the mechanism by which the parts of the system are connected', and hence followed his aim of avoiding the formulation of a mechanical ether model, the method did, however, keep 'constantly in mind the ideas appropriate to the fundamental science of dynamics', and hence satisfied his criterion for dynamical explanation (Maxwell 1873a, 2:193–4; see Harman 1982a, 93–8). Concepts of energy, momentum and velocity were fundamental to Maxwell's aim to form a 'consistent representation' of the field (Maxwell 1873a, 2:435), an expression he derived from a remark by Gauss on the need to form a 'constructible representation [*construirbare Vorstellung*]' of the manner in which the propagation of electric action takes place (Gauss 1863–1933, 5:629; see Heimann 1970a). This required the expression of 'dynamical ideas from a physical point of view', and an emphasis on the link between the mathematical formalism and the physical reality depicted, ensuring that 'we must have our minds imbued with these dynamical truths as well as with mathematical methods' (Maxwell 1873a, 2:184, 194).

In seeking to provide an exposition of dynamics from a physical point of view in the chapter 'On the equations of motion of a connected system' in the *Treatise*, Maxwell seeks to develop a new framework for dynamical theory. His argument here should be considered in relation to a draft 'On the interpretation of Lagrange's and Hamilton's equations of motion', where he observes that 'our popular dynamical ideas are far too exclusively drawn from the dynamics of a particle ... [and that] it is unfortunate that in expressing the relations between these ideas we have sometimes adopted a form of expression, which, though true for a particle, is not easily applicable to a connected system [of particles]'.[10] As he explains in the *Treatise*, in attempting to achieve this objective Maxwell favours the mode of expressing kinetic energy in terms of the velocity and momenta of the particles constituting a material system, obtaining $\dot{q} = dT_p/dp$, where 'the velocity corresponding to the variable q is the differential coefficient of T_p [the kinetic energy expressed in terms of the coordinate q and momentum p] with respect to the corresponding momentum p'. Thus 'the velocities and momenta, depend on the actual state of motion of the system at the given instant, and not on its previous history' (Maxwell 1873a, 2:189). This is the mode of expressing kinetic energy due to William Rowan Hamilton, and Maxwell obtains an equation derived by Hamilton for the impressed forces acting on the material system:

P.M. Harman

$$F_r = \frac{dp_r}{dt} + \frac{dT_p}{dq_r}$$

In the draft Maxwell notes that the first term on the right 'expresses the fact that part of the force is expended in increasing the momentum', and 'the second term indicates that if the increase of the variable q has a direct effect in increasing the kinetic energy, a force will arise from this circumstance'. Maxwell prefers this Hamiltonian form of expressing the equation of motion to the alternative form due to Lagrange. In the Lagrangian form, momenta are expressed in terms of the velocities, obtaining an expression for kinetic energy in terms of the velocities and the variables q (denoted $T_{\dot q}$), and the equation of motion

$$F_r = \frac{dp_r}{dt} - \frac{dT_{\dot q}}{dq_r}$$

Maxwell prefers the form of the equation of motion given by Hamilton because of the physical meaning of the equation, based on the momentum rather than (as with Lagrange) on velocity:

> ... it is not the velocities which obey Newton's law of persevering in their actual state, but the momenta or 'quantities of motion'. Hence if we wish to apply Newton's law we must express the kinetic energy in terms of the momenta and use Hamilton's form of the equations of motion. We then see at once that the second term indicates that if a given displacement has a direct tendency to increase the kinetic energy, the momenta remaining the same, a quantity of work, equal to this increase of kinetic energy is performed by the external force F during the displacement.

The dynamical theory of the electromagnetic field is therefore grounded on concepts of momentum, velocity and energy, and the equations are translated 'into language which may be intelligible without the use of symbols'; thus the 'language of dynamics' would express 'some property of moving bodies'. The physical meaning of the equation of motion is justified by its expression in terms of momentum and hence in its relation to Newton's first and second laws of motion (Maxwell 1873a, 2:185, 194).

The metaphysical foundations of dynamics

Maxwell thus accorded dynamics a fundamental explanatory status, and the analysis of his conception of dynamical explanation provides the necessary basis for understanding his discussion of the philosophical foundations of dynamics. Maxwell argues that dynamical theory is the fundamental mode of representing nature because 'as soon as we know what is meant by the words configuration, motion, mass, and force, we see that the ideas which they represent are so elementary that they cannot be explained by means of anything else'. For this reason a 'dynamical explanation' of a phenomenon in terms of the configuration and motion of a material system was not only a 'complete'

The natural philosophy of James Clerk Maxwell

explanation, for 'we cannot conceive any further explanation to be either necessary, desirable, or possible' (Maxwell 1890, 2:418), but (he added in a draft) 'the process by which physical science is "unified" must' he implied (breaking off abruptly) be a dynamical explanation.[11]

Maxwell does provide some hints about his view of the fundamental status of dynamics in his review of Todhunter's account of Whewell's writings in 1876, where he gives an approving glance at Whewell's discussion of the foundations of mechanics. In Whewell's view, as Maxwell expresses it, 'the fundamental doctrines of mechanics' were suggested by experiment but 'once fairly set before the mind [are] apprehended by it as strictly true' (Maxwell 1890, 2:530). He had developed this point earlier in a review of Thomson and Tait's *Elements of Natural Philosophy* (1873), where he referred to the 'fundamental ideas' of a science — the term is Whewell's (Whewell 1847, 1:66) — the 'modes of thought by which the process of our minds is brought into the most complete harmony with the process of nature' (Maxwell 1890, 2:325). As he had expressed it in his 1870 Address to the British Association, this was the 'hidden and dimmer region where Thought weds Fact' (Maxwell 1890, 2:216). These 'fundamental ideas' would only achieve their perfect form when these ideas were, as in the case of the dynamical theory of the *Treatise*, 'clothed with the imagery ... of the phenomena of the science itself, [not merely with] ... the machinery with which mathematicians have been accustomed to work problems about pure quantities' (Maxwell 1890, 2:325).

Maxwell's appeal to dynamical explanation in the *Treatise* is therefore justified by the claim that dynamical concepts and laws could be expressed as a mathematical formalism and were fundamental, elementary and strictly true. Maxwell does not, however, claim that the concepts of mechanics were innate, known *a priori*, or had the status of the propositions of geometry. In his inaugural lecture at Marischal College, Aberdeen, in 1856, he declared that 'I have no reason to believe that the human intellect is able to weave a system of physics out of its own resources without experimental labour' (Jones 1973, 57). Maxwell's view of the status of dynamics in no sense echoed Hamilton's argument on knowledge as *a priori* and grounded in our consciousness. By contrast, in *Matter and Motion* (1877) Maxwell argues that our conviction of the truth of Newton's first law of motion is strengthened by recognising that its denial could not be conceived:

> Suppose the law to be that a body, not acted on by any force, ceases at once to move. This is not only contradicted by experience, but it leads to a definition of absolute rest as the state which a body assumes as soon as it is freed from the action of external forces. It may thus be shown that the denial of Newton's law is in contradiction to the only system of consistent doctrine about space and time which the human mind has been able to form. (Maxwell 1877, 29.)

Maxwell's argument here again suggests the influence of Whewell's writings.

P.M. Harman

Whewell himself had observed that 'necessary truths are those of which we cannot distinctly conceive the contrary' (Whewell 1847, 1:59). Whewell did not regard necessary truths as logically necessary, but rather that necessary truths are those whose negation cannot be clearly and distinctly conceived. Necessary truths are not innate but are known to be fundamental because 'they give rise to inevitable convictions or intuitions (Whewell 1860, 337). In his inaugural lecture at King's College, London, in 1860, Maxwell echoes Whewell in regarding 'necessary truths' as those 'which the mind must acknowledge as true as soon as its attention has been directed to them', but he does not follow Whewell in seeking to establish the necessity of the laws of motion by an appeal to their *a priori* form.[12] In his essay on the 'Nature of the truth of the laws of motion' (1834) and in his *Philosophy of the Inductive Sciences* (1840) Whewell developed an argument for the *a priori* form of the laws of motion. Whewell's argument shows some parallels to, but also notable differences from, Kant's discussion of the laws of mechanics in the *Metaphysical Foundations of Natural Science* (1786) (Harman 1982*b*, 73–5). Whewell expresses the idea of cause in terms of three axioms of causality, which when applied to causes of motion yield three *a priori* laws of forces; when reformulated in terms of the motion of bodies, these laws in turn generate the Newtonian laws of motion (Whewell 1847, 2:473–94). Whewell's aim is to establish that the laws of motion have an *a priori* form as well as an empirical content (Butts 1965), a thesis that has no counterpart in Maxwell's discussion of the status of the laws of motion. For Maxwell, dynamics was fundamental because its concepts were elementary, and the negation of its laws could not be intelligibly conceived.

Maxwell's application of the equations of analytical dynamics is rooted in his interpretation of the physical meaning of the algebraic formalism of Hamilton's equation of motion, but he is also concerned to establish the conceptual status of his theory of dynamics. In a draft Maxwell claims that an understanding of the laws of motion leads us to form 'an idea of mass as the quantitative aspect of matter which is as necessary a part of our thoughts as the triple extension of space or the continual flux of time'. He goes on, however, to make it clear that the idea of mass is not known *a priori* but is necessary in the sense that it is a fundamental idea (in Whewell's sense) of the science of dynamics:

> I do not think it necessary to enquire whether this is the metaphysical idea of matter, or whether any metaphysician has come to the conclusion that the property commonly called inertia is the fundamental and inseparable property of matter. We may be satisfied with our dynamical reasons for asserting that the mass of a body is a measurable and constant quantity and that for all dynamical purposes a body must be measured by its mass and not by any other property such as its volume, its weight, or its chemical activity.[13]

A deeper understanding of Maxwell's theory of mass as providing the basis for a dynamical theory of matter may be gained by relating these remarks to his

The natural philosophy of James Clerk Maxwell

discussion of the foundations of dynamics in his review of the second edition of Thomson and Tait's *Treatise on Natural Philosophy* (1879). In this review Maxwell argues that the basic concepts of dynamics were to be defined in a manner which was strictly independent of any speculations about the material substratum of physical reality. He declares that the dynamical method avoids consideration of 'the relation of mass, as defined in dynamics, to the matter which constitutes real bodies'. The problem as to whether 'real bodies may or may not have such a substratum [of matter]' is simply not a question which is raised by the dynamical method. The dynamical concept of matter is not conceived as a substantial entity defined by physical properties of solidity, impenetrability and inertia (which Maxwell terms 'the innate depravity of matter'): 'what we sometimes, even in abstract dynamics, call matter, is not that unknown substratum of real bodies, against which Berkeley directed his arguments, but something as perfectly intelligible as a straight line or a sphere' (Maxwell 1890, 2:781). The dynamical concept of matter is not, however, a purely mathematical entity, for it is defined by the dynamical concept of mass, a physical not a geometrical concept.

In a letter to Tait in 1868, Maxwell also refers to Berkeley's critique of the ascription of primary qualities of extension and solidity to the unobservable corpuscles that Boyle and Newton had postulated as the constituents of material reality: 'Matter is *never* perceived by the senses. According to Torricelli quoted by Berkeley "Matter is nothing but an enchanted vase of Circe, fitted to receive Impulse and Energy [Force], essences so subtle that nothing but the inmost nature of material bodies is able to contain them"' (Knott 1911, 195; cf. Berkeley 1948–57, 4:13). While matter as the substratum of real bodies is unknowable, the dynamical theory of physics does not seek to explicate the nature of the substratum of matter but to provide a physical interpretation, grounded on concepts of energy, momentum, mass and velocity, of the equations of motion of a mechanical system (Harman 1982*b*, 140–8). In the *Treatise*, using terms which seem to echo the philosophical expressions of Whewell and Hamilton, Maxwell expounds his dynamical theory of physics. 'The fundamental dynamical idea of matter, as capable by its motion of becoming the recipient of momentum and of energy, is so interwoven with our forms of thought that, whenever we catch a glimpse of it in any part of nature, we feel that a path is before us leading, sooner or later, to the complete understanding of the subject' (Maxwell 1873*a*, 2:181). This was a conception of nature grounded in physical and mathematical practice, not determined *a priori* by regulative metaphysical principles. The chief legacy of William Hamilton's Logic class on the brilliant young James Clerk Maxwell was in Maxwell's concern to provide a justification of his physical and mathematical worldview by appeal to philosophical argument. As Maxwell remarked to Richard Litchfield in March 1858, 'the chief *philosophical* value of physics is that it gives the mind something distinct to lay hold of ...' (Campbell and Garnett 1882, 305).

P.M. Harman

Acknowledgements

I am grateful to the Council of the Royal Society for generously supporting my work on the Maxwell papers, and to the Syndics of the University Library, Cambridge, and the Keeper of Western MSS, Bodleian Library, Oxford, for their kind permission to quote from documents in their keeping.

A preliminary version of this paper was presented at a seminar in Cambridge in February 1983. I am grateful to D.T. Whiteside for his helpful comments on that occasion.

I thank Martin J. Klein for his helpful commentary on the presentation of this paper at the conference on 'Cambridge Mathematical Physics in the Nineteenth Century' in March 1984.

Notes

1. Maxwell's MS Notebook, March 1848, 'Essay: The propositions in Wallace's treatise on the ellipse demonstrated analytically', University Library, Cambridge, Add. 7655/V/d/2.
2. Maxwell's MS Notebook, 'Differential & integral calculus', University Library, Cambridge, Add. 7655/V/m/7.
3. Maxwell's MS Notebooks, University Library, Cambridge, Add. 7655/V/m/5, 8, 9, 10.
4. Maxwell, MS 'On Faraday's Lines of Force', University Library, Cambridge, Add. 7655/V/c/7.
5. MS cited in note 4.
6. Maxwell, MSS 'On the motion of "fluids"' and 'On the steady motion of an incompressible fluid', University Library, Cambridge, Add. 7655/V/c/4 and 5. Maxwell to William Thomson, 15 May 1855, University Library, Cambridge, Add. 7342/M90 (Larmor 1936, 707–10).
7. Maxwell, MS Notebook 'Hydrostatics, Hydrodynamics & Optics', University Library, Cambridge, Add. 7655/V/m/8.
8. Maxwell to Mark Pattison, 7 April 1868, Bodleian Library, Oxford, MS Pattison 56, fols 438r–441v.
9. Maxwell, MS notes on dynamics, University Library, Cambridge, Add. 7655/V/e/15.
10. Maxwell, MS 'On the interpretation of Lagrange's and Hamilton's equations of motion', University Library, Cambridge, Add. 7655/V/e/9.
11. Maxwell, MS 'On the dynamical explanation of electric phenomena', University Library, Cambridge, Add. 7655/V/c/10.
12. Maxwell, Inaugural Lecture at King's College, London, 1860, University Library, Cambridge, Add. 7655/V/h/3.
13. MS cited in note 11.

= 9 =

Modifying the continuum: methods of Maxwellian electrodynamics

JED Z. BUCHWALD

When contemporary students of physics first encounter electromagnetism, they soon learn that it consists of two distinctly different parts. One part deals with an entity — the electromagnetic field — which can transmit energy but which cannot in itself be altered in structure. The other part deals with matter, which can be altered in structure because it is assumed to consist of particles. These particles can be moved bodily by the field, and by radiating they feed energy into the field. The particles can also join in groups to produce effects which we take account of in our electromagnetic equations by forming space and time averages over the microphysical processes. Although we thereby introduce constants and vectors on the large or 'macroscopic' scale for these average actions, we realise that they do not in themselves represent attributes of the electromagnetic field proper, because the field's properties are never altered. The task of electromagnetic theory, in this modern view, is to construct appropriate models of particle groupings which, by averaging over many of them, will generate observable effects. This method — in which one alters the equations that govern matter but never alters the field equations proper — is today considered so fundamental that many students believe it to be the only physically correct way to proceed.

It was not always so. During the last quarter of the nineteenth century, British physicists elaborated a method that permitted them to explain electromagnetic phenomena without recourse to microphysics but required them instead to alter the field equations. In this chapter I shall describe the major characteristics of the method and point out what there is about it that cannot be accepted after the development of electron theory.[1]

Dynamical theory
In recent years several historians of physics have discussed the creation by Maxwell and William Thomson (later Lord Kelvin) of a 'dynamical' approach to physics. They have pointed out that the essence of this technique was to employ Lagrange's equations in ways that permitted one to generate equations for observable processes without employing unobservable or 'hidden' entities.[2]

Jed Z. Buchwald

This approach was strikingly employed by Maxwell in his *Treatise on Electricity and Magnetism* of 1873, which was utilised as their major text on electromagnetism by a group of British physicists in the 1880s and 1890s whom I shall call the 'Maxwellians'.

The 'dynamical' approach to physics relied on two basic postulates. First, it presumed that all processes can be exhaustively described in terms of the energy changes they effect. Secondly, these changes are governed by Hamilton's principle. This principle, as used by the British, asserts the following: given the kinetic (T) and potential (V) energy densities of a substance, its development is governed by the requirement that the path taken by the system between two given states and over a given time interval minimises the integral $\int_1^2 (T - V)dt$. Although neither Maxwell nor William Thomson explicitly isolated these postulates, nevertheless either they, or variant implications of them, are implicit in several of the major treatises of the day — including those by Thomson and Tait on mechanics, the *Treatise on Natural Philosophy* (1867). Moreover, the Maxwellians of the 1880s and 1890s, as we shall shortly see, tacitly presumed both assumptions in almost all of their work.

Consider, as an example, how a British dynamical theorist might generate a theory of an elastic substance which exhibits unusual properties. He would begin with the expressions for the kinetic and potential energy densities of the usual elastic body. When these expressions are fed into Hamilton's principle, two results emerge: first, one obtains the partial differential equations that govern the behaviour of the substance — these equations can also be obtained directly from Lagrange's equations, since the latter are implications of Hamilton's principle; secondly, and of equal importance, one obtains equations that describe what happens at the boundary between two bodies with different values of the constants that appear in the energy expressions. In this way one can solve such problems as the reflection and refraction of light waves.

The major point to grasp about this method is that different energy expressions necessarily lead to different differential equations and to different boundary conditions. Consequently, when one has a substance that exhibits unusual behaviour, the natural procedure is to modify the usual energy expressions and then to follow out the implications of the modifications by inserting the new expressions into Hamilton's principle.

Although this may seem an unexceptional procedure even today, the modern physicist would raise two objections. First, he would ask, from where does one obtain the energy expressions? Secondly, it is in fact possible to proceed in this way, according to modern theory, *only* in circumstances wherein one can assume that the microphysical structure of the body does not extract or emit energy which cannot be accounted for by energy expressions employed in Hamilton's principle. These two (modern) objections did not occur to Maxwellians, essentially for one reason, which marks a major divide between British dynamical theory and modern physics: the Maxwellians tacitly assumed that all

Methods of Maxwellian electrodynamics

processes can be represented by energy functions that represent *continuous* properties.

Again the example of an elastic substance will aid us in grasping this point. According to modern physics, elastic bodies are in fact composed of particles that exert forces on one another. On the large or 'macroscopic' scale we ignore this microphysical structure by averaging over the particles' effects. This enables us to construct macroscopic energy densities, which we can then use in Hamilton's principle. However, we are allowed to proceed in this way only so long as the particles do not absorb or radiate energy in ways which the averaging process would obliterate. This can occur, though it is unlikely to reveal itself in readily observable mechanical processes because these take long periods of time to occcur in comparison with the rapidity of microphysical phenomena. However, the substance's optical properties may depend quite critically upon the periodic behaviour of its microphysical constituents, and here it may not be possible to encompass its observable behaviour in a macroscopic energy function which effectively obliterates this hidden structure. In fact, one cannot in this latter way explain the dispersive behaviour of most bodies because of interactions between the natural vibrations of the body's particles and the optical frequencies that set these particles in motion. Indeed, no modern physicist would think to explain a newly discovered process simply by inventing a continuous energy function for it and then following out the implications of Hamilton's principle — he would not do so because microphysical structure would often preclude such a procedure.

The Maxwellians, by contrast, strongly felt that every process would yield to the deployment of continuous energy functions in Hamilton's principle. Those few processes that did not immediately yield to the method — like dispersion — required more intricate energy expressions, perhaps ones that depended upon frequency. The primary goal of research was therefore thought to be the creation of appropriate energy formulae.[3]

One was then committed to following out whatever the energy expressions, when inserted into Hamilton's principle, implied. Throughout the 1880s and early 1890s this was a benefit and not a liability because the procedure led to important connections between a number of electromagnetic processes that seemed to be otherwise unconnected.

The group of Maxwellians can be grossly defined as those physicists who pursued this kind of science. But we can provide a finer and more complete definition by considering briefly certain other characteristics which they shared as a group. The group existed as a fairly cohesive unit for about a quarter of a century, beginning shortly after the publication of Maxwell's *Treatise on Electricity and Magnetism* in 1873. By the early 1880s it consisted of about forty or so active members.

The majority of these Maxwellians were born between 1850 and 1860 and were educated during the mid to late 1870s. About two-thirds of them had some

Jed Z. Buchwald

connection with Cambridge, either as undergraduates or as Cavendish Laboratory students. At an early stage in their careers all of the Maxwellians studied three central texts: Thomson and Tait's *Treatise on Natural Philosophy* (1867), Maxwell's *Treatise on Electricity and Magnetism* (1873) and Rayleigh's *Theory of Sound* (1877). These three texts together embodied the 'dynamical' approach to physics; Maxwell's text applied certain aspects of that approach to electromagnetism.

Despite the overall intellectual coherence among the Maxwellians, they were not of one mind on all topics — even on such major issues as the use of Hamilton's principle. This principle was singled out as especially important by Cambridge men educated in the 1880s. For them, Hamilton's principle encapsulated in a single statement the essence of dynamics. Yet not every Maxwellian shared the Cambridge enthusiasm for the principle. Heaviside, who was not educated at Cambridge, objected quite strongly to its undisciplined use. He remarked:

> Whether good mathematicians, when they die, go to Cambridge, I do not know. But it is well known that a large number of men go there when they are young for the purpose of being converted into senior wranglers and Smith's prizemen. Now at Cambridge, or somewhere else, there is a golden or brazen idol called the Principle of Least Action. Its exact locality is kept secret, but numerous copies have been made and distributed amongst the mathematical tutors and lecturers at Cambridge, who make the young men fall down and worship the idol.
>
> I have nothing to say against the Principle. But I think a good deal may be said against the practice of the Principle ... ('The principle of least action. Lagrange's equations' (1903) in Heaviside 1893–1912, (:sec. 514.)

Heaviside went on to criticise the principle as 'unnatural' and unnecessary. But he offered in its place a method which shares fully the essential dynamical spirit of the principle. His alternative employed a generalised form of Newton's third law based on continuous energy expressions.

Rather than indicating a deviation from the 'dynamical' consensus among British physicists, Heaviside's unorthodoxy emphasises the deep unity among them. Heaviside refused to use the overly mathematical and unintuitive (as he saw it) Hamilton's principle; but he refused to do so precisely because he felt that the principle obscured dynamical foundations. Moreover, Heaviside felt himself to be a stranger among the Cambridge-educated Maxwellian mass. He remained throughout his life a critical, partly alien force among the Maxwellians. This both intensified his sense of rejection (already present by the mid-1880s) and heightened his self-image as a defender of true Maxwellian orthodoxy. Indeed, Heaviside's work more than any other illuminates the darkest corners of Maxwellian thought. From it and other Maxwellian work we can uncover the commonly accepted principles in electromagnetism.

Methods of Maxwellian electrodynamics

The Electromagnetic continuum

Maxwell's *Treatise on Electricity and Magnetism* has always been considered, even by his immediate British followers, a difficult work. One reason for this difficulty is that the *Treatise* attempts simultaneously to be a comprehensive account of methods in electromagnetic theory as well as to explain Maxwell's own theory. Although Maxwell attempted to separate these two goals from one another, he did not usually succeed in doing so, since his accounts of even quite standard methods are permeated with his theory. The result was that neophytes found the going difficult, and those who had learned electromagnetic theory elsewhere often found it impossible to understand Maxwell. Nevertheless, the *Treatise* embodies a coherent and comprehensive theory; after the first shock of encountering the book as an introduction to electromagnetism was absorbed, British students were able to transform Maxwell's ideas into an intricate, comprehensive and elegant theory.

As presented by Maxwell, electromagnetic theory was based on the assumption that the seat of electromagnetic processes is a continuous medium, or ether, which is governed by the laws of dynamics. Consequently to solve problems in electromagnetism we need only to have expressions for the energy functions of the ether; we do not need to know its true structure, nor do we need to know how changes in the ether's energy functions are brought about.

The ether has certain properties which can be altered by the presence of matter. In particular, it possesses two of immediate significance: specific inductive capacity and magnetic permeability. It is essential to understand that, for Maxwell and for the Maxwellians, these two properties are represented by continuous functions of position. The values of the functions may, and, in the presence of matter, do, change, but the changes are continuous — even though matter, the Maxwellians admit, is particulate. In, for example, a piece of iron, we find that the magnetic permeability is different from what it is outside the iron; the change occurs continuously, even though the iron is built of discrete particles. The iron particles, that is, effect *continuous* changes in ether properties.

The Maxwellians nevertheless did admit that changes in permeability and capacity are due to the effects of material molecules, and that the values we use in macroscopic equations are consequently averages over microscopic effects. However, in their view each material molecule itself effects a *continuous* alteration in the ether's properties. As a result even at the microscopic level continuity is never breached.[4]

The core of Maxwellian theory consisted in its abandonment of the conservation of charge in the previous sense of the phrase. Instead of considering charge to be a special electric substance which can accumulate in bodies, Maxwell treated charge as an epiphenomenon of the field. His concept involved the transformation of energy stored in the ether into material form (as heat). This could occur wherever matter was present — though neither Maxwell

Jed Z. Buchwald

nor the Maxwellians attempted to explain why matter could have this effect, which is represented macroscopically by electric conductivity.[5]

Imagine a region of the ether void of matter but in which an electric field — which stores potential energy in the ether — exists. Place a piece of matter in the region. Since all material substances, according to the Maxwellians, have some conductivity, the region of the ether now also occupied by matter begins to lose the energy stored in the electric field. This energy appears in the matter as heat. The result is the creation of a difference in the values of a certain quantity — the electric displacement, which is the product of inductive capacity by electric field intensity — at the boundary between matter and free ether. This difference represents, at any instant, the electric 'charge' on the boundary.

Maxwellian + and − 'charges' are not individually conserved because there might very well be no charge at all in the universe on these principles. This contrasts with electric fluid theories, which assert only that there may be no *net* charge in the universe — but the individual positive and negative particles of course continue to exist. Nevertheless, Maxwellian theory satisfies the very same charge conservation equation that particle theories satisfy, despite this difference. It can do so because of its conception of electric current.

In Maxwellian theory an electric 'current' is the rate at which a portion of the ether is moving. If the ether is quiescent, then no 'current' exists. Electric 'charge' occurs because the ether has moved or is moving through regions in which the ratio of conductivity to inductive capacity varies from point to point. To link the two processes — charge and current — we assume either that the current generates a magnetic field or else that a changing magnetic field generates an ether shift. Either assumption, well formulated, leads with other field equations to the very same equation of charge continuity satisfied by particle theories. The major point to understand is that, despite this agreement with particle theories, the Maxwellian current properly speaking is *not* the rate of change of charge with time: it only may lead to such a change.

These several ideas are present in Maxwell's *Treatise* but are there obscured by the text's novelty and comprehensive character. They and several other central concepts are clearly evident in the work of Maxwell's British followers. For our purposes six doctrines widely admitted by the Maxwellians are of particular importance:

(1) 'Charge' is nothing but a discontinuity in displacement. 'Current' is nothing but moving ether.
(2) To create a new theory, modify the energy function of the ether.
(3) The effect of matter upon ether is mysterious and must be put off until problems are solved through energy methods.
(4) Electric conductivity is particularly mysterious and has something vague to do with the particulate structure of matter.
(5) Boundary conditions are crucial analytical tools.

Methods of Maxwellian electrodynamics

(6) Mechanical models of the ether are deeply revealing as illustrations of energy exchanges but not as reflections of the ether's true structure.

One could easily expand this list of six but it will suffice for our purposes. The major differences among Maxwellians, as we saw earlier in the case of Heaviside, concerned how one generates differential equations and boundary conditions from energy expressions. Cambridge-educated Maxwellians usually employed Hamilton's principle. Heaviside and several others had alternative methods. But all of these methods presumed the six elementary doctrines outlined above. The best way to see how the system worked is through a concrete example. We shall accordingly examine how the Maxwellians dealt with several intricate phenomena.

The Hall effect

Perhaps the best illustration of the power of Maxwellian theory, and of its difference from electromagnetism after the electron, involves the 'Hall effect'.[6] Discovered in 1879 by the American physicist Edwin Hall, this effect is today thought to demonstrate that the electric current consists of negatively charged, moving particles. The experiment can be easily performed with modern equipment, and Hall's own technique was not fundamentally different from the modern one. One takes a plate, usually of copper, and sends a current across its length. Then one attaches a sensitive galvanometer across the plate's width. Place the device between the poles of an electromagnet, with the field normal to the plane of the plate. With a sensitive galvanometer — readily available today, though not in 1879 (this would be the major difference between Hall's and a modern experiment) — one detects a current across the plate's width only while the field is turned on. This, modern theory argues, directly reveals the deflection of the moving electrons in a magnetic field. Moreover, the direction of the deflection reveals their sign.

Hall was the student of Henry Rowland, who, by the late 1870s, was a convinced Maxwellian, and Hall was reading Maxwell's *Treatise* under Rowland's direction. We do not need to discuss the steps that led Hall to his discovery in order to grasp the important fact that, in his, Rowland's and many other Maxwellians' eyes, the new effect graphically demonstrated *Maxwell's* theory, which rejected electric particles.

Consider how a Maxwellian might interpret Hall's discovery. There were only two possibilities available. The least radical, which was widely received for some time until Hall empirically refuted a theory based upon it, referred the effect to an action of the magnetic field upon the material structure of the metal. According to this theory, Hall had discovered something (and possibly not something very new) about metals stressed by magnetic fields and not something about currents. Hall had discovered only that a magnetic field can alter the material structure of a metal. This interpretation had the advantage of making the effect less than fundamental, thereby avoiding major alterations in Maxwell's equations proper. Nevertheless, this view does not consider the Hall effect to involve an action of a magnetic field upon moving electric particles — as, of course, such things do not exist in Maxwellian theory.

The second possibility, which both Hall and Rowland championed from the first, makes the new discovery deeply significant for Maxwellian theory. The reasoning is quite simple. If, as Hall argued, he had discovered a new way to produce an electric current (and not a new way to stress metals), then he had necessarily also discovered a new way to produce an electric field, since in Maxwellian theory an electric current always requires an electric field. The next question was precisely what conditions generated this new field; about this there was much room for discussion.

By 1880 Hall and Rowland had come to the following understanding. Whenever an electric current exists in the presence of a magnetic field, a subsidiary electric field also exists which is at right angles to the current and to the magnetic field. Since a 'current' is simply an ether flow, this means that Hall's action should exist even in non-conductors: an ether flow in the presence of a magnetic field implies a 'Hall effect'. In fact, since ether flows are much simpler to understand in non-conductors than they are in conductors, theory can better deal with the former than with the latter situation — though Hall had found the effect only in conductors.

Let us examine what this interpretation of Hall's discovery leads to. In Maxwellian theory, field equations reflect the energy characteristics of the ether. If the ether's energy properties are changed, then the field equations are changed and vice versa. Hall had discovered an effect which he interpreted as requiring the addition of a new term to one of the field equations, though an admittedly small term. Since the Hall effect is not present in the usual field equations, and since in Hall's interpretation the effect involves a new field, the fundamental equations must somehow be modified. This necessarily means that the energy properties of the electromagnetic field must also be altered, since in their usual form they lead directly to Maxwell's original equations. Consequently Hall was proposing to alter the energy properties of the ether, and he was well aware that he was doing so.

This was a major proposal. The earliest reaction was to reject it, preferring instead to attribute the effect, as I mentioned above, to material processes. Nevertheless, Hall's proposal was entirely in keeping with the spirit of Maxwellian theory. Indeed, it epitomises that spirit by seeking to assimilate a new phenomenon to basic aspects of the theory. Moreover, we shall now see that Hall's proposal led, in the hands of Rowland, to major developments in Maxwellian theory which seemed to vindicate its dynamical structure and its understanding of the electric current.

In 1880 Rowland incorporated Hall's new field into one of Maxwell's equations, the one which we today call the Faraday law.[7] In essence Rowland

Methods of Maxwellian electrodynamics

simply added a term to the usual Faraday law: $\mathbf{B}_E$ is the applied magnetic field, $\mathbf{D}$ is the displacement, and h determines the magnitude of the Hall effect:

$$\nabla \times (\mathbf{E} - h(\mathbf{B}_E \times \partial \mathbf{D}/\partial t)) = -\partial \mathbf{B}/\partial t$$

This new equation, combined with the other, unaltered Maxwell equations, leads to a wave equation different from the usual one. In fact, this new wave equation turns out to have precisely the correct form to explain the rotation of the plane of polarisation of light passing through a magnetic field which Faraday had discovered in 1845 (Knudsen 1976).

This was a major result. Rowland wrote that it '... raises Maxwell's theory almost to the realm of fact' (Rowland, 1881). Nor was he alone in his excitement. Soon a number of Maxwellians were actively engaged in following out the implications of this new, 'Hall' term in the field equations.

One of them was Richard Tetley Glazebrook, who turned to the energy questions raised by the new term (Glazebrook, 1881). Through an intricate and careful examination Glazebrook was able to show that the energy term which Maxwell had himself used in the *Treatise* to explain the Faraday rotation (and which Fitzgerald had, in 1880, given electromagnetic significance) leads through Hamilton's principle to the 'Hall' term in the Faraday law which Rowland had introduced.

This amounted to a deep vindication of Maxwellian principles. First, the very idea of adding the 'Hall' term to a field equation was required by the Maxwellian conception of the electric current as an ether flow. Secondly, the dynamical underpinnings of Maxwellian theory required that the term be derived from altered energy expressions. When this is done, the new term is seen to be a consequence of the very same energy expressions long known to be necessary to explain the Faraday rotation. One could hardly ask for a more tightly knit analysis.

But this was not all. Thus far we have considered only the ways in which the new energy terms affect the differential equations of the Maxwell theory. Hamilton's principle, as I remarked earlier, provides more. It also provides boundary conditions — and if one alters energy expressions, then one alters the boundary conditions as well as the differential equations. This was at once seen in the case of the 'Hall' term because Fitzgerald had already examined what the alterations must be, given the energy expression Maxwell had used to generate the Faraday rotation (Fitzgerald, 1880). Since Glazebrook proved that this same expression led to the 'Hall' term, it was obvious to all concerned that the usual boundary conditions in electromagnetisem are altered fundamentally wherever the Hall coefficient (h in the equation) is non-zero.

This did not at all bother the Maxwellians. They were as willing to alter boundary conditions as they were to alter Maxwell's equations, as long as the changes derived — as they did in the Hall case — from deeper Maxwellian principles. How well did the new boundary conditions work?

Jed Z. Buchwald

To ask this last question is to ask about a new experimental situation, namely, one in which light is reflected in a strong magnetic field. Moreover, since the Hall coefficient was clearly much greater in metals than in non-conductors (given the comparative magnitudes of the Faraday rotation in thin metal films and in liquids), the experiment had to involve the reflection of light from metal surfaces. John Kerr had discovered some of the major characteristics of this situation (Kerr 1878). Fitzgerald had been able to capture some, but not all, of Kerr's facts in the 1880 paper which Glazebrook later related to the Hall effect. Consequently it was already known that the Hall term did not lead to a full explanation of what happens in magneto-optic reflection.

However, Fitzgerald was also quick to point out that he had ignored conductivity in his theory, as Glazebrook did after him. And it did seem that those of Kerr's facts which the theory could not capture were related to known properties of metallic reflection.

The events that occurred in this area between 1881 and 1891 are intricate. We need only note that Fitzgerald's deduction of the boundary conditions was found to have been flawed, necessitating a new one, which was first done by A.B. Basset (Basset 1891). Basset's deduction bypassed Hamilton's principle (though Basset was a Cambridge man he came late to electromagnetism), but it did not lead to markedly improved results since Basset did not incorporate conductivity. This was done, using the same conditions as Basset, by J.J. Thomson two years later (Thomson 1893). Thomson discovered that, when conductivity is incorporated into the theory, then all of Kerr's observations can be successfully captured.

This result was, if anything, even more striking than the earlier successes, because it provided for the first time a theory of magneto-optic reflection. The modern reader should be remarkably puzzled by Thomson's success, because it was achieved at the price of violating the usual electromagnetic boundary conditions. We today explain the effect by retaining the usual field equations and boundary conditions and modifying only the differential equation for the microscopic charges.

In fact Thomson's theory does not work empirically, but to know that required more refined observations than were readily available at the time, observations ultimately provided by Pieter Zeeman (Zeeman 1898-99). The difficulty Maxwellian theory faced in 1893 was not empirical failure. The difficulty was to understand the nature of conductivity.

That difficulty was not of immediate significance for the magneto-optic problem. There conductivity posed an analytical problem, but it did not directly raise other problems. Although neither Basset nor Thomson attempted to incorporate conductivity into Hamilton's principle to deduce the boundary conditions for magneto-optics, nevertheless they were quite convinced that, were one to so, then one would obtain the very conditions they used. This turned out to be quite correct. Here, then, conductivity posed only calculational difficulties. But even here it raised serious questions of understanding, and these questions could be answered either by creatively ignoring them or by grappling with them directly. Most Maxwellians preferred to ignore the problem. But not all of them did so, with the result that, by the mid-1890s, Maxwellian theory was in deep difficulty, not with experiment, but with problems of understanding it had long sought to avoid. Once enough Maxwellians became convinced that these problems could not be avoided (and this occurred after about 1895), the foundations of Maxwellian electrodynamics were seriously threatened and even irremediably eroded. This was pre-eminently the work of Joseph Larmor.

Bypassing conductivity
Thoughout the 1880s most Maxwellians did not consider conductivity to pose a problem. This was not because they were able to incorporate it in their dynamical field equations. They were not able to do so — and in the early 1890s, as we shall see, Heaviside conclusively demonstrated that conductivity cannot be inserted directly into dynamical equations. Rather, Maxwellians avoided the entire question by relegating conductivity to an area about which, they were willing at once to admit, they knew little — the unknown mechanism which links matter and ether.

In 1885 J.H. Poynting deduced from Maxwell's equations an expression which can be interpreted as the rate at which energy flows through the electromagnetic field ('On the transfer of energy in the electromagnetic field', 1885: Poynting 1920, 175-93). (Some such expression had necessarily to exist because Maxwellian theory presumes the idea that force arises from local inhomogeneities in the field's energy.) The major impact of Poynting's theorem at the time was his, and, later, J.J. Thomson's, use of it to put conductivity to the side.

The manner in which this was done is rather intricate, but the central idea is simple. One can use either Poynting's theorem (as Poynting himself did), or consequences of it (as J.J. Thomson did), to show that, as far as energy is concerned, conduction need not involve any kind of transfer *along* the conducting path. (The reason is that the energy which dissipates as Joule heat in conduction flows radially from the field into the conducting path.) In Poynting's and J.J. Thomson's views this made it possible to dispense altogether with a theory of conduction. To do so they (in different but related ways) reinterpreted Maxwell's equations to trace the energy flow through the non-conducting medium up to the conducting region. They did this by linking the energy to flows of electric displacement (Poynting 1920, 224-34). Then the function of the conducting region was to destroy the displacement, a process which involved the obscure and unknown mechanism by which matter influences ether processes.

J.J. Thomson went so far as to offer a diagrammatic representation of this destruction *via* molecular dissociation (Thomson 1891, 1893). But all he did was to carry the process down to the microscopic level: he envisaged the molecules, in forming ions, to destroy inflowing displacement. He offered no explanation for how, or why, this occurred, other than that, since the process is microscopic and intermittent, one should expect it to be dissipative on a macroscopic scale.

Poynting's and Thomson's theories were well received among Maxwellians. They justified putting off for the present the difficulties posed by conductivity. Nevertheless, neither theory excluded the possibility of incorporating conductivity analytically by discovering an appropriate dynamical representation for it. Indeed, Maxwellians felt that one would ultimately be able to bring in conductivity in much the same way that viscosity can be brought into continuum theory — that is, an adequate dynamical interpretation of the field energies could have a term for conduction which was analogous to Rayleigh's dissipation function for mechanical continua. However, in 1893 Oliver Heaviside demonstrated that no such term is allowable under any otherwise tenable dynamical representation for the field (Heaviside 1893–1912: 1, secs. 146–59). His argument was complex and mathematically intricate. In essence, he showed that none of the several possible dynamical interpretations of the field equations is compatible with a representation for conductivity that involves terms which are linear functions of velocity. But only terms of this kind can produce dissipative effects, so Heaviside had effectively shown that conductivity cannot be fitted analytically into any dynamical representation of the field.

There was little immediate reaction in print to Heaviside's argument, though correspondence among Maxwellians shows that it was well known. Most probably Maxwellians felt Heaviside had shown that a theory like Poynting's or J.J. Thomson's — in which conductivity is bypassed by ascribing it to unknown molecular processes — was not merely convenient but necessary. The answer to the enigma posed by conductivity lay exclusively, therefore, in the greater enigma of the ether–matter link. Since few Maxwellians seemed to think that the latter problem would soon be solved, one is not surprised to find little discussion among them of conduction in the early 1890s.

This kind of attitude was easily adopted only by those who had been deeply involved in Maxwellian activity for a decade or more. Joseph Larmor, though he attended Cambridge with J.J. Thomson, came somewhat later to the depths of Maxwellianism. Unlike Thomson or Poynting, Larmor, without intending at first to do so, ultimately tried to tackle the ether–matter nexus, not fully understanding the grave difficulties conductivity posed for any such undertaking.

Larmor's vortices

In late 1893 Joseph Larmor began his attempt to explain the link between ether and matter. His reaults were published over several years in 'A dynamical theory of the electric and luminiferous medium' (1893–97). (I will not here give

Methods of Maxwellian electrodynamics

complete references to Larmor's material because the dating of events is a delicate matter.)

Larmor based his theory upon energy expressions originally due to James MacCullagh in the 1840s in a paper 'An essay towards a dynamical theory of crystalline reflexion and refraction' (1839) (MacCullagh 1880, 145–84). Larmor chose these expressions because in 1880 George Fitzgerald had shown that they led directly to the usual field equations, and that they could be modified to yield the Faraday effect as well. It was already quite well known that such a scheme was adequate in many areas, but that, like other Maxwellian theories, it did not encompass conductivity.

In Larmor's (as in Fitzgerald's and MacCullagh's) scheme, the ether stores energy reversibly only for absolute differential rotations of its elements: it does not offer any elastic resistance to ordinary strains. By identifying the magnetic intensity with the velocity of an ether motion, one can build a scheme which works very well for the usual field equations.

Larmor was first interested in the model primarily for optical purposes. But he felt early on that it could also be used to provide a representation for conduction. Most of his Maxwellian contemporaries would not have agreed with him because they were already well aware of the pitfalls posed by conduction for any dynamical field theory. But, unlike them, Larmor came to the depths of field theory a decade after graduating from Cambridge; he had not been carefully following the more intricate contemporary developments in the subject. He was willing to attempt something his more experienced contemporaries sought to avoid. Larmor argued that one can incorporate conductivity, and explain magnetic substances, by assuming that the rotational elasticity of the ether can, under certain circumstances, literally vanish.

What happens in and near regions where the elasticity vanishes? Imagine a toroid in which the ether elasticity is at first finite. Assume that the ether circulates around the toroid, passing through its central aperture. In Larmor's theory this cannot go on forever because the equations show that such a process will necessarily produce differential rotation along the toroid's axis, and this involves energy storage. But suppose that the elasticity within the toroid suddenly vanishes. Then the circulation through the central aperture can continue indefinitely without storing ever-increasing quantities of elastic energy. In fact, not only can the circulation continue forever, it must do so because there is no way to alter the circulation as long as the ethereal elasticity within the toroid is zero.

In Larmor's first theory (late 1893 to early 1894) such regions of null elasticity represent the Ampèrean currents which, most Maxwellians agreed, exist in magnetisable bodies. These currents persist indefinitely. But what of ordinary conduction? Here, Larmor argues, the ether in the conductor is not permanently inelastic. At this point Larmor begins radically to depart from his Maxwellian contemporaries.

Jed Z. Buchwald

In order to incorporate conductivity into the scheme, Larmor argues that, in conductors, the ether's elasticity continually vanishes and reappears. This was a radical assertion because it was so extreme: the ether's elasticity did not just become very small under certain circumstances, it entirely vanished, only to reappear later.

This idea differed greatly from Thomson's and Poynting's treatments of conduction. In their theories conductivity does not require the ether's properties *per se* to be changed. Rather, they assume only that matter can somehow extract energy from the ether. By contrast, in Larmor's theory the ether elasticity must itself undergo rapid and drastic changes in extremely short time intervals. In Larmor's theory the ether experiences radical structural changes during conduction; in Poynting's and Thomson's theories changes in ether structure are unknown and so play only indirect roles.

Despite this difference, Larmor's theory offerred no better an explanation for the transfer of energy from ether to matter which occurs in conduction than did the Thomson–Poynting theories. The exchanges remained mysterious. But now there were two mysteries: the vanishing of the ether's elasticity and the conversion of ethereal to material energy. Moreover, in Larmor's theory it is extremely difficult, and probably impossible, to envisage a situation in which displacement decays *in situ* without producing magnetic intensity. The correlate to displacement decay in the Larmor scheme is the vanishing of elasticity, but, if displacement exists in the regions where the elasticity is soon to vanish, then magnetic intensity will always appear for some time after the elasticity vanishes since the external ether must readjust its equilibrium. Yet the process of displacement decay, wherein no magnetic intensity arises, was central to Maxwellian theory, and, in various forms, it appears in Maxwell's *Treatise* and the work of Thomson and Poynting. Larmor's theory ignores the process.

In Larmor we have, then, an incomplete Maxwellian. He was by no means as thorough a student of the *Treatise* as were Thomson, Poynting, Oliver Heaviside and several others. Unlike them, he had not fully absorbed the Maxwellian understanding of charge and current. Had he done so, he would not have run roughshod over the delicate Maxwellian constructions based upon displacement decay.

The collapse of Maxwellian theory

Larmor's theory was unfortunately incapable of accounting correctly for two cardinal electromagnetic processes: the forces between permanent magnets, and the induction of a conduction current in a wire by motion through a magnetic field. The reasons for these failures are complex. Suffice it to say that the first difficulty derives from the fact that the vorticities of Larmor's Ampèrean currents are unalterable. The second problem is common to most continuum theories and derives from the fact that, in such theories, field equations are the medium's partial differential equations of motion. To obtain electromagnetic

Methods of Maxwellian electrodynamics

induction due to motion in the magnetic field one needs to employ a total or convective derivative. However, the field equations yield only the derivative's local or stationary part. Hence one has to bring in other considerations to obtain the phenomenon.

These difficulties puzzled Larmor for many weeks. He discussed them at length in his correspondence with Fitzgerald, who frequently and strongly criticised Larmor's attempted solutions. In the end Larmor could offer only an admittedly weak account, one which both he and others found unsatisfactory because of its vagueness. Nevertheless, one finds here the germ of the idea that soon led him to break fundamentally with Maxwellian theory: that processes which involve the apparent vanishing of the ether's elasticity always require the presence of what Larmor termed 'monads' of charge. These monads soon evolved into fully fledged replacements for Maxwellian continuum theory.

According to Larmor the Ampèrean vortices that explain magnetism are not purely hydrodynamical. The surface of the vortex contains discrete loci of 'charge', i.e. of differential rotation. As a result the vortices are not governed entirely by fluid pressure. This, Larmor argued, can be shown to overcome the problem of the sign of magnetic action: like conduction currents, the Ampèrean vortices can be affected by electromotive force, though not in the same way.

To explain induction by motion through a magnetic field, Larmor, rather obscurely, relied on 'statistics'. He argued that even a macroscopically steady magnetic field is in reality intermittent (as a result, for example, of rapidly rotating Ampèrean atoms). This yields a process of growth and decay of induction in neighbouring circuits which will have a net result of zero only if the atoms and circuits are not in relative motion.

Larmor's monads did not possess mass, nor were they protoelectrons in any other respect. They did not, for example, yield by motion the magnetic field of conduction currents. Nevertheless, they could be transferred from 'atom' to 'atom', and the process was supposed to occur pre-eminently during conduction, whether electrolytic or metallic. For the first time an avowedly Maxwellian theory was trying to incorporate discrete 'charges' at a fundamental level.

But Larmor's scheme was very weak. Not only was it obscure in concept, but it was also essentially qualitative: one could hardly calculate the behaviour of Larmor's monad-studded vortices when even ordinary vortices presented almost insuperable difficulties. Larmor himself recognised the inadequacy of the scheme, as did J.J. Thomson (who refereed the work for the *Philosophical Transactions*) and, most importantly, George Fitzgerald.

Larmor was in constant correspondence with Fitzgerald. Indeed, many of the changes in Larmor's initial work were prompted directly by Fitzgerald's criticisms. Fitzgerald was especially dissatisfied with Larmor's vague use of 'statistics' and with his monad-ridden vortex. In the course of thinking about Larmor's several makeshift proposals, Fitzgerald realised that Larmor did not need vortices *per se* for these purposes at all. Instead, all Larmor needed to accomplish the same goals was to assume that a monad or monads circulate through the core of the vortex region. This will yield external hydrodynamic circulation (the magnetic field), and it will also permit electromotive force to have an effect — and without placing monads obscurely on vortex surfaces.

The interaction between Larmor and Fitzgerald was remarkably intricate. Larmor would throw out an undeveloped idea; Fitzgerald would try to understand it and would then suggest ways in which it could be clarified. This process of creation, criticism and modification led Larmor to introduce monads into his theory quite early on. But Larmor used the monads only to strengthen weak elements in the theory; he did not use them for fundamental purposes.

Larmor, then, tended to see Fitzgerald's suggestions as helpful hints for fixing weak spots. Fitzgerald himself seems to have thought of his own suggestions in the same way. Fitzgerald did not accordingly think of his suggested replacement of vortices with circulating monads as extraordinarily significant. But at some point Larmor realised that the idea was very important — one which he could use to recast the entire theory to remove nearly every obscurity and contradiction that plagued it in a consistent way.

Larmor rapidly replaced the vortices with Fitzgerald's circulating monads. His theory achieved thereby a degree of unity and consistency which had eluded him since 1893. But Larmor did not at first go much further with the new, circulating monads. He continued to think in starkly Maxwellian terms even of his modified theory. Circulating monads made his theory tenable, but they were not supposed to overturn fundamental Maxwellian assumptions. Larmor did not conceive of these new entities as replacements for Maxwellian macroscopic analysis. He still thought of the field as a Maxwellian entity — as a dynamical structure which can be modified.

One sees this strikingly illustrated in Larmor's insistence at this time on continuing to employ current elements in the same way Maxwell had. This procedure leads to an extra tension in current-bearing circuits; one, that is, which is not implied by the Ampère law. Larmor asked Oliver Lodge to look for the tension. However, even before Lodge reported his negative results, Larmor had abandoned the current elements.

The reason was that he had in the interim seen a copy of Lorentz's 1895 *Versuch einer Theorie der electrischen und optischen Erscheinungen in bewegten Körpern*. Lorentz's theory relied directly upon electrons and made current elements illegitimate objects of inquiry. Seeing this work evidently convinced Larmor that his own theory had to break even more fundamentally with Maxwellian tradition.

Over the next few years Larmor proposed a reconstruction of electrodynamics along the same sorts of lines sketched by Lorentz, though he worked out the practical details in greater depth than Lorentz did. In this new theory the Maxwellian dynamical field no longer exists, because the field's properties cannot be affected by matter. Instead of modifying field energy

Methods of Maxwellian electrodynamics

functions, the new theory relied on models of material bodies built out of charged particles.

By $c.$ 1901 few if any articles employed Maxwellian conceptions. The shift from the old to the new physics took, therefore, about six years. Some physicists, like Oliver Heaviside, had great difficulty making the transition, and in a deep sense Heaviside himself never did make it. But Larmor engaged in a powerful drive to force the new theory's acceptance. Not only did he write many articles applying it to new phenomena, but he also had his student, J.G. Leathem, demonstrate unequivocally that Maxwellian theory, carried out in full detail, actually cannot explain many of the features of magneto-optic reflection — though that had been one of its great successes in the early 1890s. By the turn of the century Maxwellian theory was a rapidly fading memory even in Britain, where it had motivated highly successful research for a quarter of a century.

Notes

1. I have given a detailed, technical account of these issues in Buchwald (1979) and Buchwald (1981–82). See also my forthcoming book *From Maxwell to Microphysics — Aspects of Electromagnetic Theory in the Last Quarter of the Nineteenth Century.* Full bibliographical material can be found there.
2. Moyer (1977 and 1978); Siegel (1981); Simpson (1970); Topper (1971 and 1980).
3. The first full example of this procedure was Fitzgerald's 'On the electromagnetic theory of the reflection and refraction of light' (1880). This was one of the last papers refereed by Maxwell before his death.
4. Poynting's 'Molecular electricity' (1895) shows how the Maxwellians could accommodate their ideas to the admitted discreteness of matter (Poynting 1920, 224–34).
5. Maxwell's fullest discussions of the process are in the *Treatise*, Part I, art. 111, Part II, chapter 10 (Maxwell 1873*a*, 1:132–4, 374–87).
6. Most of Hall's articles appeared in the *Philosophical Magazine* in 1880 and 1881. His laboratory notebooks are preserved at Houghton Library, Harvard University.
7. Rowland 1881: this article was not included in Rowland's *Physical Papers* of 1902.

Bibliography

Achinstein, P. 1968. *Concepts of Science: A Philosophical Analysis*. Baltimore.
Addison, W.I. 1898. *A Roll of the Graduates of the University of Glasgow*. Glasgow.
―― 1902. *Prize Lists of the University of Glasgow from Session 1778-9 to Session 1832-33*. Glasgow.
―― 1913. *The Matriculation Albums of the University of Glasgow from 1727 to 1858*. Glasgow.
Airy, G.B. 1826. *Mathematical Tracts on Physical Astronomy, the Figure of the Earth, Precession and Nutation, and the Calculus of Variations*. Cambridge.
―― 1840. On the regulator of the clock-work for effecting uniform motion of equatoreals. *Memoirs of the Astronomical Society*, 11:249-67.
―― 1842. *Mathematical Tracts on the Lunar and Planetary Theories, the Figure of the Earth, Precession and Nutation, the Calculus of Variations, and the Undulatory Theory of Optics*. Cambridge.
―― 1896. *Autobiography*, ed. W. Airy. Cambridge.
Ampère, A.-M. 1827. Mèmoire sur la théorie mathématique des phénomènes électrodynamique uniquement déduite de l'éxperience. *Mémoires de l'Académie royale des Sciences*, année 1823, 6:175-388. Printed separately: Paris, 1826; and reprinted: Paris 1958.
Anderson, J. 1837. *Extracts from the Latter Will and Codicil of Professor John Anderson*. Glasgow.
Anderson, R.D. 1983. *Education and Opportunity in Victorian Scotland: Schools and Universities*. Oxford.
Antropova, V.I. 1957. [History of Ostrogradsky's Integral theorem.] *Trudy Instituta Istorii Estestvoznanii i Techniki A.N.S.S.S.R.*, 17:226-69.
―― 1965. [Remarks on the 'Memoir on the propagation of heat inside solid bodies' by M.V. Ostrogradsky.] *Istoriko-mathematicheskie Issledovanie*, 16:97-126.
Auroux, S. 1981. Condillac ou la vertu des signes. In *Condillac, la langue des signes*, ed. A.M. Chouillet, pp. i-xxvii. Lille.
Babbage, C. 1830. *Reflections on the Decline of Science in England*. London.
―― 1864. *Passages from the Life of a Philosopher*. London.
Bachelard, G. 1928. *Etude sur l'évolution d'un problème de physique*. Paris.
Bailhache, P. 1975. *Louis Poinsot. La théorie générale de l'équilibre et du mouvement des systèmes*. Paris.
Ball, W.W.R. 1889. *History of the Study of Mathematics at Cambridge*. Cambridge.

Bibliography

Basset, A.B. 1891. On the reflection and refraction of light at the surface of a magnetized medium. *Philosophical Transactions of the Royal Society*, 182:371–96.

Becher, H. 1980*a*. William Whewell and Cambridge mathematics. *Historical Studies in the Physical Sciences*, 11:1–48.

—— 1980*b*. Woodhouse, Babbage, Peacock, and modern algebra. *Historia Mathematica*, 7:389–400.

Berkeley, G. 1948–57. *The Works of George Berkeley*, ed. A.A. Luce and T.E. Jessop, 9 vols. London.

Bernkopf, M. 1968. A history of infinite matrices. *Archive for History of Exact Sciences*, 4:308–58.

Blake, R.M., Ducasse, C.J. and Madden, E.H. 1960. *Theories of Scientific Method: the Renaissance through the Nineteenth Century*. Seattle.

Bowler, P.J. 1976. *Fossils and Progress. Palaeontology and the Idea of Progressive Evolution in the Nineteenth Century*. New York.

Bowley, R.M. 1976. *George Green*. Nottingham.

Boyer, C.B. 1956. *History of Analytic Geometry*. New York.

Bromberg, J. 1967*a*. Maxwell's displacement current and his theory of light. *Archive for History of Exact Sciences*, 4:218–34.

—— 1967*b*. Maxwell's electrostatics. *American Journal of Physics*, 36:142–51.

Brooke, J.H. 1979. The natural theology of the geologists: some theological strata. In *Images of the Earth*, eds L.J. Jordanova and R.S. Porter, pp. 39–64. Chalfont St. Giles.

Brush, S.G., ed. 1965. *Kinetic Theory, Vol. 1, The Nature of Gases and Heat*. Oxford.

—— 1974. The development of the kinetic theory of gases. VIII. Randomness and irreversibility. *Archive for History of Exact Sciences*, 12:1–88.

—— 1976. *The Kind of Motion we Call Heat: A History of the Kinetic Theory of Gases in the Nineteenth Century*, 2 vols. Amsterdam.

Buchwald, J.Z. 1977. William Thomson and the mathematisation of Faraday's electrostatics. *Historical Studies in the Physical Sciences*, 8:102–36.

—— 1979. The Hall effect and Maxwellian electrodynamics in the 1880s. *Centaurus*, 23:51–99, 118–62.

—— 1980. Optics and the theory of the punctiform ether. *Archive for History of Exact Sciences*, 21:245–78.

—— 1981. The quantitative ether in the first half of the nineteenth century. In *Conceptions of Ether*, eds G.N. Cantor and M.J.S. Hodge, pp. 215–37. Cambridge.

—— 1981–82. The abandonment of Maxwellian electrodynamics: Joseph Larmor's theory of the electron. *Archives internationales d'histoire des sciences*, 31:135–80, 373–438.

Burkhardt, K.H. 1908. Entwicklung nach oscillirenden Functionen. *Jahresbericht der Deutschen Mathematiker-Vereinung*, 10: part 2.

Bibliography

Büttner, F. 1900. *Studien über die Greensche Abhandlung.* Leipzig.

Butts, R.E. 1965. Necessary truth in Whewell's theory of science. *American Philosophical Quarterly*, 2:1–21.

Cambridge University Calendar. 1835–54. Cambridge.

Campbell, L. and Garnett, W. 1882. *The Life of James Clerk Maxwell.* London. Reprinted: New York, 1969.

Cannon, W.F. 1960. The uniformitarian–catastrophist debate. *Isis*, 51:38–55.

―― 1961. John Herschel and the idea of science. *Journal of the History of Ideas*, 22:215–39.

―― 1964. Scientists and broad churchmen: an early Victorian intellectual network. *Journal of British Studies*, 4:65–88.

Cantor, G.N. 1975. The reception of the wave theory of light in Britain. *Historical Studies in the Physical Sciences*, 6:109–32.

Cantor, G.N. and Hodge, M.J.S., eds 1981. *Conceptions of Ether. Studies in the History of Ether Theories 1740–1900.* Cambridge.

Catalogus togatorum in academia Glasquensi. 1836–54. Glasgow.

Cauchy, A.L. 1825. *Mémoire sur les intégrales définies, prises entre des limites.* Paris.

―― 1836. *Sur la dispersion de la lumière.* Prague.

―― 1882–. *Oeuvres complètes d'Augustin Cauchy.* Paris.

Challis, J. 1834. Report on the present state of the analytical theory of hydrostatics and hydrodynamics. *Report of the British Association*, pp. 131–51. London.

―― 1838. *Syllabus of a Course of Experimental Lectures on the Equilibrium and Motion of Fluids, and on Optics.* Cambridge.

―― 1878. *Cambridge Mathematical Studies and their Relation to Modern Physical Science.* Cambridge.

Channell, D.F. 1982. The harmony of theory and practice: the engineering science of W.J.M. Rankine. *Technology and Culture*, 23:39–52.

Chrystal, G. and Tait, P.G. 1879. The Rev. Professor Kelland. *Proceedings of the Royal Society of Edinburgh*, 10:321–9.

Church, R.W. 1891. *The Oxford Movement. Twelve Years 1833–1845.* London.

Clark, J.W. and Hughes, T. McK. 1890. *The Life and Letters of the Reverend Adam Sedgwick.* 2 vols. Cambridge.

Conybeare, W.D. 1832. Report on the progress, actual state, and ulterior prospects of geological science. *Report of the British Association for the Advancement of Science*, pp. 365–414.

Cotter, C.H. 1976. George Biddell Airy and his mechanical correction of the magnetic compass. *Annals of Science*, 33:263–74.

Crosland, M.P. 1967. *The Society of Arcueil.* London.

Crosland M.P. and Smith C.W. 1978. The transmission of physics from France to Britain, 1800–1840. *Historical Studies in the Physical Sciences*, 9:1–61.

Bibliography

Cross, J.J. 1983. Euler's contributions to potential theory 1730–1755. In *Leonhard Euler 1707–1783*, pp. 331–43. Basle.

Crowe, M.J. 1967. *A History of Vector Analysis*. London.

Davie, G.E. 1961. *The Democratic Intellect. Scotland and her Universities in the Nineteenth Century*. 2nd edn, 1964. Edinburgh.

Davies, G.L. 1969. *The Earth in Decay. A History of British Geomorphology 1578–1878*. London.

Deakin, M. 1981–82. The development of the Laplace transform, 1737–1937. *Archive for History of Exact Sciences*, 25:343–90; 26:351–81.

Dealtry, W. 1810. *The Principles of Fluxions. Designed for the Use of Students in the University*. Cambridge.

Delambre, J.B.J. 1806. *Tables du soleil*. Paris.

—— ed. 1806–10. *Base du système métrique décimal*, 3 vols. Paris.

[De Morgan, A.] 1835. [Review article on *Ecole Polytechnique*.] *Quarterly Journal of Education*, 10:330–40.

—— 1856. Robert Murphy. *Supplement to the Penny Cyclopaedia*. Vol. II, pp. 337–8. London.

Dirichlet, P.G.L. 1889–97. *Werke*. Berlin.

Dubbey, J. 1978. *The Mathematical Work of Charles Babbage*. Cambridge.

Duhamel, J.M.C. 1833. Mémoire sur la méthode générale rélative au mouvement de la chaleur dans les corps solides plonges dans les milieux dont la temperature varie avec le temps. *Journal de l'Ecole polytechnique*, cahier 22, 14:20–77.

Duhem, P. 1954. *The Aim and Structure of Physical Theory*. Transl. P.P. Wiener. Princeton.

Edinburgh Academy Register. 1914. Edinburgh.

Edinburgh University Almanack. 1833–34. Edinburgh.

Ellis, R.L. 1845a. Notes on magnetism. *Cambridge Mathematical Journal*, 4:90–5, 139–43.

—— 1845b. Memoir of the late D.F. Gregory. *Cambridge Mathematical Journal*, 4:145–52.

Engelsman, S.B.1982. *Families of Curves and the Origins of Partial Differentiation*. Amsterdam.

Enros, P.C. 1983. The Analytical Society (1812–1813). *Historia Mathematica*, 10:24–47.

Euler 1954–55. *Leonhardi Euleri Opera Omnia*. Series II, Vols XII and XIII, ed. C. Truesdell. Zurich.

Everitt, C.W.F. 1975. *James Clerk Maxwell: Physicist and Natural Philosopher*. New York.

Evidence, oral and documentary, taken and received by the Commissioners appointed [...] for visiting the universities of Scotland, Vol. II: University of Glasgow. 1837. *Parliamentary Papers*, Vol. 36.

Bibliography

Faraday, M. 1839–55. *Experimental Researches in Electricity*. 3 vols. London.
Ferguson, E.S. 1962. Kinematics of mechanisms from the time of Watt. *Bulletin of the U.S. National Museum*, 228:185–230.
Ferguson, J. 1908. Lord Kelvin: a recollection and an impression. *Glasgow University Magazine*, 20:276–82.
Fitzgerald, G. 1880. On the electromagnetic theory of the reflection and refraction of light. *Philosophical Transactions of the Royal Society*, 171:691–711.
Forbes, J.D. 1836a. On the refraction and polarization of heat. *Transactions of the Royal Society of Edinburgh*, 13:131–68.
―― 1836b. Note respecting the undulatory theory of heat and on the circular polarization of heat by total reflexion. *Philosophical Magazine*, 8:246–9.
―― 1860. Dissertation sixth: exhibiting a general view of the progress of mathematical and physical science, principally from 1775 to 1850. In *The Encyclopaedia Britannica*, 8th edn, I, pp. 795–996. Edinburgh.
Forman, P., Heilbron, J.L. and Weart S. 1975. Physics *circa* 1900: personnel, funding and productivity of the academic establishments. *Historical Studies in the Physical Sciences*, 5.
Fourier, J. 1878. *Analytical Theory of Heat*, transl. A. Freeman. London.
Fuller, A.T. 1976. The early development of control theory. *Journal of Dynamic Systems, Measurement and Control*, 98:109–18.
Garland, M.M. 1980. *Cambridge before Darwin: the Ideal of a Liberal Education 1800–1860*. Cambridge.
Gauss, C.F. 1863–1933. *Werke*, 12 vols. Göttingen.
Gibson, G.A. 1889–90. Green's and allied theorems: an historical sketch. *Proceedings of the Edinburgh Mathematical Society*, 8:2–5.
Gillispie, C.C. 1951. *Genesis and Geology*. Cambridge, Mass.
Glasgow University Calendar. 1826–29, 1833, 1844. Glasgow.
[Glasgow University Library.] 1977. *Kelvin Papers: Index to the Manuscript Collection of William Thomson, Baron Kelvin in Glasgow University Library*. Glasgow.
Glasgow University Prize and Degree Lists, 1834–1863.
Glazebrook, R.T. 1881. On the molecular vortex theory of electromagnetic action. *Philosophical Magazine*, 11:397–413.
Gnedenko, B.V. and Pogrebysski, I.B. 1963. [*Mikhail Vasil'evich Ostrogradskii 1801–1862: Life and Work, Scientific and Pedagogical Legacy.*] Moscow.
Gooding, D. 1980. Faraday, Thomson and the concept of the magnetic field. *British Journal for History of Science*, 13:91–120.
―― 1982. Convergence of opinion on the divergence of lines: Faraday and Thomson's discussion of diamagnetism. *Notes and Records of the Royal Society*, 36:243–59.
Goursat, E. 1884. Demonstration du théorème de Cauchy. *Acta Mathematica*, 4:197–200.

Bibliography

——— 1900. Sur la définition générale des fonctions analytiques d'après Cauchy. *Transactions of the American Mathematical Society*, 1:14–16.

Grattan-Guinness, I. 1979. Babbage's mathematics in its time. *British Journal for History of Science*, 12:82–8.

——— ed. 1980. *From the Calculus to Set Theory, 1630–1910*. London.

——— 1981. Mathematical physics in France, 1800–1840: knowledge, activity and historiography. In *Mathematical Perspectives*, ed. J.W. Dauben, pp. 95–138. New York.

——— 1984. Work for the workers: advances in engineering mathematics and its instruction in France, 1800–1830. *Annals of Science*, 41:1–33.

Gray, A. 1908. *Lord Kelvin: An Account of his Scientific Life and Work*. London.

Green G. 1828. *An Essay on the Application of Mathematical Analysis to the Theories of Electricity and Magnetism*. Nottingham.

——— 1871. *Mathematical Papers*, ed. N.M. Ferrers. London.

Green, H. 1947. A biography of George Green. In *Studies and Essays in the History of Science and Learning*, ed. M.F. Ashley Montague, pp. 545–94. New York.

Greenberg, J.L. 1982. Alexis Fontaine's integration of ordinary differential equations and the origins of the calculus of several variables. *Annals of Science*, 39:1–36.

Gregory, D.F. 1839. Notes on Fourier's Heat. *Cambridge Mathematical Journal*, 1:104–7.

Hall, A.R. 1969. *A History of the Cambridge Philosophical Society*. Cambridge.

Hamilton, W. 1859. *Lectures on Metaphysics and Logic*, eds H.L. Mansel and J. Veitch, 4 vols. Edinburgh.

Hamilton, W.R. 1931–67. *The Mathematical Papers of Sir William Rowan Hamilton*, 3 vols. Cambridge.

Hankel, H. 1861. *Zur allgemeine Theorie der Bewegung der Flüssigkeiten*. Göttingen.

Hankins, T.L. 1980. *Sir William Rowan Hamilton*. Baltimore.

Harman, P.M. 1982a. *Energy, Force, and Matter. The Conceptual Development of Nineteenth Century Physics*. Cambridge.

——— 1982b. *Metaphysics and Natural Philosophy. The Problem of Substance in Classical Physics*. Brighton.

Heaviside, O. 1893–1912. *Electromagnetic Theory*, 3 vols. London.

Heimann [Harman], P.M. 1970a. Maxwell and the modes of consistent representation. *Archive for History of Exact Sciences*, 6:171–213.

——— 1970b. Molecular forces, statistical representation and Maxwell's demon. *Studies in History and Philosophy of Science*, 1:189–211.

Helmholtz, H. 1882–95. *Wissenschaftliche Abhandlungen von Hermann von Helmholtz*, 3 vols. Leipzig.

Herschel, J.F.W. 1830. *A Preliminary Discourse on the Study of Natural Philosophy*. London.

Bibliography

_____ 1841. Whewell on inductive sciences. *The Quarterly Review*, 68:177–238.

Hertz, H. 1893. *Electric Waves. Being Researches on the Propagation of Electric Action with Finite Velocity Through Space*. London.

Hilken, J.J.N. 1967. *Engineering in Cambridge University 1783–1965*. Cambridge.

Historical Manuscripts Commission. 1982. *The Manuscript Papers of British Scientists*. London.

Hopkins, W. 1835. Researches in physical geology. *Transactions of the Cambridge Philosophical Society*, 6:1–84.

_____ 1836. An abstract of a memoir on physical geology; with a further exposition of certain points connected with the subject. *Philosophical Magazine*, 8:227–36, 272–81, 357–66.

_____ 1840. Researches in physical geology — second series. *Philosophical Transactions of the Royal Society*, 130:193–208.

_____ 1841. *Remarks on Certain Proposed Regulations Respecting the Study of the University*. Cambridge.

_____ 1842a. Researches in physical geology — third series. *Philosophical Transactions of the Royal Society*, 132:43–56.

_____ 1842b. On the elevation and denudation of the District of the Lakes of Cumberland and Westmorland, *Proceedings of the Geological Society of London*, 3:757–66.

_____ 1847. Report on the geological theories of elevation and earthquakes. *Report of the British Association for the Advancement of Science*, pp. 33–92.

_____ 1848. On the elevation and denudation of the District of the Lakes of Cumberland and Westmorland. *The Quarterly Journal of the Geological Society of London*, 4:70–98.

Howson, A.G. 1982. *A History of Mathematical Education in England*. Cambridge.

Hutchison, K. 1981. W.J.M. Rankine and the rise of thermodynamics. *The British Journal for the History of Science*, 14:1–26.

Jackson, T. 1827. *Elements of Theoretical Mechanics*. Edinburgh.

Jones, R.V. 1973. James Clerk Maxwell at Aberdeen, 1856–1860. *Notes and Records of the Royal Society of London*, 28:57–81.

Kelland, P. 1836a. On the dispersion of light, as explained by the hypothesis of finite intervals. *Transactions of the Cambridge Philosophical Society*, 6:153–84.

_____ 1836b. On the motion of a system of particles, considered with reference to the phenomena of sound and heat. *Transactions of the Cambridge Philosophical Society*, 6:235–88.

_____ 1837a. On the laws of the transmission of light and heat in uncrystallized media. *Philosophical Magazine*, 10:336–42.

_____ 1837b. *Theory of Heat*. Cambridge.

Bibliography

——— 1840. On the conduction of heat. *Report of the British Association*, pp. 15–116.

——— 1841. On the present state of our theoretical and experimental knowledge of the laws of conduction of heat. *Report of the British Association*, pp. 1–25.

Kerr, John. 1878. On reflection of polarized light from the equatorial surface of a magnet. *Philosophical Magazine*, 5:161–77.

King, A.G. 1925. *Kelvin the Man: A Biographical Sketch by his Niece*. London.

Kirchhoff, G. 1857. Ueber die Bewegung der Elektrizität in Drähten. *Annalen der Physik und Chemie*, 100:193–217.

Klein, F. 1926. *Vorlesungen uber die Entwicklung der Mathematik im 19 Jahrhundert*. Teil I. Berlin.

Klein, M. J. 1970. Maxwell, his demon and the second law of thermodynamics. *American Scientist*, 58:84–97.

——— 1972. Mechanical explanation at the end of the nineteenth century. *Centaurus*, 17:58–82.

Knott, C.G. 1911. *Life and Scientific Work of Peter Guthrie Tait*. Cambridge.

Knudsen, O. 1972. From Lord Kelvin's notebook: ether speculations. *Centaurus*, 16:41–53.

——— 1976. The Faraday effect and physical theory, 1845–73. *Archive for History of Exact Sciences*, 15:235–81.

——— 1978. Electric displacement and the development of optics after Maxwell. *Centaurus*, 22:53–60.

Koppelman, E. 1972. The calculus of operations and the raise of abstract algebra. *Archive for History of Exact Sciences*, 8:155–242.

Lacroix, S.F. 1814. *Traité de calcul différentiel et du calcul intégral*. 2nd edn, 2 vols. Paris.

Lagrange, J.L. 1760–61. Nouvelles recherches sur la nature et la propagation du son. *Miscellanea Taurinensia*, 2:11–172.

——— 1973. *Mécanique analytique*. In *Oeuvres*, Vols 11 and 12, eds J.-A. Serret and G. Darboux. Reprinted: Hildesheim.

Lamé, G. 1833. Mémoire sur la propagation de la chaleur dans les polyedres, et principalement dans la prisme triangulaire regulier. *Journal de l'Ecole polytechnique*, cahier 22, 14:194–251.

Larmor, Joseph 1927–29. *Mathematical and Physical Papers*, 2 vols. Cambridge.

——— 1936. The origins of Clerk Maxwell's electric ideas, as described in familiar letters to William Thomson. *Proceedings of the Cambridge Philosophical Society*, 32:695–750.

Lawrence, P. 1978. Charles Lyell versus the theory of central heat. *Journal of the History of Biology*, 11:101–28.

Leslie, J. 1804. *An Experimental Inquiry into the Nature and Propagation of Heat*. London.

Bibliography

Love, A.E.H. 1952. *A Treatise on the Mathematical Theory of Elasticiy*, 4th edn. Cambridge.
Lyell, C. 1830–33. *Principles of Geology*, 3 vols. London.
MacCullagh, J. 1880. *The Collected Works of James MacCullagh*, eds J.H. Jellett and S. Haughton. Dublin.
Maclaurin, C. 1742. *A Treatise of Fluxions*, 2 vols. Edinburgh.
Maxwell, J.C. 1873a. *A Treatise on Electricity and Magnetism*, 2 vols. Oxford.
_____ 1873b. Quaternions [review of P. Kelland and P.G. Tait, *Introduction to Quaternions* (London, 1873).] *Nature*, 9:137–8.
_____ 1877. *Matter and Motion*. London.
_____ 1890. *The Scientific Papers of James Clerk Maxwell*, ed. W.D. Niven, 2 vols. Cambridge.
Mayr, O. 1971. Maxwell and the origins of cybernetics. *Isis*, 62:425–44.
Morrell, J. and Thackray, A. 1981. *Gentlemen of Science. Early Years of the British Association for the Advancement of Science*. Oxford.
Morse, P.M. and Feshbach, H. 1953. *Methods of Theoretical Physics*. New York.
Moseley, H. 1843. *The Mechanical Principles of Engineering and Architecture*. London.
Moyer, D.F. 1977. Energy, dynamics, hidden machinery: Rankine, Thomson and Tait, Maxwell. *Studies in History and Philosophy of Science*, 8:251–68.
_____ 1978. Continuum mechanics and field theory: Thomson and Maxwell. *Studies in History and Philosophy of Science*, 9:35–50.
Muir, J. 1950. *John Anderson, Pioneer of Technical Education, and the College he Founded*, ed. J.M. Macaulay. Glasgow.
Murphy, R. 1830. On the general properties of definite integrals. *Transactions of the Cambridge Philosophical Society*, 3:429–33.
_____ 1883a. On the resolution of algebraical equations. *Transactions of the Cambridge Philosophical Society*, 4:125–54.
_____ 1833b. On the inverse method of definite integrals, with physical applications. *Transactions of the Cambridge Philosophical Society*, 4:353–408.
_____ 1833c. *Elementary Principles of the Theories of Electricity, Heat and Molecular Actions, Part 1. On Electricity*. Cambridge.
_____ 1835a. Second memoir of the inverse method of definite integrals. *Transactions of the Cambridge Philosophical Society*, 5:113–48.
_____ 1835b. Third memoir on the inverse method of definite integrals. *Transactions of the Cambridge Philosophical Society*, 5:315–93.
Murray, D. 1927. *Memories of the Old College of Glasgow. Some Chapters in the History of the University*. Glasgow.
Navier, C.M.D.H. 1827. Mémoire sur les lois du mouvement des fluides. *Mémoires de l'Académie royale des Sciences*, année 1823, 6:389–440.
Neuenschwander, E. 1978. Der Nachlass von Casorati (1835–1890) in Pavia. *Archive for History of Exact Sciences*, 19:1–89.

Bibliography

Newton, I. 1972. *Isaac Newton's Philosophiae Naturalis Principia Mathematica: The Third Edition (1726) with Variant Readings*, eds A. Koyré and I.B. Cohen, 2 vols. Cambridge.

Olson, Richard G. 1975. *Scottish Philosophy and British Physics 1750–1880. A Study in the Foundations of the Victorian Scientific Style*. Princeton.

Ostrogradsky, M.V. 1831*a*. Note sur la théorie de la chaleur. *Mémoires de l'Académie impériale des Sciences de St. Petersbourg*, 1:129–38.

―― 1813*b*. Deuxième note sur la théorie de la chaleur. *Mémoires de l'Académie impériale des Sciences de St. Petersbourg*, 1:123–6.

―― 1838*a*. Mémoire sur le calcul des variations des intégrales multiples. *Mémoires de l'Académie impériale des Sciences de St. Petersbourg*, 3:35–58.

―― 1838*b*. Sur la transformation des variables dans les intégrales multiples. *Mémoires de l'Académie impériale des Sciences de St. Petersbourg*, 3:401–8.

―― 1959–61. *Polnoe Sobranie Trudov* [Complete Collected Works], 3 vols. Kiev.

―― 1965*a*. [Proof of a theorem in the integral calculus.] *Istoriko-mathematicheskie Issledovanie*, 16:49–64.

―― 1965*b*. [Memoir on the propagation of heat inside solid bodies.] *Istoriko-mathematicheskie Issledovanie*, 16:65–96.

Outline of the Plan of Education to be Pursued in the Bristol College. 1830. Bristol.

Petrova, S.S. 1978. [The conception of the theory of linear operators in the works of Servois and Murphy.] *Istoriya Metologiya Estestvoznaniya Nauk*, 20:122–8.

Playfair, J. 1860. Dissertation fourth: exhibiting a general view of the progress of mathematical and physical science, since the revival of letters in Europe. In *The Encyclopaedia Britannica*, 8th edn, I, pp. 548–688. Edinburgh.

Poisson, S.D. 1811. *Traité de mécanique*, 2 vols. Paris.

―― 1812. Mémoire sur la distribution de l'électricité a la surface des corps conducteurs. *Mémoires de la Classe des Sciences mathématiques et physiques de l'Institut de France*, 12:1–92, 163–274.

―― 1826*a*. Mémoire sur la théorie du magnétisme. *Mémoires de l'Académie royale des Sciences*, 5:247–338.

―― 1826*b*. Seconde mémoires sur la théorie du magnétisme. *Mémoires de l'Académie royale des Sciences*, 5:488–33.

―― 1827. Mémoire sur la théorie du magnetisme en mouvement. *Mémoires de l'Académie royale des Sciences*, 6:441–570.

―― 1829. Mémoire sur l'équilibre et le mouvement des corps élastiques. *Mémoires de l'Académie royale des Sciences*, 8:357–570, 623–7.

―― 1830. Mémoire sur l'équilibre des fluides. *Mémoires de l'Académie royale des Sciences*, 9:1–89.

―― 1835. *Théorie de la chaleur*. Paris.

Bibliography

____ 1838. Mémoire sur les déviations de la boussole, produits par le fer des vaisseaux. *Mémoires de l'Académie royale des Sciences*, 16:479–555.

____ 1842. *A Treatise of Mechanics*. transl. H.H. Harte. London.

Porter, R.S. 1977. *The Making of Geology. Earth Science in Britain 1660–1815*. Cambridge.

Poynting, J.H. 1920. *Collected Scientific Papers*. Cambridge.

Pycior, H. 1981. George Peacock and the British origins of symbolical algebra. *Historia Mathematica*, 8:23–45.

____ 1983. Augustus de Morgan's algebraic work. *Isis*, 74:211–26.

Rankine, W.J. Macquorn, 1881. *Miscellaneous Scientific Papers of W.J. Macquorn Rankine*, ed. W.J. Millar. London.

Richards, J. 1980. The art and the perception of British algebra. *Historia Mathematica*, 7:343–65.

Riemann, G.F.B. 1892. *Bernhard Riemann's Gesammelte Mathematische Werke*. Herausg. H. Weber. 2nd edn. Leipzig.

Ross, B. 1977. The development of fractional calculus 1695–1900. *Historia Mathematica*, 4:75–89.

Rowland, H. 1881. On the theory of magnetic attractions, and the magnetic rotation of polarised light. *Philosophical Magazine*, 11:254–61.

____ 1902. *Physical Papers*. Baltimore.

Royal Society of London. 1867–79. *Catalogue of Scientific Papers*, Vols 1–8 (1800–1874). London.

____ 1940. *Record of the Royal Society of London for the Promotion of Natural Knowledge*. 4th edn. London.

Rudwick, M.J.S. 1970. The strategy of Lyell's *Principles of Geology*. *Isis*, 61:4–33.

____ 1971. Uniformity and progression: reflections on the structure of geological theory in the age of Lyell. In *Perspectives in the History of Science and Technology*, ed. D.H.D. Roller, pp. 209–37. Norman, Oklahoma.

____ 1972. *The Meaning of Fossils. Episodes in the History of Palaeontology*. London.

____ 1974. Poulett Scrope on the volcanoes of Auvergne: Lyellian time and political economy. *British Journal for the History of Science*, 7:205–42.

Rupke, N.A. 1983. *The Great Chain of History. William Buckland and the Oxford School of Geology (1814–1849)*. Oxford.

Ruse, M. 1976. Charles Lyell and the philosophers of science. *British Journal for History of Science*, 9:121–31.

Saint-Venant, B. de., ed. 1864. [Appendicial material to his edition of] C. Navier, *De la résistance des corps solides*, pp. 512–852. Paris.

Schlissel, A. 1977. The development of asymptotic solutions of linear ordinary differential equations. *Archive for History of Exact Sciences*, 16:307–78.

[Scrope, G.P.] 1830. Lyell's *Principles of Geology*. *The Quarterly Review*, 43:411–69.

Bibliography

Sedgwick, A. 1821. *A Syllabus of a Course of Lectures on Geology.* Cambridge.

—— 1822. On the physical structure of those foundations which are immediately associated with the primitive ridge of Devonshire and Cornwall. *Transactions of the Cambridge Philosophical Society,* 1:89–146.

—— 1825. On diluvial formations. *Annals of Philosophy,* 10:18–35.

—— 1830. Address of the President. *Proceedings of the Geological Society of London,* 1:187–212.

—— 1831. Address of the President. *Proceedings of the Geological Society of London,* 1:281–316.

—— 1834. *A Discourse on the Studies of the University.* 2nd edn. Cambridge.

Shairp, J.C., Tait, P.G. and Adams-Reilly, A. 1873. *Life and Letters of James David Forbes.* London.

Siegel, D.M. 1975. Completeness as a goal in Maxwell's electromagnetic theory. *Isis,* 66:361–8.

—— 1981. Thomson, Maxwell and the universal ether in Victorian physics. In *Conceptions of Ether: Studies in the History of Ether Theories 1740–1900,* eds G.N. Cantor and M.J.S. Hodge, pp. 239–68. Cambridge.

—— 1983. The energy concept: a historical overview. *Materials and Society,* 7:411–24.

Simpson, T.K. 1970. Some observations on Maxwell's *Treatise on Electricity and Magnetism. Studies in History and Philosophy of Science,* 1:249–63.

Smith, C.W. 1976a. Natural philosophy and thermodynamics: William Thomson and 'The Dynamical Theory of Heat'. *British Journal for History of Science,* 9:293–319.

—— 1976b. 'Mechanical philosophy' and the emergence of physics in Britain: 1800–1850. *Annals of Science,* 33:3–29.

—— 1976c. William Thomson and the creation of thermodynamics. *Archive for History of Exact Sciences,* 16:231–88.

—— 1978. A new chart for British natural philosophy: the development of energy physics in the nineteenth century. *History of Science,* 16:231–79.

—— 1979. From design to dissolution: Thomas Chalmers' debt to John Robison. *British Journal for History of Science,* 12:59–70.

Sologub, V.S. 1975. [*The Development of the Theory of Elliptic Equations in the 18th and 19th Centuries.*] Kiev.

Stokes, G.G. 1880–1905. *Mathematical and Physical Papers,* 5 vols. Cambridge.

—— 1907. *Memoir and Scientific Correspondence of the late Sir George Gabriel Stokes,* ed. J. Larmor. 2 vols. Cambridge.

Sturm, J.C.F. 1828. [Review of various papers on spheroids.] *Bulletin universel des sciences et de l'industrie, sciences et mathématiques,* 9:150–8.

Sviedrys, R. 1970. The rise of physical science at Victorian Cambridge. *Historical Studies in the Physical Sciences,* 2:127–45.

—— 1976. The rise of physics laboratories in Britain. *Historical Studies in the Physical Sciences,* 7:405–36.

Bibliography

Tait, P.G. 1860. *The Position and Prospects of Physical Science.* Edinburgh.

_____ 1862. Formulae connected with continuous displacements of the particles of a medium. *Proceedings of the Royal Society of Edinburgh,* 4:617–23.

_____ 1863. Note on a quaternion transformation. *Proceedings of the Royal Society of Edinburgh,* 5:115–19.

_____ 1879. James Clerk Maxwell. *Proceedings of the Royal Society of Edinburgh,* 10:331–9.

Tanner, J.P., ed. 1917. *The Historical Register of the University of Cambridge.* Cambridge.

Thompson, S.P. 1910. *The Life of William Thomson, Baron Kelvin of Largs,* 2 vols. London.

Thomson, J. 1831. *An Introduction to the Differential and Integral Calculus.* Belfast.

Thomson, J.J. 1891. On the illustration of the properties of the electric field by means of tubes of electrostatic induction, *Philosophical Magazine,* 31:149–71.

_____ 1893. *Notes on Recent Researches in Electricity and Magnetism.* Oxford.

Thomson, W. 1847. Notes on hydrodynamics. On the equation of continuity. *Cambridge and Dublin Mathematical Journal,* 2:282–6.

_____ 1856. Dynamical illustrations of the magnetic and heliocoidal rotatory effects of transparent bodies on polarized light. *Proceedings of the Royal Society,* 8:150–8.

_____ 1872a. *Reprint of Papers on Electrostatics and Magnetism.* London.

_____ 1872b. On the motion of free solids through a liquid. *Proceedings of the Royal Society of Edinburgh,* 7:384–90.

_____ 1882–1911. *Mathematical and Physical Papers,* 6 vols. Cambridge.

_____ 1891. *Popular Lectures and Addresses,* 3 vols. London.

[Thomson, W. =] Lord Kelvin. 1904. *Baltimore Lectures on Molecular Dynamics and the Wave Theory of Light.* Cambridge.

Thomson, W. and Tait, P.G. 1867. *Treatise on Natural Philosophy.* Oxford.

Todhunter, I. 1876. *William Whewell, D.D. An Account of his Writings with Selections from his Literary and Scientific Correspondence.* 2 vols. London.

_____ 1886. *A History of the Theory of Attractions and the Figure of the Earth,* 2 vols. Cambridge.

Topper, D. 1971. Commitment to mechanism: J.J. Thomson, the early years. *Archive for History of Exact Sciences,* 7:393–410.

_____ 1980. 'To reason by means of images': J.J. Thomson and the mechanical picture of nature. *Annals of Science,* 37:31–57.

Truesdell, C.A. 1966. *Continuum Mechanics I: The Mechanical Foundations of Elasticity and Fluid Dynamics.* New York.

Venn, J.A. 1940–54. *Alumni Cantabrigienses, 1752–1900,* 6 vols. Cambridge.

Vince, S. 1816. *The Elements of Astronomy: designed for the Use of Students of the University,* 4th edn. Cambridge.

Whewell, W. 1819. *An Elementary Treatise on Mechanics.* Cambridge.

Bibliography

_____ 1823. *A Treatise on Dynamics*. Cambridge.

_____ 1828. On the principles of dynamics, as stated by French writers. *Edinburgh Journal of Science*, 8:27–38.

_____ 1831*a*. Herschel's *Preliminary Discourse*. *Quarterly Review*, 45:374–407.

_____ 1831*b*. Lyell — *Principles of Geology*. The British Critic, Quarterly Theological Review, and Ecclesiastical Record, 9:180–206.

_____ 1831*c*. Progress of geology. *The Edinburgh New Philosophical Journal*, 11:242–67.

_____ 1832. Lyell's Geology, Vol. 2–changes in the organic world now in progress. *Quarterly Review*, 47:103–32.

_____ 1833. *Astronomy and General Physics Considered with Reference to Natural Theology*. London.

_____ 1835. *Thoughts on the Study of Mathematics as Part of a Liberal Education*. 2nd edn, 1836. Cambridge.

_____ 1837. *History of the Inductive Sciences*, 3 vols. London.

_____ 1840. *The Philosophy of the Inductive Sciences, Founded upon their History*, 2 vols. London.

_____ 1841. *Mechanics of Engineering, Intended for Use in Universities and in Colleges of Engineers*. Cambridge.

_____ 1845. *Indications of the Creator*. London.

_____ 1847. *Philosophy of the Inductive Sciences, Founded upon their History*, 2nd. edn, 2 vols. London.

_____ 1860. *On the philosophy of Discovery*. London.

Whiteside, D.T. 1967–81, ed. *The Mathematical Papers of Isaac Newton*, 8 vols. Cambridge.

Whittaker, E.T. 1951–53. *A History of the Theories of Aether and Electricity*, 2 vols. London.

Willis, R. 1841. *Principles of Mechanism*. Cambridge.

Wilson, C.A. 1980. Perturbations and solar tables from Lacaille to Delambre. *Archive for History of Exact Sciences*, 22:53–304.

Wilson, D.B. 1972. George Gabriel Stokes on stellar aberration and the luminiferous ether. *British Journal for the History of Science*, 6:57–72.

_____ 1974. Herschel and Whewell's version of Newtonianism. *Journal of the History of Ideas*, 35:79–97.

_____ 1976. *A Catalogue of the Manuscript Collections of Sir George Gabriel Stokes and Sir William Thomson, Baron Kelvin of Largs in Cambridge University Library*. Cambridge.

_____ 1982. Experimentalists among the mathematicians: physics in the Cambridge Natural Sciences Tripos, 1851–1900. *Historical Studies in the Physical Sciences*, 12:325–71.

Winstanley, D.A. 1955. *Early Victorian Cambridge*. Reprint of 1940 edn. Cambridge.

Wise, M.N. 1979. William Thomson's mathematical route to energy conserv-

Bibliography

ation: a case study of the role of mathematics in concept formation. *Historical Studies in the Physical Sciences*, 10:49–83.

―――― 1981. The flow analogy to electricity and magnetism, Part 1: William Thomson's reformulation of action at a distance. *Archive for History of Exact Sciences*, 25:19–70.

―――― 1982. The Maxwell literature and British dynamical theory. *Historical Studies in the Physical Sciences*, 13:175–205.

Wolf, C.J.E., ed. 1889. Introduction historique. In *Mémoires sur le pendule*, part 1, pp. i–xliii. Paris.

Woodhouse, R. 1821–23. *A Treatise on Astronomy Theoretical and Practical*. Cambridge.

Woodruff, A.E. 1976. Weber, Wilhelm Eduard. *Dictionary of Scientific Biography*, Vol. 14. New York.

Youshkevitch, A.P. 1965. [On the unpublished early papers of M.V. Ostrogradsky.] *Istoriko-mathematicheskie Issledovanie*, 16:11–48.

―――― 1967. *Michel Ostrogradsky et le progrès de la science au XIXe siècle*. Paris.

Zeeman, P. 1898–99. Comparison des mésures relatives a la réflexion de la lumière sur la surface polaire d'un aimant avec les théories de Goldhammer et Drude. *Archives Néerlaindaises*, 1–2:354–7.

Index

Airy, George Biddell, 6, 15–18, 19, 23, 29, 40, 43, 87, 90, 101–2, 103–4, 108, 109
Alembert, Jean Lerond d', 112
Ampère, André Marie, 6, 19, 25, 34, 67, 86, 116, 120, 162, 165, 191, 194, 237, 239
Analytical Society, 6, 95, 207
Andrews, Thomas, 42

Babbage, Charles, 6, 20, 95, 100
Bacon, Sir Francis, 23, 25–6, 27, 41
Basset, A.B., 234
Beaumont, Elie de, 57–8
Berkeley, George, 223
Bernoulli, Daniel, 112
Biot, Jean Baptiste, 129, 151
Blackburn, Hugh, 33
Brewster, Sir David, 19, 43, 89, 206
Boole, George, 84, 99

calculus (foundations of), 6, 17, 21, 30, 33, 95, 97, 100, 207, 214
Cambridge and Dublin Mathematical Journal, 14–15, 99, 161
Cambridge Philosophical Society, 14, 74, 90–5, 122
Cambridge, University of,
 Board of Mathematical Studies, 2, 16
 Mathematical Tripos, 1–2, 10, 13, 14–19, 20–1, 33, 36–8 39–45, 61, 207–8
 Natural Sciences Tripos, 2, 10
 Smith's Prize, 7, 15, 18, 144, 145, 208
 Wranglers, 1–2, 10, 14–19, 50

see Cavendish Laboratory, 'mixed mathematics'
Campbell, Lewis, 202
Cauchy, Augustin Louis, 6, 95, 100–1, 104, 105, 116–17, 138, 207
Cavendish Laboratory, 11
Cayley, Arthur, 2, 39, 84
Challis, James, 15, 18, 105
Clairaut, Alexis Claude, 112
Clausius, Rudolph, 195
common sense philosophy, 9, 181, 202, 203, 204, 212–14
continuity equation, 151–2, 209, 210
Coulomb, Charles Augustin, 129

Dealtry, W., 95
Delambre, J.B.J., 101
De Morgan, Augustus, 89, 100
Dirichlet, P.G.L., 117, 132, 140, 141, 144, 145, 159
Duhem, Pierre, 8–9
dynamical explanation, 8–10

Ecole Polytechnique, 85, 89
Earnshaw, Samuel, 19, 30
Edinburgh, University of, 4–5, 19–26, 205
electromagnetism, 3, 25, 41–2, 105–6, 180, 225, 229–30, 231–6
 dynamical theory of, 3, 164–8, 171–6, 188–99, 217–18, 225
 field theory of, 7–10, 157–61, 188–99, 217–18
 Maxwell, 181–3, 184–7, 188–91, 191–4, 195–9, 208–10, 217–18
 Thomson, 149, 151–6, 161–4, 164–8, 171–6

Index

see field, concept of, Maxwellian physics
Ellis, R.L., 39
encyclopaedias, mathematical 87–9, 100
ether, 3, 4, 6–10, 17–18, 164–8, 169–71, 171–6, 188–91, 197–9, 217–18, 229–31, 236–8
Euler, Leonhard, 112, 113, 119, 205

Faraday, Michael, 1, 19, 25, 29, 41, 42, 154–5, 158, 164, 165, 181, 183, 185, 191, 196, 205, 208–9
  Faraday effect, 33, 158, 164–8, 184–5, 194, 217, 234, 237

field, concept of, 3, 7, 8–10, 154–6, 157–61, 163–4, 171–6, 208–10, 225, 229–31, 236–8
Fitzgerald, George Francis, 169, 171, 237, 239–40
Forbes, James David, 4–5, 13–14, 19–26, 27, 32–3, 34, 41, 42, 43, 204–5, 207, 212
Fourier, Joseph, 4, 54, 66, 86, 106, 113, 140, 150–2, 156–7, 206, 209
French mathematics, 85–7, 115–21; see Ecole polytechnique
Fresnel, Augustin, 4, 24, 104

Gauss, Carl Friedrich, 6, 24, 114, 123, 126, 130, 140, 153, 219
geology
  Cambridge school, 50, 64
  catastrophists, 59
  directionalism, 55–6, 63
  dynamics, 5, 54–5, 68–72, 75, 78–83
  geological structures, 52
  geometry, 50, 52, 74–5, 76–7
  uniformitarians, 53, 56–7, 71, 79
Glasgow, University of, 5, 13–14, 26–33

Glazebrook, Richard Tetley, 233
Green, George, 1, 3, 6, 7, 12, 33, 40, 43, 104, 105, 106, 118, 121, 124, 129–35, 137, 139, 140, 145, 153
Gregory, D.F., 39, 84

Hall, Edwin, 231–3
  effect, 231
Hamilton, Sir William, 202, 203, 204, 212, 221, 223
Hamilton, Sir William Rowan, 10, 42, 100, 214–16, 219–20, 231, 234
Hankel, Hermann, 144
Heaviside, Oliver, 228, 231, 235, 236, 238, 241
Helmholtz, Hermann von, 163
Herschel, Sir John, 3, 6, 20, 23–4, 33–4, 54, 55, 67, 69–70, 87, 95, 100
Hertz, Heinrich Rudolf, 176, 177, 178
Hopkins, William
  as mathematics coach, 4–5, 15–18, 33–4, 43, 61, 77, 106, 136
  on physical geology, 5, 50, 66, 67, 73–6, 78–83, 103, 108
hydrodynamics, 3, 7, 15, 17, 40, 102, 136–9, 157–61, 165–8, 182–3, 209–11

integral theorems, 6–7, 112–15
  French mathematics, 115–21
  Green, 129–35
  Maxwell, 144–5
  Murphy, 121–9
  Stokes, 136–9
  Stokes theorem, 7, 144–5
  Thomson, 139–44

Joule, James Prescott, 1, 25, 43

Kant, Immanuel, 101, 212, 222
Kelland, Philip, 33, 39, 104, 105, 106, 207

258

Index

Kepler, Johann, 24, 28, 67
Kerr, John, 43, 234
Kelvin, Lord, *see* Thomson, Sir William
Kirchhoff, Gustav, 174
Kohlrausch, Rudolph, 190

Lacroix, S.F., 95, 99, 101, 129
Lagrange, Joseph Louis, 6, 29, 95, 97, 113, 115, 119, 129, 133, 135, 197, 207, 218–19, 225–8
Laplace, Pierre Simon de, 19, 29, 30, 54, 106, 129, 131, 132, 145, 151, 152, 159, 206
Larmor, Sir Joseph, 1, 3, 10, 235, 236–41
Leathem, J.B., 241
Leslie, John, 19–20, 43
Liouville, Joseph, 131, 139, 141, 154
Lorentz, Henrik Antoon, 240
Lyell, Sir Charles, 53–5, 57, 59, 60, 65, 69, 71, 74, 79

MacCullagh, James, 138, 237
Maclaurin, Colin, 204, 207, 213–14
Mathematical Tripos, *see* Cambridge
Maxwell, James Clerk, 1–2, 4–5, 10, 12, 15, 23, 39, 42, 44, 77, 119, 149, 226
  analogy, 4, 8–9, 153, 164, 181–4, 209–12
  Cambridge mathematics, 3, 9, 33, 207–8, 212
  common sense philosophy, 9, 203, 204, 212–14
  dynamical explanation, 3, 8–10, 167–8, 169, 171–6, 197–9, 217–20, 220–3, 225
  electromagnetic theory of light, 9, 168, 170–1, 173–5, 196
  electromagnetism 8–9, 41, 164, 167, 171–6, 181–99, 208–10, 229
  elastic solid, 206, 212
  ether 8–9, 167, 188–9, 197–9, 209
  field concept, 3, 8–10, 164, 167, 171–6, 208–10
  geometry, 9, 204–6, 208
  hydrodynamics, 164, 169, 210–11
  kinetic theory of gases, 3, 195
  laws of motion, 10, 219, 220, 221–3
  lines of force, 8, 41, 164, 208–10
  matter, 10, 199, 223
  magneto-optic rotation, 8, 167, 184, 194
  mechanical models, 8–10, 171–6, 180, 185–91, 195–7, 198–9
  metaphysics, 9, 212–14, 214–15, 220–3
  quaternions, 215–16
  Stokes' theorem, 7, 145
  vortices (molecular), 8–9, 167, 185–7, 188–90, 191–9, 217
Maxwellian physics, 10
  dynamics, 225, 226–7
  electric charge, 229–31, 231–5, 238–41
  electromagnetic continuum, 229–31, 236–8, 239–41
  Hamilton's principle, 10, 226–8
  Poynting's theorem, 235–6
  mechanical explanation, 2–4, 7–10, 16–18, 23–4, 42, 141, 164–8, 169–71, 171–6, 177–8, 184–5, 195–7, 198–200, 217–20, 220–3, 225
Meikleham, William, 13–14, 26–30, 32, 33–4, 41, 43
Mill, John Stewart, 24
Miller, William Hallows, 19, 102, 136
'mixed mathematics', 1–2, 5–6, 8–10, 15–19, 21, 33–4, 39–45, 77, 95–9, 101–6, 107–9, 207–8, 212; *see* Cambridge, Mathematical Tripos
Moseley, Henry, 21, 90
Murchison, Roderick, 51, 52

Index

Murphy, Robert, 7, 15, 106, 121–9

natural theology, 3, 5, 60–1, 64, 69
Navier, C.M.D.H., 137, 138
Newton, Sir Isaac, 9, 15, 17, 24–6, 28, 29, 30, 40, 60, 61, 75, 99, 104, 205, 207, 214, 220

optics, 2, 3, 4, 5, 7, 16, 87, 103–4
  elastic solid, theory of, 171–6
  electromagnetic theory of, 8–9, 168, 170–1, 173–5, 190
  wave theory of, 4, 6, 16–18, 20, 24, 104
Ostrogradsky, M., 117–18, 119, 120, 129

Peacock, George, 6, 100
physics
  Cambridge, 1–2, 3, 4, 6, 7–11, 12, 14–19, 33–4, 36–8, 39–45, 207–12
  Scottish, 4–5, 12–14, 21–5, 26–32, 33–4, 35–6, 39–45, 204–6, 212
Playfair, John, 21, 23, 26
Poisson, Siméon Denis, 6, 23, 30, 66, 86, 106, 115, 118–20, 123, 129, 131, 132, 137, 138, 140, 145, 151, 154, 162, 165, 167, 206
Poynting, John Henry, 235, 238

quaternions, 215–17

Rankine, W.J.M., 33, 39, 166, 184
Rayleigh, Lord, 11, 44, 228
Riemann, Bernhard, 141, 144
Rowland, Henry A., 231–3

Scottish philosophy, *see* common sense philosophy
Scottish physics, *see* Edinburgh, Glasgow (Universities); physics
Sedgwick, Adam, 49, 51, 60, 73, 74, 75, 79, 83

  on education, 60–2
  geometrical style of geology, 50, 52
  on Lyell, 54, 56–7
  natural theology, 61–2, 64
  progressionist geology, 63–4
Siemens, C.W., 217
Smith, Archibald, 39, 73
Smith, William, 51
Smith's prize, *see* Cambridge
Steele, W.J., 12, 15, 31, 43
Stewart, Balfour, 39, 43, 44
Stewart, Dugald, 212
Stokes, Sir George Gabriel, Bt., 12, 15, 84, 115, 118, 135, 136–7, 143, 144
  Cambridge mathematics, 1–3, 9, 10, 41, 43–4, 203, 206, 207, 212, 214
  ether, 6, 40–1
  hydrodynamics, 15, 40–1, 206
  integral theorems, 136–9, 144
  Stokes' theorem, 7, 143–4, 145
Sylvester, J.J., 2, 39

Tait, Peter Guthrie, 12, 15, 23, 42, 144, 145, 177, 216, 217, 223, 228
Taylor series, 6, 95, 207
Thomson, David, 26, 29–30, 41
Thomson, James, 30, 33
Thomson, Joseph John, 11, 44, 234, 235–6, 238, 239
Thomson, Sir William (Lord Kelvin), 1–2, 12, 14, 15, 26, 33, 39, 42–3, 73, 84, 106, 118–19, 137, 139, 196, 203, 208–9, 226, 228
  analogy, 7, 32–3, 41, 150, 152, 155, 157–61, 181, 186, 210–11
  Cambridge mathematics, 1–3, 8, 33, 43–4, 203, 212
  continuity equation, 151–2, 210
  elastic solid theory, 7, 40, 157, 169–71, 171–6
  electrostatics, 7, 43, 141, 152–7, 209–10

260

Index

on electromagnetic theory of light, 173–5
ether, 7, 40, 164–8, 169–71, 171–6
field theory, 154, 157–61, 163–4, 171–6
hydrodynamics, 3, 7, 141, 157, 165–7, 168–9
integral theorems, 7, 140, 143
lectures, 30–32
magnetism, 3, 141–2, 161–4
magneto-optic rotation, 33, 164–8, 184–5, 217
on Maxwell, 149, 167, 171–6
mechanical explanation, 7–8, 42, 141, 164–8, 169–71, 171–6, 177–8, 184–5
Stokes' theorem, 7, 143
vortex motion, 3, 43, 164–8, 184–5, 200, 217
Tyndall, John, 170

vortex motion, 3, 8–9, 43, 164–8, 184–7, 188–90, 191–9, 200, 217

Wallace, William, 33, 103

Warren, John, 100
Weber, Wilhelm Eduard, 164, 182–3, 190, 194
Whewell, William
on Cambridge education, 2–3, 15, 20, 43
on dynamics, 23–4, 29, 68, 101
geology, 5, 49–50, 54–5, 59, 60, 64–9, 72–3, 75, 79
laws of motion, 24, 101, 204, 222
mathematics, 2–3, 6, 101
metaphysics, 19, 23, 54, 68, 72, 82, 203, 213, 221, 223
theology, 60–1, 64
Whittaker, Sir E.T., 1
Willis, Robert, 67–8, 205, 207
Woodhouse, R., 100, 108
Wordsworth, Christopher, 49
Wordsworth, William, 49–50, 83
Wranglers, 1–2, 10, 14–19, 50; *see* Cambridge

Young, Thomas, 19, 34

Zeeman, Pieter, 234